SURVEYS OF THE SOUTHERN GALAXY

ASTROPHYSICS AND SPACE SCIENCE LIBRARY

A SERIES OF BOOKS ON THE RECENT DEVELOPMENTS OF SPACE SCIENCE AND OF GENERAL GEOPHYSICS AND ASTROPHYSICS PUBLISHED IN CONNECTION WITH THE JOURNAL SPACE SCIENCE REVIEWS

VOLUME 105

PROCEEDINGS

SURVEYS OF THE SOUTHERN GALAXY

PROCEEDINGS OF A WORKSHOP HELD AT
THE LEIDEN OBSERVATORY, THE NETHERLANDS, AUGUST 4–6, 1982

Edited by

W. B. BURTON
Sterrewacht Leiden, The Netherlands

and

F. P. ISRAEL
Space Science Department, ESTEC, European Space Agency, Noordwijk, The Netherlands

D. REIDEL PUBLISHING COMPANY

A MEMBER OF THE KLUWER ACADEMIC PUBLISHERS GROUP

DORDRECHT / BOSTON / LANCASTER

Library of Congress Cataloging in Publication Data

Main entry under title:

Surveys of the southern galaxy.

(Astrophysics and space science library ; v. 105)
Includes index.
1. Galaxies–Congresses. I. Burton, W. B. (William Butler), 1940– . II. Israel, F. P. (Frank Pieter), 1946– . III. Series.
QB857.S94 1983 523.1′12 83-16054
ISBN 90-277-1651-X

Published by D. Reidel Publishing Company,
P.O. Box 17, 3300 AA Dordrecht, Holland.

Sold and distributed in the U.S.A. and Canada
by Kluwer Academic Publishers,
190 Old Derby Street, Hingham, MA 02043, U.S.A.

In all other countries, sold and distributed
by Kluwer Academic Publishers Group,
P.O. Box 322, 3300 AH Dordrecht, Holland.

Printed in The Netherlands

TABLE OF CONTENTS

3. LARGE-SCALE GALACTIC STRUCTURE

4. THE GALACTIC CENTER

5. DETAILED SURVEYS

1. J. Albinson
2. T.H. Trolund
3. R.N. Manchester
4. E. Bajaja
5. H.S. Liszt
6. F.M. Olnon
7. C.P. de Vries
8. M. Gottwald
9. W.B. Burton
10. F.P. Israel
11. A. Leene
12. D. Despois
13. J. Brand
14. F.J. Kerr
15. C.J. Mayer
16. F.N. Bash
17. M.L. Kutner
18. A.W. Strong
19. M. van der Bij
20. J.B. Whiteoak
21. J.S. Young
22. W. Hermsen
23. E. Brinks
24. P.A. Riley
25. H. Bloemen
26. J.H. Oort
27. L. Blitz
28. F. Lebrun
29. D. Emerson
30. R.S. Cohen
31. R.S. Cohen
32. R.A.M. Walterbos
33. A.R. Gillespie
34. D. Sanders
35. A.I. Sargent
36. R.C. Kennicutt
37. K. Bennett
38. R.N. Martin
39. C. Heiles
40. H.J. Habing
41. K. Ruf
42. R.J. Cohen
43. B. Baud
44. F. Combes
45. J. Graham
46. T. de Graauw
47. A. Baudry
48. L.R. de Leeuw

LIST OF PARTICIPANTS

J. Albinson, Foundation Radio Astronomy, Dwingeloo, Netherlands
E. Bajaja, Sterrewacht Leiden, Netherlands
F. Bash, University of Texas, Austin, TX, USA
B. Baud, University of Groningen, Netherlands
A. Baudry, Observatoire de Bordeaux, France
K. Bennett, ESTEC, Noordwijk, Netherlands
A. Blaauw, Sterrewacht Leiden, Netherlands
L. Blitz, University of Maryland, College Park, MD, USA
J. Brand, Sterrewacht Leiden, Netherlands
H. Bloemen, Huygens Laboratory, Leiden, Netherlands
M.A. Braz, University of Sao Paolo, Brazil
E. Brinks, Sterrewacht Leiden, Netherlands
W.B. Burton, Sterrewacht Leiden, Netherlands
R. Cohen, Columbia University, New York, NY, USA
R.J. Cohen, Jodrell Bank, UK
F. Combes, Observatoire de Meudon, France
Th. de Graauw, ESTEC, Noordwijk, Netherlands
D. Despois, Observatoire de Bordeaux, France
C. de Vries, Sterrewacht Leiden, Netherlands
D. Emerson, IRAM, Grenoble, France
N. Epchtein, Observatoire de Meudon, France
A.R. Gillespie, MPI Radioastronomie, Bonn, West Germany
M. Gottwald, MPI Extraterrestrische Physik, Garching, West Germany
J. Graham, Cerro Tololo Observatory, La Serena, Chile
H.J. Habing, Sterrewacht Leiden, Netherlands
C. Heiles, UC Berkeley, CA, USA
L. d'Hendecourt, Huygens Laboratory, Leiden, Netherlands
W. Hermsen, Huygens Laboratory, Leiden, Netherlands
F.P. Israel, ESTEC, Noordwijk, Netherlands
R.C. Kennicutt, University of Minnesota, Minneapolis, MN, USA
F.J. Kerr, University of Maryland, College Park, MD, USA
M. Kessler, ESTEC, Noordwijk, Netherlands
G. Knapp, Princeton University, NJ, USA
M. Kutner, Rensselaer Polytechnic Institute, Troy, NY, USA
F. Lebrun, Saclay, France
H.S. Liszt, NRAO, Charlottesville, VA, USA
J. Lub, Sterrewacht Leiden, Netherlands
R.N. Manchester, CSIRO, Epping, Australia
R.N. Martin, IRAM, Grenoble, France
C.J. Mayer, University of Durham, UK
J.H. Oort, Sterrewacht Leiden, Netherlands
P.A. Riley, University of Durham, UK

B.J. Robinson, CSIRO, Epping, Australia
K. Ruf, MPI Radioastronomie, Bonn, West Germany
D. Sanders, University of Massachusetts, Amherst, MA, USA
A.I. Sargent, California Institute of Technology, Pasadena, CA, USA
A.W. Strong, Instit. Fisica Cosmica CNR, Milano, Italy
T. Troland, University of Kentucky, Lexington, KY, USA
Th. van der Hulst, Foundation Radio Astronomy, Dwingeloo, Netherlands
J.B. Whiteoak, CSIRO, Epping, Australia
A.W. Wolfendale, University of Durham, UK
J.S. Young, University of Massachusetts, Amherst, MA, USA

FOREWORD

Problems associated with a general scarcity of observations of the southern sky have persisted since the present era of galactic research began some sixty years ago. In his 1930 Halley Lecture A.S. Eddington commented on the observational support given to J.H. Oort's theory of galactic rotation by the stellar radial velocities measured by Plaskett and Pearce: " ... out of 250 stars only 4 were between 193° and 343° galactic longitude [ℓ^{I}: $225^{o} < \ell^{II} < 15^{o}$]; a stretch of one-third of the whole circuit was unrepresented by a single star. This is the operation which Kapteyn used to describe as "flying with one wing". By mathematical dexterity the required constants of rotation have been extracted from the lopsided data; but no mathematical dexterity can avert the possibility that the neglected part of the sky may spring an unpleasant surprise. As a spectator I watch the achievements of our monopterous aviators with keen enthusiasm; but I confess to a feeling of nervousness when my turn comes to depend on this mode of progression."

During the past few years substantial gains have been made in securing fundamental data on the southern sky. Interpretations based on combined southern and northern surveys are producing a balanced description of galactic morphology. These matters were discussed at a Workshop held at the Leiden Observatory, August 4-6, 1982, attended by some 60 astronomers from 9 countries. Support for the Workshop was provided by the University of Leiden and by ESTEC of the European Space Agency. We are grateful to Lenore de Leeuw for secretarial help during the meeting, to Luc Zuyderduin for photographic work, and to Lena Cijntje for secretarial help during the preparation of the proceedings.

In an appendix to these proceedings we have included some maps based on combined northern and southern data from what are, or will become, standard reference of surveys. Among the maps are comparable ℓ,v maps of the HI and CO distribution in the first and fourth quadrants, as well as sky maps of the γ-ray emission over selected energy ranges, HI integrated emission over selected velocity ranges, and radio continuum flux at 408 MHz. Also included is a map of the HI surface density in M31, made with linear resolution comparable to that of the standard single-dish surveys of our own galaxy; we have no doubt that the complete solution of many of the problems originally posed for our own galaxy will be provided to an important extent by studies of some of the nearby external galaxies. We are grateful to the many authors of the various surveys, referenced in the appendix, for providing their material in machine readable form. We are especially grateful to H.S. Liszt, J.B.G.M. Bloemen, C.G.T. Haslam, and E. Brinks for preparing the maps for publication. This preparation was a time-consuming chore, involving such matters as reading unfamiliar tape formats, re-griding disparate data sets, and adjusting

W. B. Burton and F. P. Israel (eds.), Surveys of the Southern Galaxy, xiii–xiv.

intensity scales. The additional costs accrued by publishing these reference maps in large formats have been partly carried by the Leiden Kerkhoven-Bosscha Foundation.

W.B. Burton
Sterrewacht, Leiden

F.P. Israel
ESTEC, Noordwijk

CO DISTRIBUTION ALONG THE SOUTHERN GALACTIC PLANE

B.J. Robinson, W.H. McCutcheon,* R.N. Manchester and
J.B. Whiteoak
Division of Radiophysics, CSIRO, Sydney, Australia

ABSTRACT

Results are presented for a well-sampled survey of J = 1-0 $^{12}C^{16}O$ emission along the southern galactic plane in the range $294° \leqslant \ell \leqslant 358°$, $-0.075° \leqslant b \leqslant 0.075°$ with an effective angular resolution of about 8' arc. The variation of radial velocity with longitude shows well-defined terminal velocities whose locus matches closely the rotation curves determined from atomic hydrogen and CO observations along the northern galactic plane. Over particular ranges of longitude, absence of CO emission near the tangential velocity is most apparent in the southern observations, and provides compelling evidence for arm-like structures in the CO distribution. The variation of CO column density with longitude suggests a marked contrast in the CO emission between arm and inter-arm regions.

The radial distribution of CO from the galactic centre (averaged over all azimuths) displays two pronounced peaks - a sharp peak near R = 3.5 kpc and a broader peak centred near R = 7 kpc. Northern CO data show only a broad peak, centred near R = 6 kpc.

A preliminary interpretation of the combined southern and northern CO data is consistent with a four-arm spiral structure with a pitch angle of 12° ± 1°.

1. INTRODUCTION

Millimetre-wave emission from the CO molecule is an extremely useful probe of the cold, dense clouds of molecular gas in our galaxy. Giant molecular clouds are closely related to the interstellar dust and recognized as the sites of current star formation in the Galaxy. Thus CO emission is likely to be a good tracer for concentrations of dust and massive young stars which delineate the spiral arms in external galaxies.

*On leave from Dept. of Physics, University of British Columbia.

W. B. Burton and F. P. Israel (eds.), Surveys of the Southern Galaxy, 1–15.

Early studies of galactic CO using telescopes in the northern hemisphere (Scoville and Solomon, 1975; Burton et al., 1975; Bash and Peters, 1976; Burton and Gordon, 1978; Solomon et al. 1979) concentrated on longitudes $0° < \ell < 80°$. The galactic plane was undersampled, which hindered the detection in the CO data both of continuous features which might correspond to spiral arms and of gaps in the CO emission which could be associated with interarm regions.

More recently Cohen et al. (1980) have published a well-sampled survey of the region $12° \leqslant \ell \leqslant 60°$ and $-1° \leqslant b \leqslant 1°$ which provides clear evidence that large-scale structures are present in the CO emission and that there are gaps over significant ranges of longitude and radial velocity.

We have completed a well-sampled survey of J = 1-0 CO emission at intervals of 3' arc over the range $294° \leqslant \ell \leqslant 358°$, $-0.075° \leqslant b \leqslant 0.075°$. The results are discussed in terms of the distribution of emission with longitude and velocity, the variation of CO column density with longitude and the radial distribution of CO from the galactic centre.

2. OBSERVATIONS

The observations were made during the periods September-December 1980 and June-November 1981 with a 4-m Cassegrain telescope at the CSIRO Division of Radiophysics. The beamwidth at 115 GHz was 2'.8 arc.

Details of the receiving system are described by Robinson, McCutcheon and Whiteoak (1982). In brief, the cryogenically-cooled Schottky-barrier mixer receiver had a single-sideband noise temperature on the sky of 1000 K. The CO spectra were measured with a 512-channel acousto-optical spectrograph with radial velocity coverage of 244 km s^{-1} and an effective resolution of 0.6 km s^{-1}.

For a given observation a grid of nine positions was observed, covering $\Delta\ell$ = 0', ±3' and b = 0', ±3'. Adjacent observations were spaced by 9' arc in longitude. To remove the instrumental baseline from the spectra the observations were interspersed with measurements at reference positions 10° above or below the galactic plane. For direct comparison with the results of Cohen et al. (1980), taken with a 1.2-m telescope, the nine spectra in a grid were summed to form a single spectrum which was then smoothed to an effective velocity resolution of 1.6 km s^{-1}.

The intensity calibration is based on a value of T_A^* of 65 K for the peak of the J = 1-0 CO emission from OMC-1 (Ulich and Haas, 1976; Davis and Vandenbout 1973). For our double-sideband receiver there is a pronounced difference in the atmospheric opacity of O_2 at the signal frequency (ν_s = 115.271 GHz) and the image frequency (ν_i = 106.2 GHz). An analysis similar to that of Davis and Vandenbout [see their equation (4)] led to a correction factor

$$W = \frac{T_{atm}}{2}\left[1 + \exp\left(\tau_s - \tau_i\right) \sec Z\right] \quad , \tag{1}$$

where Z is zenith angle, to be applied to the usual absorbing paddle calibration. For our site we have assumed that the atmospheric temperature T_{atm} is the same as that of the calibration paddle. From observations of OMC-1 over a 50° range of zenith angle Z we determined that $(\tau_s - \tau_i) = 0.60$, and all profiles have been corrected with the corresponding W.

3. ROTATION CURVE

Typical CO line profiles at different southern longitudes (in McCutcheon et al. (1981) and Robinson et al. (1982)) show a sharp fall-off in the CO emission at the highest negative velocity, giving a well-defined "terminal velocity". This can be seen clearly in the lower part of Figure 1, where the CO emission for $294° \leqslant \ell \leqslant 358°$ is shown on a logarithmic grey scale of intensity as a function of galactic longitude and radial velocity. The terminal velocity increases steadily from $\ell \approx 300°$ to $\ell \approx 341°$, the tangential point of the "3 kiloparsec arm".

The CO observations of Cohen et al. (1980) are shown in the upper part of Figure 1. These also show a steady increase in terminal velocity from $\ell \approx 60°$ to $\ell \approx 20°$.

For gas in circular orbits with velocity V(R) at galactocentric radius R, the tangential point at longitude ℓ is at $R_T = R_\odot \sin \ell$ ($R_\odot$ being the Sun-centre distance, assumed to be 10 kpc). The radial velocity V_T at the tangential point is

$$V_T = V(R_\odot \sin \ell) - V(R_\odot) \sin \ell \quad . \tag{2}$$

Figure 2 shows the locus of V_T for the rotation curves of Schmidt (1965), Burton and Gordon (1978) and Sinha (1978) with $V(R_\odot) = 250$ km s^{-1}; these curves are based on 21 cm data for the northern hemisphere. Over the longitude range covered by the CO data the differences in the three models are small. The Sinha curve is the best fit to the southern CO terminal velocities, while the Burton and Gordon curve is a somewhat better fit to the northern CO velocities.

4. TANGENTIAL POINTS

One of the striking features of the southern CO observations in Figures 1 or 2 is the absence of CO near the tangential velocity over significant stretches of longitude. There are pronounced holes in CO

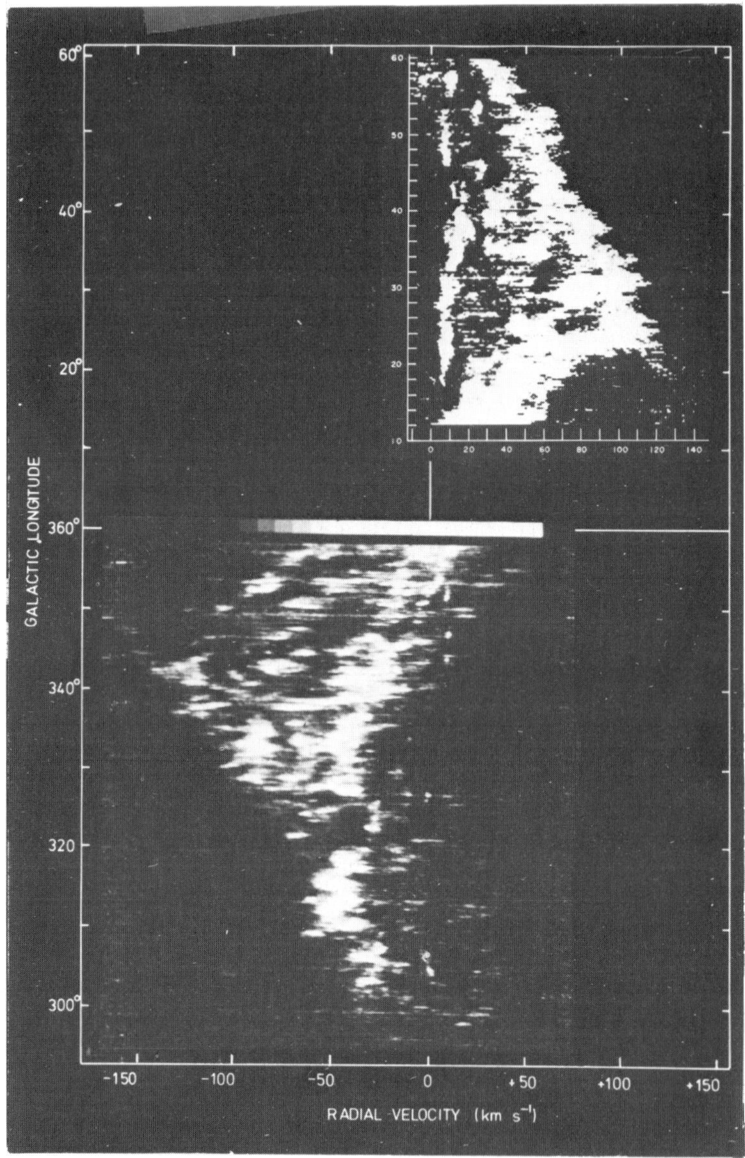

Fig. 1 - Longitude-velocity plot of CO emission with grey-scale of intensity for $294° \leqslant \ell \leqslant 358°$ (this paper) and $12° \leqslant \ell \leqslant 60°$ (Cohen et al., 1980).

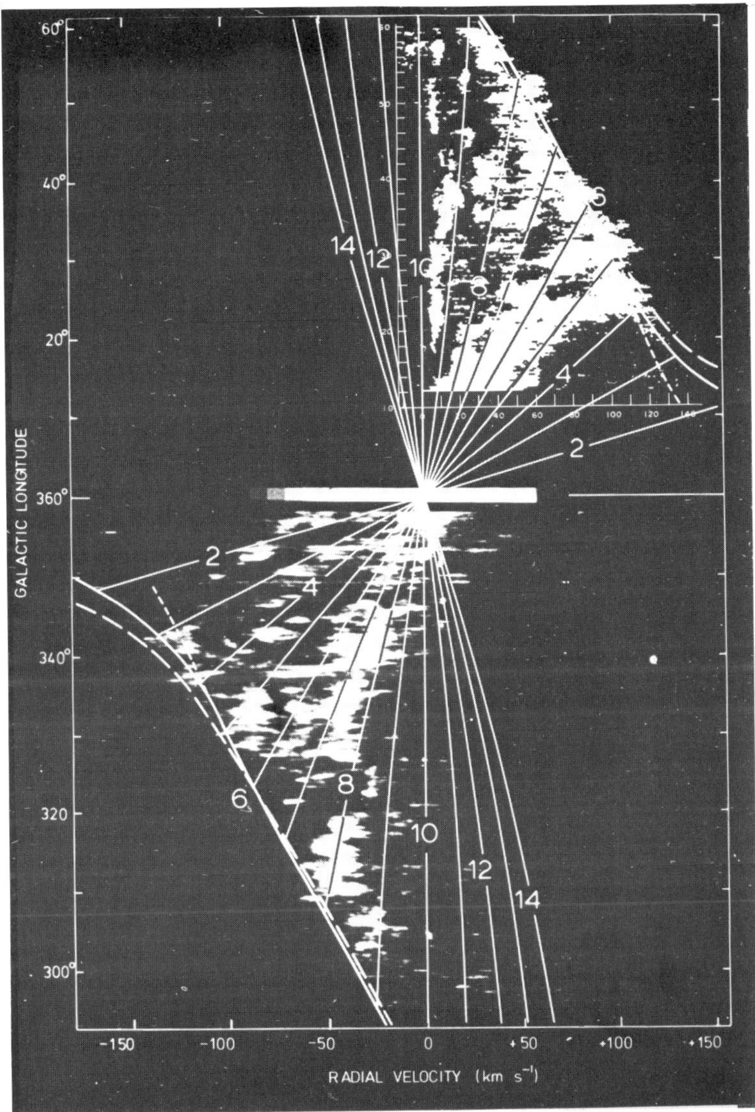

Fig. 2 - Longitude-velocity plot of CO emission showing locus of tangential velocity for the rotation curves of Schmidt (1965) (------), Burton and Gordon (1978) (——) and Sinha (1978) (— — —). The lines radiating from $\ell = 0°$, $v = 0$ km s^{-1} are the loci of annuli with galactocentric radius 2, 3, ... 14 kpc (see text, equation (2)).

for $305° < \ell < 308°$, $316° < \ell < 323°$, $332° < \ell < 336°$ and $338° < \ell < 340°$, which suggests that we are observing tangential points near longitudes of 309°, 327°, 337° and 341°.

The CO seen near $\ell = 341°$ with velocity close to -150 km s^{-1} is associated with the "3 kpc expanding arm", well known from observations of HI and other lines. The velocity of the gas in this arm decreases with increasing longitude to a value of -53 km s^{-1} at $\ell = 0°$.

For low optical depths, or isolated, non-overlapping clouds, the column density of CO is proportional to $\int_0^\infty T_{CO}(v)\,dv$. This quantity is shown as a function of longitude in Figure 3. The lack of CO for

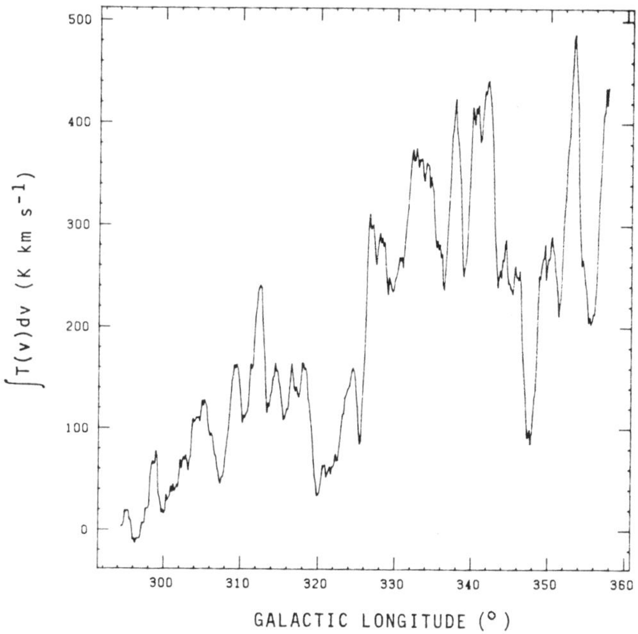

Fig. 3 - Variation of $\int T_{CO}(v)\,dv$ with galactic longitude.

$319° < \ell < 325°$ is a pronounced feature of this figure. Also notable are sharp increases in column density at $\ell = 309°$, 327°, 331°, 337°, 340° and 352°. As noted above, the increases at 309°, 327°, 337° and 340° are produced mainly by CO near the tangential velocity.

For the northern data published by Cohen et al. (1980) and reproduced in Figure 1, CO near the tangential velocity begins abruptly near $\ell = 22°$ but extends fairly continuously at all longitudes up to $\ell = 60°$. There are strong concentrations near $\ell = 31°$ and 52°, and a minimum for $33° < \ell < 37°$. However, there is no strong clumping near particular tangential velocities which marks the southern CO data. The results of the latitude survey by Manchester et al. (1983) show that the holes in the southern CO emission in Figure 1 are not a result of our limited latitude coverage compared to that of Cohen et al. In addition, the northern CO results for b near 0° (Cohen, private communication) are very similar to the published results (integrated over latitude).

5. RADIAL DISTRIBUTION OF CO

For gas in a ring of galactocentric radius R, the radial velocity seen from the Sun is

$$v(\ell) = \left[\frac{R_\odot}{R} V(R) - V(R_\odot)\right] \sin \ell \quad . \tag{2}$$

The loci of $v(\ell)$ for R = 2, 3, ..., 14 kpc have been superimposed on Figure 2 with the Burton and Gordon (1978) model of V(R) within $R = R_\odot$ and a flat rotation curve beyond $R_\odot$. These loci will move only slightly if the Sinha model is used within $R = R_\odot$.

The radial distribution of CO emissivity for longitudes 294° to 350° was determined for concentric galactocentric annuli of width 0.5 kpc, using the Burton and Gordon rotation curve. For each annulus the integrated CO temperature along each line of sight was normalized by the path length through the annulus, and averaged over all lines of sight. Figure 4 shows the derived distribution (full line) and the

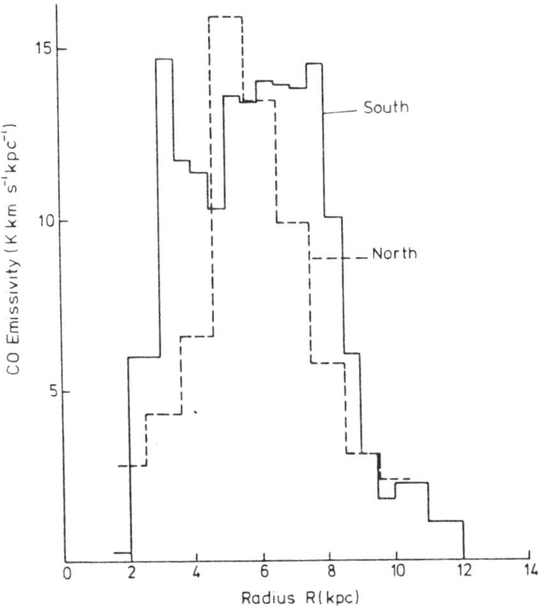

Fig. 4 - Radial distribution of CO emissivity in annuli 0.5 kpc wide. —— southern data; ---- northern data.

corresponding distribution from northern hemisphere data (dashed line) determined by Scoville and Sanders (private communication).

The southern CO radial distribution shows two distinct peaks, a narrow peak near 3.5 kpc and a broader peak centred near 7 kpc, whereas the northern CO radial distribution shows one broad peak, centred on 6 kpc. The difference in the radius of the broad peak between the two sides of the Galaxy will be interpreted below as a result of trailing spiral arms which wind closer to the centre on the northern side. On the southern side the CO emissivity is much higher than the corresponding northern values for $2 < R < 4.5$ kpc.

It is worth noting that the low CO emissivity outside $R = 9$ kpc will reduce the effectiveness of the survey in delineating structure in the gas distribution outside that radius.

6. COMPLETENESS OF THE CO SURVEY

The sizes and volume density of CO clouds are relevant to the completeness of the CO surveys discussed in this paper, which have an effective angular resolution of about 8' arc, corresponding to a linear dimension of 23 pc at a distance of 10 kpc.

(a) Beam dilution for distant clouds

The data given in Solomon et al. (1979) on cloud sizes and numbers suggest that two-thirds of the CO clouds would be unresolved by an 8' arc beam at a distance of 13 kpc, while 90% would be unresolved at a distance of 22 kpc. When the effect of beam dilution is combined with the decrease in CO emissivity outside $R = 9$ kpc, the effectiveness of this survey and the Cohen et al. survey in detecting CO clouds on the far side of the Galaxy is quite low. Only the very large CO complexes will be seen right across the Galaxy. On the other hand, the resolution used in radial velocity (1.6 km s^{-1}) is adequate to resolve the velocity width of CO clouds (typically 9.5 km s^{-1}).

(b) Effect of limited latitude coverage

For $23° \leqslant \ell \leqslant 30°.5$ Solomon et al. (1979) found a scale height for the CO of 60 pc, with about 360 clouds per kpc^3. Manchester et al. (1983) find a similar width in z for $303° \leqslant \ell \leqslant 342°$. With the limited latitude coverage of the southern CO survey we would expect to miss many clouds at distances less than 10 kpc from the Sun. Cohen et al. (1980) observed to ±1° in latitude and hence have sampled more clouds much closer to the Sun. This can be seen in Figure 1, where there is much more CO observed with $0 < v < 20$ km s^{-1} in the northern survey. However, the northern latitude coverage exceeds a scale height for distances more than 3.5 kpc from the Sun.

A further question concerns the departures of the mean CO plane from b = 0° for R < 10 kpc. Manchester et al. (1983) find that $\bar{z}$, the mean distance of CO from the galactic plane, can vary by up to ±60 pc. In some cases this would result in reduced observed intensities.

7. EVIDENCE FOR SPIRAL STRUCTURE IN THE GALAXY

Interpretation of CO surveys of the northern part of the Galaxy has led to controversy as to whether the observations reveal continuous structures which could be regarded as spiral arms or sections of arms. We discuss here what the observations of CO on the southern side of the galactic centre add to the resolution of this controversy.

(a) Holes in the ℓ-v distribution

In the ℓ-v plane there are holes in the CO emissivity at velocities well short of the tangential velocity which extend over longitude spans up to 40° in length. As may be seen in Figure 2 there is, for example, lack of CO with $R \approx 5$ kpc for $338° < \ell < 352°$, while for $328° < \ell < 337°$ there are minima in the CO emission centred on v = -55 and -75 km s^{-1}.

There are also pronounced holes in the CO centred on ℓ = 25°, v = 80 km s^{-1} and ℓ = 31°, v = 60 km s^{-1}, as well as the extended gap in CO near $v \approx 20$ km s^{-1} for $14° < \ell < 55°$.

There are opposed points of view in the interpretation of the clumpiness of the northern CO data. The view of Burton and Gordon (1978) and Solomon et al. (1979) is that the CO is found in clouds and cloud complexes which are randomly distributed in ℓ and v and reveal no large-scale structure in the gas. The opposite view, expressed by Cohen et al. (1980) is that there is significant continuity in the CO emission and the holes over wide ranges of the (ℓ,v) plane, and that these translate into quasi-continuous structures when transformed into the galactic plane.

(b) Kinematic distance ambiguities

In transforming from the (ℓ,v) plane to the (R,θ) galactocentric system there are difficulties associated with non-circular motions and kinematic distance ambiguities for gas with $R < R_\odot$. The distance ambiguity problem will be handled by drawing on independent observational data of two kinds.

(i) The latitude distribution of CO determined from cuts across the southern galactic plane each 3° in longitude to b = ±1°. The CO latitude survey is discussed by Manchester et al. (1983).

(ii) Absorption measurements, mainly atomic hydrogen and formaldehyde, on HII regions associated with the CO.

For circular motion the velocities in the (ℓ,v) plane would approach zero as $\ell\to 0°$. Figure 2 shows that this is not the case. The figure suggests that there are systematic departures from circular motion for $350° < \ell < 358°$. These will be discussed in a later paper with similar data taken for $2° < \ell < 12°$.

(c) (R,θ) to (ℓ,v) transformation

We assume model spiral features in the (R,θ) plane and transform these to the (ℓ,v) plane for direct comparison with the observations of CO and HII regions. The transformation assumes circular motions and we adopt the rotation curve of Burton and Gordon (1978) within R = 10 kpc and a flat rotation curve for $R > 10$ kpc.

The spiral arms are modelled by

$$R = R_s \exp\left[(\theta-\theta_s) \tan \mu\right] ,$$

where R_s is taken as 4 kpc and the starting phase θ_s and pitch angle μ are variable.

(d) Tangential point near $\ell = 327°$

The radius of a spiral arm seen tangentially at $\ell \approx 327°$ is $R_T = R_\odot \sin 327° = 5.45$ kpc (for $R_\odot = 10$ kpc). A variety of models were explored with $5° \leqslant \mu \leqslant 20°$, and the best fit was obtained for $\mu = 12°$, $\theta_s = 25°$. This model is shown as locus 1 superimposed on the CO (ℓ,v) plot in Figure 5. The choice of this model was largely dictated by the requirement to pass through near-side material on the upper side of the locus, and through identifiable far-side material on the lower side of the locus.

A spiral arm model with these best-fit parameters would reach R = 10 kpc at $\ell \approx 20°$, and would then have negative radial velocity in the upper part of Figure 5. The CO observations of Cohen et al. (1980) do not extend to negative velocities; however, the HI observations of Westerhout et al. (1983 - see the atlas section of this volume) show a clearly defined locus of hydrogen emission at negative velocities starting from near $\ell = 20°$. This HI locus fits surprisingly well with the CO locus 1 on Figure 5.

(e) Tangential point near $\ell = 309°$

For the tangential point near $\ell = 309°$ the radius $R_T = R_\odot \sin 309° = 7.8$ kpc. Models were tried with $5° \leqslant \mu \leqslant 20°$, and the best fit was obtained with $\mu = 12°$ and $\theta_s = 135°$. This model has also been superimposed as locus 2 on the observations in Figure 5. The choice of the parameters was largely based on the requirement to pass

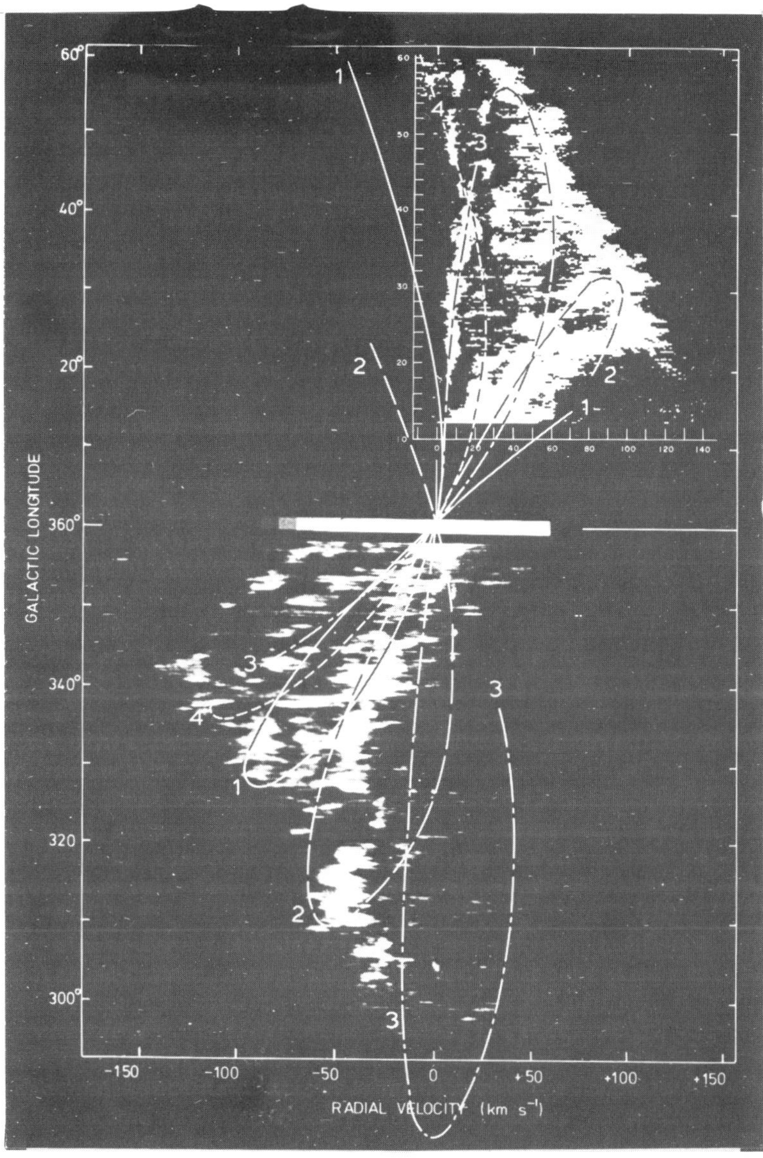

Fig. 5 - CO data from Figure 1 with loci of proposed spiral arm features.

(1) Pitch angle 12°, starting phase 25°.
(2) " " 12°, " " 135°.
(3) " " 11°, " " 247°.
(4) " " 13°, " " 300°.

through near-side material on the upper side of the locus. Beyond the tangential point the gas would have $R > 8$ kpc and the CO emissivity would be falling rapidly. Thus the locus has been fitted to the few patches of far-side CO material which occur in the appropriate (positive) radial velocity range from $320° < \ell < 355°$. The far side of the model arm also lies between 15 and 22 kpc from the Sun and the effects of beam dilution will make undetectable all but the giant CO complexes at these distances. However, for $0° < \ell < 25°$ model locus 2 predicts distant gas at negative velocities which can be seen clearly in the HI observations of Westerhout et al.

The closest point of approach to the Sun of this arm is R = 6.5 kpc. It would be seen tangentially again near $\ell = 30°$, corresponding to $R_T = 5$ kpc. As shown by locus 2 on the upper part of Figure 5, the model encloses the well-defined hole in the northern CO observations centred at $\ell = 25°$, $v = 80$ km s^{-1}.

(f) Tangential point in Carina (near $\ell = 282°$)

Both HI and continuum observations show a tangential point to a spiral feature near $\ell = 282°$ in Carina. This longitude has not been covered in the current CO survey near $b = 0°$; because of the warp in the galactic plane the HI, CO and HII regions in Carina lie nearly 1° below the plane. A CO survey of this region is under way.

We adopt a tangential direction for the Carina arm of $\ell = 282°$, based on HI and HII observations (see Caswell and Haynes, 1983). The locus of a spiral arm with pitch angle 11° and $\theta_S = 247°$ has been drawn on Figure 5 (locus 3).

Beyond the tangential point the velocity of the gas is between +30 and +50 km s^{-1} for $295° < \ell < 320°$. Material at these velocities (assuming a flat rotation curve beyond R = 10 kpc) lies in the range $11 < R < 12$ kpc [cf. equation (2)]. Weak CO is detected in the current $b = 0°$ survey at radial velocities near those corresponding to the model.

The near side of the model arm comes very close to the Sun, with $R \approx 9.2$ kpc at the sub-solar point. For $294° < \ell < 340°$ locus 3 at negative velocities is defined by only occasional patches of CO emission. This is to be expected from the limited latitude coverage of the southern survey, as discussed in Section 6(b). In contrast, CO emission corresponding to the sub-solar section of locus 3 is much more continuous for $18° < \ell < 54°$, reflecting the wider latitude coverage of the northern survey.

On the northern side of the Galaxy locus 3 is tangential again near $\ell = 55°$. The northern CO survey does not clearly define a tangential point, but CO is observed along the far side of the locus from $\ell \approx 30°$ to $\ell \approx 60°$. HII region data would place the tangential point of this arm nearer to 50°, being dominated by the W51 complex; but the CO data show gas extending significantly higher in longitude.

The adopted model starts at $R_S = 4$ kpc, $\ell \approx 345°$, $v = -90$ km s^{-1}. There is a significant amount of CO close to this locus.

(g) Does the Galaxy have a fourth spiral arm?

In this preliminary discussion of the CO observations we have presented arguments for three spiral features with pitch angle near 12° and having θ_S values of 25°, 135° and 247°. Three arm spirals are rare, but Georgelin and Georgelin (1976) have argued for four spiral arms in our galaxy with pitch angle near 12° on the basis of radio and optical observations of HII regions. Henderson (1977) fitted a four-arm spiral with 13° pitch angle to HI data from the Maryland-Green Bank and Parkes surveys. Recently Blitz (1983) has used the Weaver and Williams (1973) HI survey to derive a four-arm spiral with a 20° pitch angle.

Do the CO data show evidence for a fourth arm? We note that for the three arms discussed so far the values of θ_S are not symmetrically displaced around the galactic centre. If a fourth arm with $\theta_S \approx 310°$ were present the pattern of arms would be more symmetric. The locus of such an arm with $\mu = 13°$, $\theta_S = 300°$ is sketched in Figure 5 (locus 4). It is tangential near $\ell = 337°$, corresponding to the tangential point suggested earlier at this longitude.

On the northern side of the centre such an arm would include some of the CO observed inside the loop of locus 3 (see Fig. 5). This fourth arm does not have a tangential point on the northern side of the centre. However, we note that it crosses R = 10 kpc ($v = 0$ km s^{-1}) near $\ell = 55°$ and would thereafter have negative velocities. The HI observations of Westerhout et al. show precisely such an HI feature.

Within the range $335° < \ell < 40°$ the material in this postualted arm lies on the far side of the galactic centre. Its maximum distance from the Sun is 16 or 17 kpc. Beam dilution effects in the CO surveys would restrict the detection of CO to the larger complexes (at a distance of 17 kpc 80% of the CO clouds would be smaller than the beam area). The CO at $\ell = 43°$, $v = +9$ km s^{-1} associated with W49 (at a distance of 14 kpc) lies right on locus 4.

The higher-resolution (3' arc) data available in the southern CO survey need to be examined for further evidence of this fourth arm. Higher-resolution observations in the north (Sanders, 1983) may also provide supporting evidence.

The four-arm spiral models of Georgelin and Georgelin (1976), Henderson (1977) and Blitz (1983) are compared with the CO model in Table 1 in terms of observable parameters:

(a) the longitudes of tangential points;

(b) the longitudes where the arms cross $R = R_\odot$ and change the sign of their radial velocity.

TABLE 1. Comparison of four-arm spiral models of the Galaxy

Model	Longitudes of tangential points (°)						Longitudes of $R=R_{\odot}$ crossing (°)			
This paper	282	309	328	337	33	55	282	326	22	55
Georgelin & Georgelin (1976)	285	310	329	-	35	49	285	323	14	55
Henderson Model II (1977)	290	314	329	-	28	46	290	340	20	65
Blitz (1983)	296	-	324	339	29	-	296	338	29	70

There is excellent agreement between the model presented in this paper and the Georgelin and Georgelin model, and fairly good agreement with Henderson's model II (pitch angle 13°), which included density-wave perturbations. The Blitz model (pitch angle 20°) fails to predict the tangential points near $\ell = 310°$ and $\ell \approx 50°$.

Acknowledgments

Many people have contributed to the success of the CO survey. In particular we thank R.A. Batchelor and M.G. McCulloch for constructing the 115 GHz receiver and L.W. Simons and P.T. Rayner for the on-line software. Isobel Goddard, Anne Manefield, Robina Otrupcek, C.J. Rennie, Betty Siegman and J.S. Wang have assisted with the observations and data analysis.

REFERENCES

Bash, F.N., and Peters, W.L.: 1976, *Astrophys. J.* 205, p. 786.
Blitz, L.: 1983, paper in these proceedings.
Burton, W.B., and Gordon, M.A.: 1978, *Astron. Astrophys.* 63, p. 7.
Burton, W.B., Gordon, M.A., Bania, T.M., and Lockman, F.J.: 1975, *Astrophys. J.* 202, p. 30.
Caswell, J.L., and Haynes, R.F.: 1983, in preparation.
Cohen, R.S., Cong, H., Dame, T.M., and Thaddeus, P.: 1980, *Astrophys. J.* 239, p. L53.
Davis, J.H., and Vandenbout, P.: 1973, *Astrophys. Lett.* 15, p. 43.
Georgelin, Y.M., and Georgelin, Y.P.: 1976, *Astron. Astrophys.* 49, p. 57.
Henderson, A.P.: 1977, *Astron. Astrophys.* 58, p. 189.
McCutcheon, W.H., Robinson, B.J., and Whiteoak, J.B.: 1981, *Proc. Astron. Soc. Aust.* 4, p. 243.

Manchester, R.N., Whiteoak, J.B., Robinson, B.J., Otrupcek, R.E., and Rennie, C.J.: 1983, paper in these proceedings.

Robinson, B.J., McCutcheon, W.H., and Whiteoak, J.B.: 1982, *Int. J. Infrared Millimeter Waves* 3, p. 63.

Sanders, D.: 1983, paper in these proceedings.

Schmidt, M.: 1965, in "Stars and Stellar Systems", Vol. V (A. Blaauw and M. Schmidt eds.) p. 513 (University of Chicago Press).

Scoville, N.Z., and Solomon, P.M.: 1975, *Astrophys. J.* 199, p. L105.

Sinha, R.P.: 1978, *Astron. Astrophys.* 69, p. 227.

Solomon, P.M., Sanders, D.B., and Scoville, N.Z.: 1979, in "The Large Scale Characteristics of the Galaxy" (W.B. Burton ed.) p. 35 (Reidel, Dordrecht).

Ulich, B.L., and Haas, R.W.: 1976, *Astrophys. J. Suppl.* 30, p. 247.

Weaver, H., and Williams, D.R.W.: 1973, *Astron. Astrophys. Suppl. Ser.* 8, p. 1.

Westerhout, G., Kerr, F.J., and Bowers, P.F.: 1983 (the HI reference map in the appendix to these proceedings; comprises the southern data of Kerr and Bowers and the northern data of Westerhout).

A FIRST CO(J=2-1) SURVEY OF THE SOUTHERN HEMISPHERE*.

Th. de Graauw and F.P. Israel
Estec, European Space Agency, Noordwijk (NL)

C.P. de Vries
Sterrewacht Leiden (NL)

on behalf of the Dutch CO Group.

ABSTRACT

The fourth galactic quadrant ($l = 270^{\circ} - 355^{\circ}$) was surveyed in the CO(J=2-1) transition at a wavelength of 1.3 mm. The CO distribution is found to be clumpy; the radial distribution of CO in the fourth quadrant differs from that in the (northern) first quadrant. The cloud-cloud velocity dispersion is, however, very similar to that found in the northern hemisphere.

1. INTRODUCTION

A variety of observational work over the last decade has shown that the CO molecule is an excellent tracer of molecular clouds in the galaxy. For this reason, we have embarked upon a survey of southern hemisphere molecular clouds using the telescopes of the European Southern Observatory at La Silla (Chile). The survey consists of three parts: a survey of molecular clouds associated with HII regions, a survey of globules, dark clouds and Herbig-Haro type objects, and a survey of the galactic plane at $b = 0^{\circ}$ from $l = 270^{\circ}$ to $l = 355^{\circ}$ The results of the first two parts are summarized elsewhere (Brand, these proceedings). This contribution reports on our observational setup and the results obtained in the third part. This is the first galactic survey in the CO(J=2-1) transition; the results are therefore complementary to e.g. the southern hemisphere CO(J=1-0) survey reported by Robinson et al (page 1 ff, these proceedings).

2. OBSERVATIONS AND REDUCTION.

The observations were made with the Estec-Utrecht submillimeter line receiver and the 1.4 m Coude Auxiliary Telescope (CAT) at ESO La

* Based on Observations Made at the European Southern Observatory.

W. B. Burton and F. P. Israel (eds.), Surveys of the Southern Galaxy, 17–24.

Silla in the period May 1981 to April 1982. This optical telescope, designed and planned to feed the High Resolution Spectrograph located at the Coude floor of the 3.6 m telescope building, has a three-mirror Coude arrangement with an f/120 beam. One of the four secondary mirrors in the turret was replaced to obtain an f/30 beam suitable to feed the receiver frontend in the CAT dome. An extra polyethylene lens was used to match the receiver f/9 beam to the f/30 optics.

The peculiar optical arrangement of the CAT with a rotating third mirror introduced considerable baseline ripple due to LO and mixer noise power reflected by the telescope and dome structure. With absorbing material at the center of the secondary the ripple amplitude could be reduced; however, the greatest reduction came from using a reciprocating mirror (Gustincic, 1976) which continuously varied the optical pathlength by half a wavelength during integration. This had the effect of making the noise power reflected into the mixer frequency independent. The receiver system (Lidholm and De Graauw, 1979) used uncooled Schottky diode mixers with noise temperatures (SSB) between 3500 K (1981) and 2500 K (1982). The backend consisted of 256 channels of 1 MHz bandwidth and 256 channels of 250 kHz bandwidth. Consequently, the resolving powers at 230 GHz were 1.3 and 0.3 km s^{-1} respectively, covering velocity ranges of 333 and 88 km s^{-1}. Calibration of the receiver was done by measuring hot and ambient loads; beam profile and beam efficiency were determined by scanning the Moon. HPBW beamsize was 5.5 arcmin. Overall system efficiency and atmospheric transmission were determined by measuring apparent sky temperatures at different airmasses (see also Brand, 1982). All data were taken with respect to reference positions generally several degrees away; these were checked for emission. We use temperatures T_A^* defined such that Orion A has an observed T_A^* = 36 K, following the definitions by Kutner and Ulich (1981).

For the plane survey, we sampled essentially a grid with 30 arcmin spacing, although sometimes a denser grid was sampled. In order to obtain sufficient signal to noise we generated new scans by averaging gridpoints over typically two degrees. Residual ripple patterns were removed by Fourier transforming each of these scans and eliminating the spatial frequency corresponding to the ripple pattern. A very slow baseline variation was removed by eye estimate. Removal of the slow baseline variation had, however, the unfortunate tendency of also removing or decreasing weak and broad (typically 50 km s^{-1} or more) CO emission whenever present. This effect becomes increasingly significant at longitudes $l > 320^o$. Consequently, at these longitudes a large fraction of low-level diffuse emission is removed, thereby enhancing the contrast of the remaining clumpy structure which is still correctly represented. Obviously, determinations of CO emission integrated over velocity suffer most strongly from this effect, to the extent that in profiles with emission extended over a wide velocity range we may underestimate integrated strengths by a factor of two.

3. RESULTS

In Figure 1, we show a longitude-velocity map of the CO(J=2-1) distribution of the fourth galactic quadrant ($b=0^{o}$) with a velocity resolution of 5 km s^{-1} and a longitude resolution of 2° (as compared with a beam resolution of 5.5' and a sampling interval of 30'). We have also produced a longitude-velocity map over a more limited range ($l=300-335^{o}$), but with a longitude resolution of 1°, based on the highest quality spectra obtained in the survey. On the basis of this map we have listed the intensities and sizes of the brightest CO(J=2-1) maxima (Table 1).

TABLE 1. BRIGHTEST CO(2-1) MAXIMA IN THE FOURTH QUADRANT.

Longitude (^{o})	Size (^{o})	V_{LSR} (km/s)	Vel. Width (km/s)	T_A^* (K)	Assoc. HII Regions
287.0	1.5	-27	10	1.7	RCW 53 Carina
305.5	3.1	-38	25	2.2	RCW 74
311.5	2.6	-59	15	2.3	G 311.5+0.3
315.0	2.5	-58	40	3.5	
323.0	1.6	-72	25	3.5	
327.0	1.9	-50	20	2.3	RCW 97
334.0	2.1	-92	30	2.2	RCW 106 Norma
	1.5	-52	30	2.8	
336.5	1.7	-115	20	2.0	G 336.4-0.3
	1.7	-79	15	2.0	
340.5	2.7	-41	10	2.6	
342.8	2.7	-32	20	2.5	G 343.5+0.0
	1.6	-21	15	2.0	
344.4	2.3	-115	25	2.0	
	2.3	-22	10	1.8	

The CO distribution is very clumpy; few cloud complexes extend over more than two degrees. In the map, the molecular cloud complexes associated with Carina, RCW74 and RCW87 stand out. We note that the general lack of CO shortwards of $l=300^{o}$ is due to the tilt of the galactic plane, placing most of the molecular material below $b=0^{o}$. However, as mentioned before, the observed clumpiness in our maps has an enhanced contrast as a consequence of our baseline fitting method discriminating against broad, low-intensity CO emission.As is shown below, about one half of the total CO emission appears to be missing in Figure 1. At the same time it should be stressed, however, that the overall distribution of CO in the fourth galactic quadrant is well-represented in Figure 1 as is borne out by the excellent detailed agreement between the CO(J=2-1) map shown here, and the CO(J=1-0) map obtained by the Australian group (Robinson et al, page 1 ff, these proceedings). This very good agreement also indicates that, on a large scale, the two lowest CO transitions behave very similarly,i.e. on the scales relevant here optical depth effects seem to be negligible.

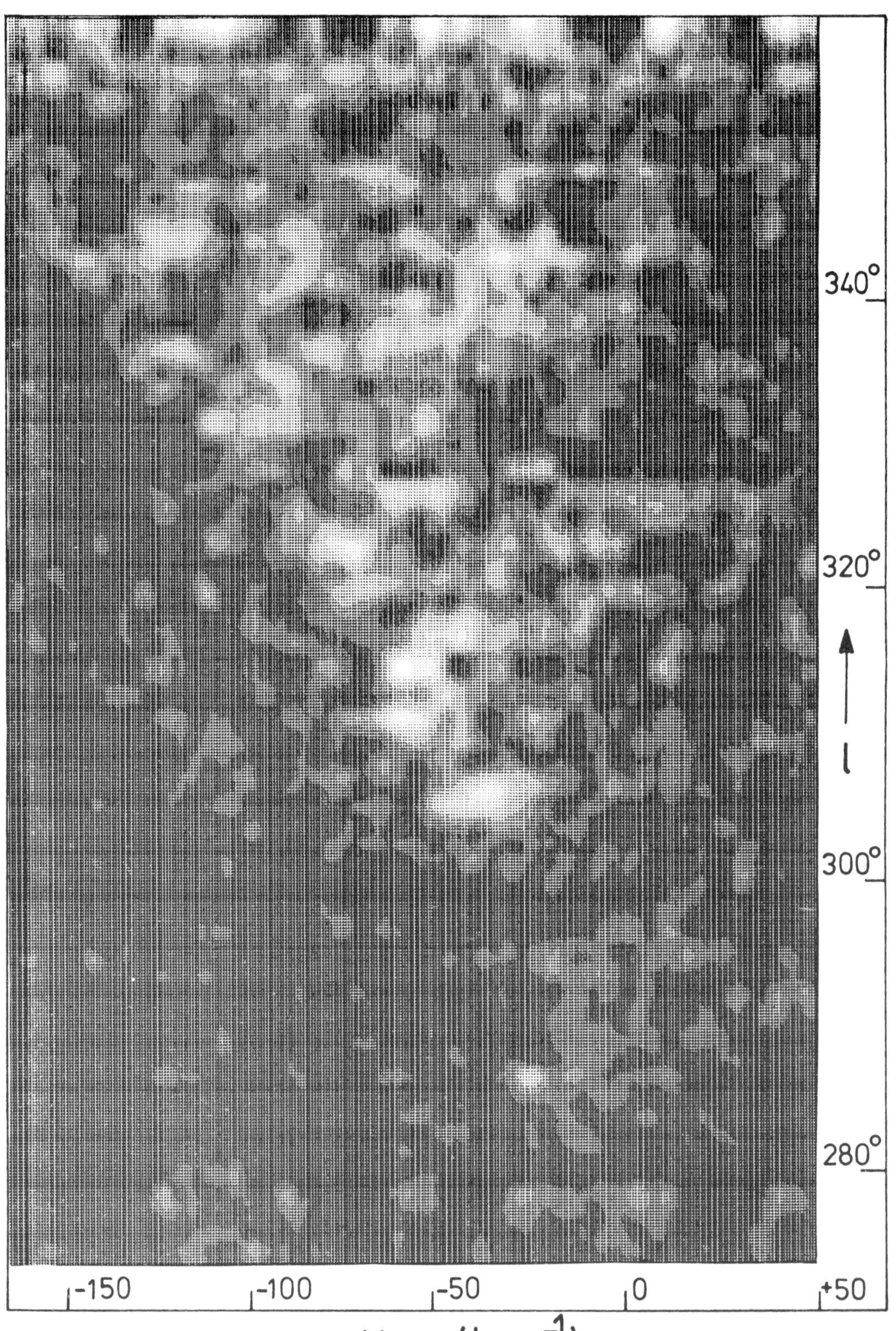

340°
320°
300°
280°
l
-150
-100
-50
0
+50
V_{LSR} (km s^{-1})

Figure 1. Map of the CO(J=2-1) distribution in the galactic plane, between longitudes 270° and 355°. Sampling was at intervals of 30 arcmin, but the map has been smoothed to an effective resolution of two degrees, and 5 km s^{-1}.

In Figure 2 we show the longitude distribution of CO(J=2-1) integrated over the full velocity range. Again the lack of CO below l=300° is largely due to the tilt of the galactic plane. In the longitude distribution, the minima around l=310° and l=330° are remarkable; the former is also found in the southern HI distribution. This particular minimum is accentuated by the presence of the strong CO maximum at l=306° associated with the HII region complex RCW74. Below l=300° integrated CO(J=2-1) intensities are comparable to those found for CO(J=1-0) in both the southern and northern hemisphers at corresponding longitudes. Above l=300° the increase in integrated CO strength is readily seen, but it is not as rapid as in the CO(J=1-0) survey. As mentioned above, this is due to our baseline fitting method which effectively removes low-intensity CO emission in spectra that have emission over a large velocity range. Comparison of Figure 2 with the result obtained by the Australian group for CO(J=1-0) suggests that up to 50 per cent of the total CO emission may have been missed in this way. Nevertheless, the CO(J=2-1) and

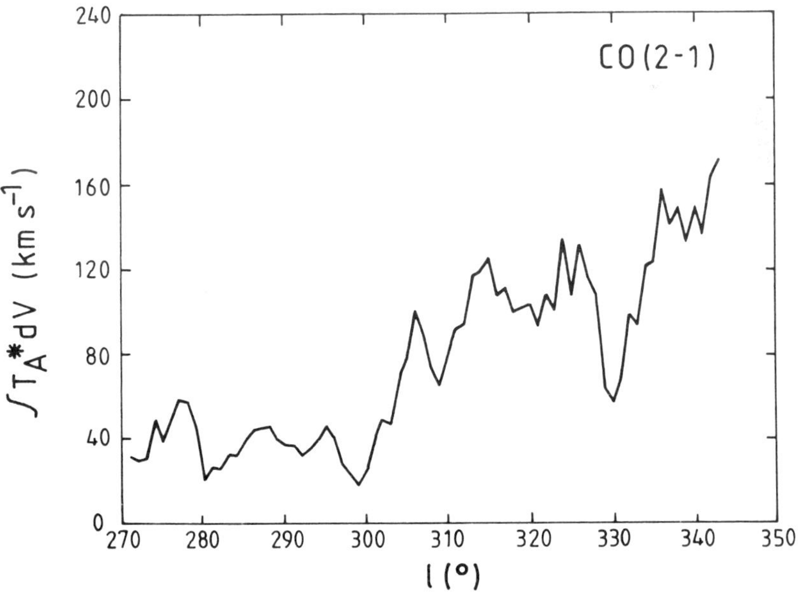

Figure 2. Longitude distribution of CO(J=2-1) integrated over the velocity range -170 to +60 km s^{-1}.

CO(J=1-0) longitude distributions are very similar. The major difference is the deep minimum seen in CO(J=2-1) around $l=330^o$, that has no counterpart in the CO(J=1-0) survey.

In order to study the kinematics of the CO(J=2-1) cloud ensemble we have plotted the longitudinal distribution of CO(2-1) terminal velocities in Figure 3. Also, we have plotted the southern

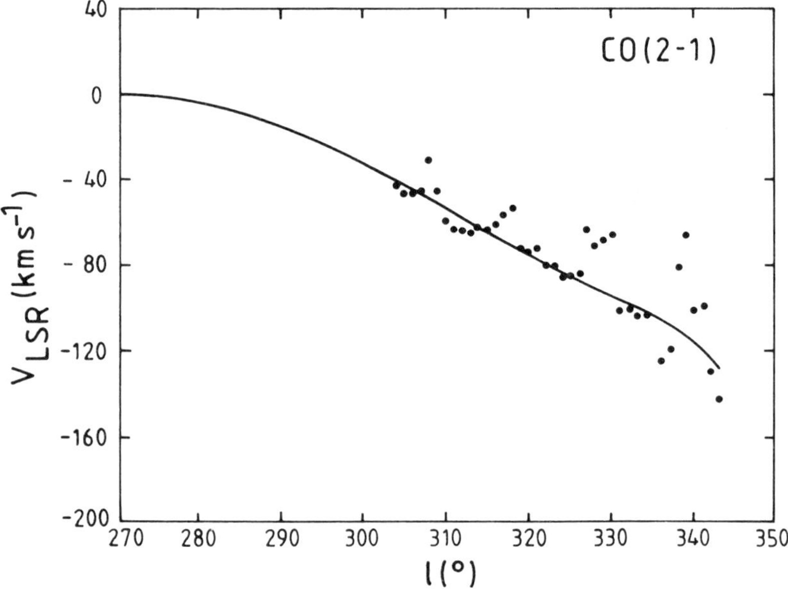

Figure 3. CO(J=2-1) terminal velocities as a function of galactic longitude. The solid line indicates velocities expected on the basis of the Burton and Gordon (1978) rotation curve.

hemisphere galactic rotation curve, derived from HI observations by Kerr, Bowers and Kerr, 1982). As is clear from Figure 3, there is very good general agreement between the observed terminal velocities and the model curve. Lack of CO -- 'CO holes' -- is seen at $l=307-309^o$, $l=315-319^o$, $l=326-331^o$, $l=338-341^o$. These longitudes agree roughly, but not precisely with those found by the Australian group. Only the second 'CO hole' corresponds to an 'HI hole'. From the scatter of points outside these 'holes' around the model curve, we derive an r.m.s velocity dispersion of 4.5 ± 0.5 km s^{-1} for the CO(J=2-1) clouds at the tangential points. At the same time, the dispersion of CO tangential velocities with respect to HI tangential velocities is 3.7 ± 0.5 km/s. It should be noted that these values refer to an effective resolution of 1° to 2°; if clouds were to show significant velocity structure on smaller scales, the velocity dispersion could be somewhat higher. Nevertheless, there is excellent agreement with the result obtained for the northern hemisphere (4.5 km s^{-1}, Dame, 1983).

Finally, we have used the model rotation curve to derive the CO(J=2-1) emissivity as a function of galactic radius R in annuli of 0.5 kpc (Figure 4). Since the inner part of the Galaxy was poorly

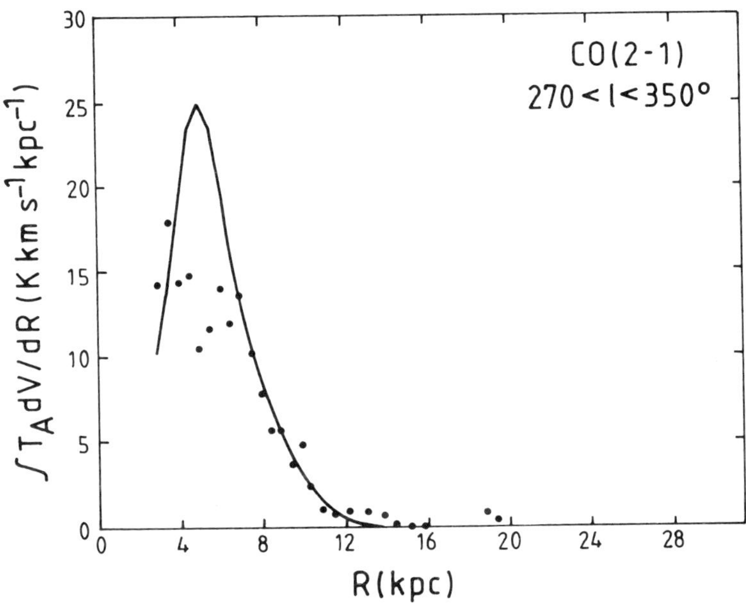

Figure 4. Integrated CO(J=2-1) emissivities as a function of distance with respect to the galactic center. Dots indicate the southern hemisphere 2-1 data, whereas the solid line indicates northern hemisphere 1-0 observations.

sampled, this procedure is meaningful only for radii R>3 kpc. As Figure 4 shows, the CO(J=2-1) emissivity has peaks at R=3.5 kpc and R=7 kpc, and beyond 8 kpc drops smoothly to zero at R=20 kpc. This is in good agreement with the CO(J=1-0) behavior, but in contrast to what is seen in HI. The HI emissivity has a flat peak between R=2.5 and R=11 kpc, and only then drops to zero at R=20 kpc.

We conclude that in the fourth galactic quadrant the distribution of CO(J=2-1) is in almost all respects very close to that of CO(J=1-0) We confirm the presence of 'holes' in the CO distribution, and its clumpy nature. CO terminal velocities follow very closely the curve predicted by the rotation curve adopted by Burton and Gordon (1978) for their CO survey, and r.m.s cloud velocity dispersions are practically identical to those in the northern first galactic quadrant. We confirm the differences between northern and southern radial CO emissivity distribution noted already for CO(J=1-0).

A more detailed version of this paper is submitted to Astronomy and Astrophysics. We are indebted to Dr. W.B. Burton for providing the computer program necessary to generate Figures 3 and 4.

REFERENCES.

Brand, J., 1982, ESO Messenger, No. 29, page 20 ff.
Burton, W.B., Gordon, M.A., 1978 Astron. Ap. 63, 7.
Dame, T., 1983, Ph.D. Thesis Columbia University
Gustincic, J.J., 1976, Sec. Int. Conference on Submm Waves and Their Application, San Juan, P.R., Dec. 1976.
Kerr, F.J, Bowers, P.F., Kerr, M., 1982 in preparation
Kutner, M.L., Ulich, B., 1981, Ap. J. 250, 341
Lidholm, S., De Graauw, Th., 1979, Fourth Int. Conference on Infrared and Submm Waves and Their Applications, Dec. 1979.

NOTE:

The Dutch CO group consists, apart from the authors, of the following persons: J. van Amerongen (3) J. van der Biezen (1), J. Brand (2), H.J. Habing (2), A. Leene (2), I. Nagtegaal (3), F. Selman (5), H. van de Stadt (3), J. Wouterloot.

(1) Estec, Noordwijk; (2) Sterrewacht Leiden; (3) Sterrewacht Utrecht; (4) ESO Garching; (5) Universidad de Chile, Santiago de Chile.

SOUTHERN GALACTIC PLANE SURVEYS OF OH AND H_2O MASERS

J.L. Caswell and R.F. Haynes
Division of Radiophysics, CSIRO, Sydney, Australia

ABSTRACT

Systematic searches for OH and H_2O masers are being made over much of the southern galactic plane using the Parkes radio telescope. The status of the surveys and the results to date are described.

1. INTRODUCTION

Surveys on all four ground-state OH transitions were first conducted from galactic longitude 326° to 340°; these and all major subsequent surveys are listed in Table 1. The latitude coverage is at least ±0°.3, and larger where the continuum radiation was seen to extend significantly beyond this range. In the following sections we discuss the major results from the various surveys.

2. 1612 MHz OH MASERS

The initial 1612 MHz survey from ℓ = 326° to ℓ = 340° was later enlarged to reach the galactic centre and in conjunction with northern surveys of comparable sensitivity provides extensive galactic plane coverage. The maser source population on this transition is dominated by late-type (OH/IR) stars; the radio maser surveys now allow detailed study of the kinematics and galactic distribution of these stars. Figure 1 (taken from Caswell et al., 1981) shows the longitude-velocity plot for those OH/IR stars with $|b| < 0°.6$. The dominant characteristic is the approximate antisymmetry about the galactic centre, as expected from the effects of galactic rotation. Within 5° of the galactic centre there is some departure from antisymmetry, with a slight overall preponderance of negative velocities. Caswell and Haynes (1982a) point out that this might result if these stars are in the central elliptical disk (or bar); it will be important to increase the latitude coverage of the OH maser survey near the galactic centre to see whether any "tilt" in the distribution is evident, comparable to

W. B. Burton and F. P. Israel (eds.), Surveys of the Southern Galaxy, 25–30.

TABLE 1 - Parkes surveys of masers in the southern galactic plane.

Maser transition	Galactic longitudes	Principal results or objectives	Reference
OH 1612 MHz	326°-340°	15 OH/IR stellar masers plus weak IIc emission	Caswell and Haynes (1975)
OH 1720 MHz	326°-340°	Weak IIa emission	Haynes and Caswell (1977)
OH 1665 and 1667 MHz	326°-340°	40 Type I masers	Caswell et al. (1980)
OH 1612 MHz	340°-360°	50 OH/IR stellar masers plus 28 other masers	Caswell et al. (1981)
OH 1665 MHz	340°-360°-02°	49 OH masers, mostly Type I	Caswell and Haynes (1982b)
H_2O 22 GHz	250°-360°	68 masers, mostly coinciding with OH masers	Batchelor et al. (1980)
H_2O 22 GHz	250°-360°-50°	A search at new OH maser positions	In progress (see text)
OH 1665 MHz	233°-326°	Survey analysis continuing	In preparation

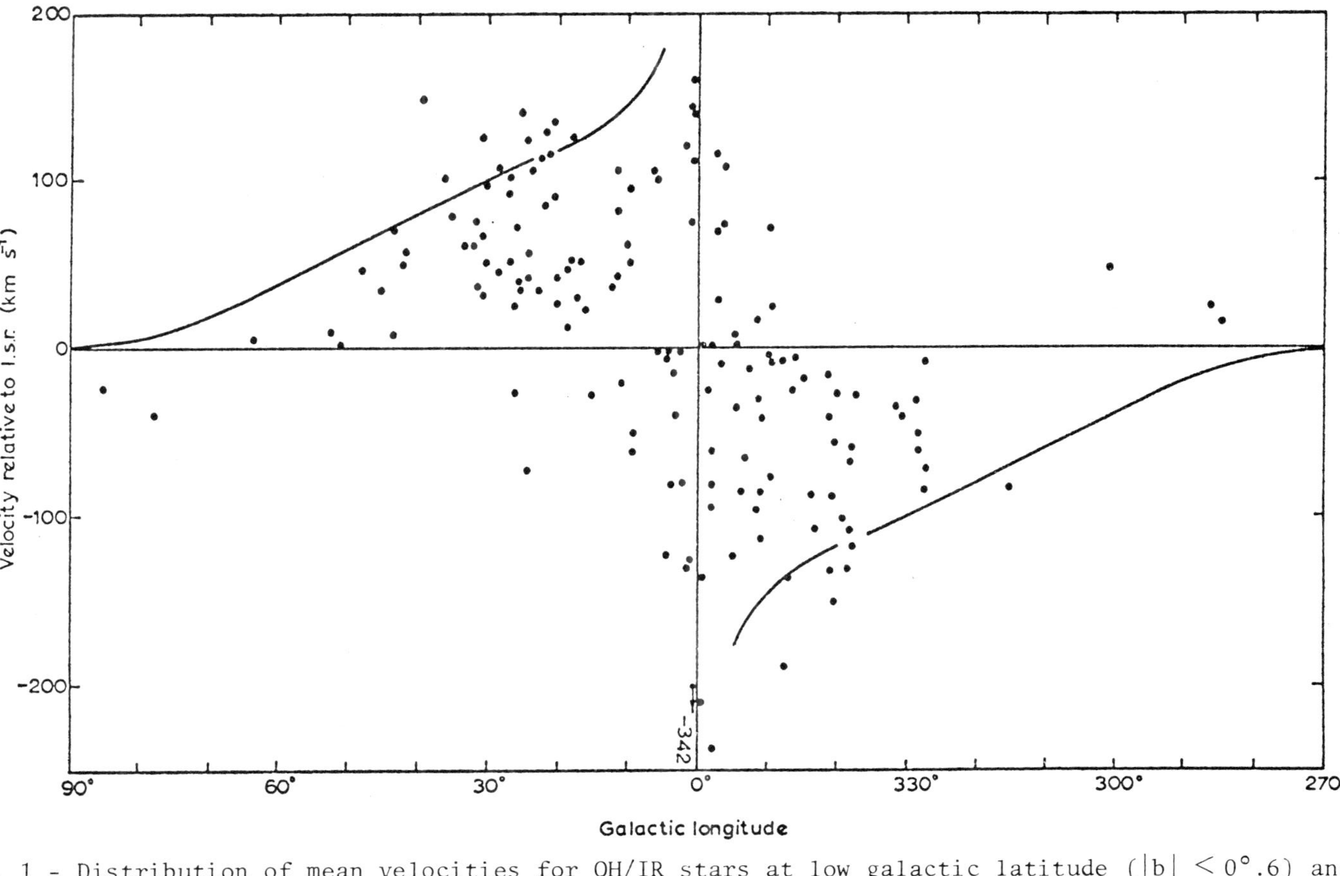

Fig. 1 - Distribution of mean velocities for OH/IR stars at low galactic latitude ($|b| < 0°.6$) and with galactic longitude within 90° of the galactic centre. The curves show the maximum "allowed" velocity assuming strict circular motion defined by Schmidt's (1965) galactic rotation model.

that found in the bar distribution of HI and CO by Liszt and Burton (1980).

3. 1665 AND 1667 MHz OH MASERS

The 1665 MHz survey has been extended from $\ell = 340°$ to the galactic centre. With this recently completed work (Caswell and Haynes, 1982b) and additional current work we will soon have completed the longitude range 270° to 360°. The emphasis is on the detection of Type I masers associated with regions of star formation (some results on the distribution of clouds of OH seen in absorption are obtained as a by-product). The surveys have been made at a uniform grid of positions and some of the OH masers have been discovered where there is negligible continuum radiation. At these locations the OH masers are especially valuable in tracing star formation sites which would have been missed had we studied only the well-developed HII regions.

4. H_2O MASERS

An extensive catalogue of southern H_2O masers (at 22 GHz) was published by Batchelor et al. (1980); this included many new detections, together with a summary of earlier discoveries in the longitude range 250° to 360°. With recent improvements in the performance of receivers at 22 GHz (notably a maser receiver with system temperature below 100 K installed on the Parkes telescope) it has been possible to detect much weaker H_2O masers than before. We are conducting a search for H_2O masers at all the Type I OH maser positions and have been able to detect H_2O counterparts to approximately 80% of the OH masers (Caswell, Batchelor, Forster and Wellington - in preparation). Indeed it now appears that sites of star formation may be just as readily detectable by their H_2O maser emission as by their OH maser emission; the major obstacle is the time-consuming nature of searching large areas with a beamsize of about 1' arc. Searches with the Itapetinga telescope (with a 4' arc beam) have discovered a number of H_2O masers with no OH counterpart (Scalise and Braz, 1980; Braz and Scalise, 1982). The difficulty of detecting any corresponding OH masers must be partly due to their intrinsic weakness but is aggravated by the presence of OH absorption in the vicinity (a problem not encountered in the H_2O case); the latter difficulty is partly compensated for by the high degree of circular polarization found in the OH masers.

5. THE GALACTIC DISTRIBUTION OF TYPE I OH MASERS

Preliminary results of the distribution of Type I OH masers are shown in Figure 2, taken from Caswell and Haynes (1982b). An analysis of this distribution shows the southern hemisphere masers to be concentrated between galactocentric radii 5 and 9 kpc but there is also a significant number near the galactic centre itself. We expect that

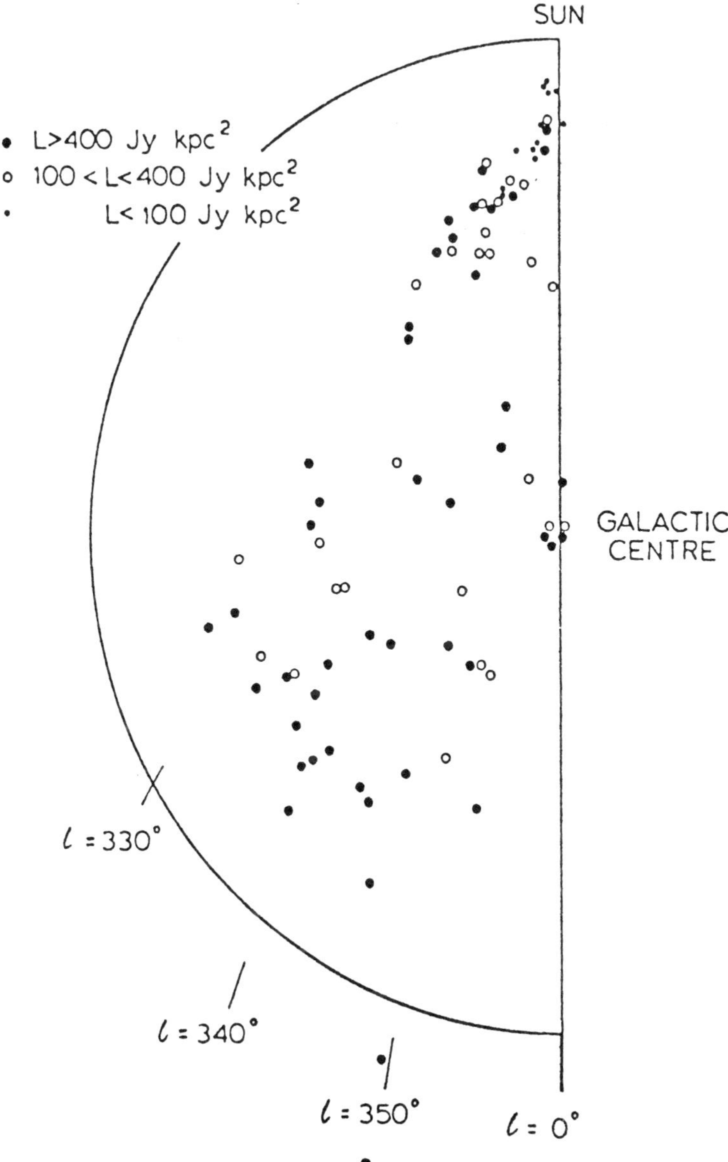

Fig. 2 - Distribution of Type I OH masers showing the 85 masers in the longitude range 326° (through 360°) to 2°. The sample of sources with L > 400 Jy kpc^2 is complete to approximately 20 kpc from the Sun and the sample with L > 100 Jy kpc^2 is complete to approximately 10 kpc from the Sun.

the combination of the longitude/velocity diagrams for the OH masers with that for the recombination line emission from HII regions (our recent studies greatly expand the number of HII regions available) will cast considerable light on the spiral structure problem. It will then be possible to assess whether this structure is similar to that derived from CO or HI studies.

REFERENCES

Batchelor, R.A., Caswell, J.L., Goss, W.M., Haynes, R.F., Knowles, S.H., and Wellington, K.J.: 1980, *Aust. J. Phys.* 33, p. 139.
Braz, M.A., and Scalise, E.: 1982, *Astron. Astrophys.* (in press).
Caswell, J.L., and Haynes, R.F.: 1975, *Mon. Not. R. Astron. Soc.* 173, p. 649.
Caswell, J.L., Haynes, R.F., and Goss, W.M.: 1980, *Aust. J. Phys.* 33, p. 639.
Caswell, J.L., Haynes, R.F., Goss, W.M., and Mebold, U.: 1981, *Aust. J. Phys.* 81, p. 333.
Caswell, J.L., and Haynes, R.F.: 1982a, *Astrophys. J. Lett.* 254, p. L31.
Caswell, J.L., and Haynes, R.F.: 1982b, *Aust. J. Phys.* (in press).
Haynes, R.F., and Caswell, J.L.: 1977, *Mon. Not. R. Astron. Soc.* 178, p. 219.
Liszt, H.S., and Burton, W.B.: 1980, *Astrophys. J.* 236, p. 779.
Scalise, E., and Braz, M.A.: 1980, *Astron. J.* 85, p. 149.
Schmidt, M.: 1965, in *Galactic Structure* (A. Blauuw and M. Schmidt, eds.) p. 513 (Univ. Chicago Press).

SOUTHERN LINE SURVEYS WITH THE PARKES 64-M AND EPPING 4-M RADIO TELESCOPES

J.B. Whiteoak
Division of Radiophysics, CSIRO, Sydney, Australia

SUMMARY

Line surveys of the Southern Milky Way have played a major role in the research carried out by the CSIRO Division of Radiophysics, providing a firm foundacion for future southern radio astronomy. The various surveys are listed. The surveys (particularly CO) towards HII regions are discussed and compared.

1. INTRODUCTION

Line surveys of the Southern Milky Way have played a major role in the research carried out by the CSIRO Division of Radiophysics, Australia, using its 64-m radio telescope at Parkes and 4-m radio telescope at Epping. It seems that a substantial fraction of the radio astronomy community is unaware of many of these surveys, presumably because most radio astronomers are located in the northern hemisphere and use northern telescopes with access only to the northern part of the galactic plane. It is probably not surprising that the following statement concerning southern hemisphere line reserach has recently appeared in a refereed journal: '.....there are extensive maps in radio continuum, extensive searches for water and hydroxyl masers, and preliminary CO and infrared surveys. Othern than this, there has been no information about the molecular clouds associated with the HII regions other than the formaldehyde results from Whiteoak and Gardner (1974).' In this talk I will endeavour to show that southern spectral line research, although lagging behind that in the north, is not as neglected as the remarks quoted would suggest.

2. SOUTHERN LINE SURVEYS

The line surveys of the Southern Milky Way that have been carried out at Parkes and Epping take several forms: observations along the galactic equator with a fixed longitude interval, observations concentrated towards the galactic centre, and observations of discrete objects

W. B. Burton and F. P. Israel (eds.), Surveys of the Southern Galaxy, 31–42.

such as HII regions, dark clouds etc. Table 1 lists the latest published surveys. It does not contain all the surveys that have been made - for some lines there have been earlier surveys with less sensitive equipment; for others there exist unpublished surveys (for CH and H_2O for instance). Hydrogen recombination line surveys have been omitted; the most extensive of these was carried out at 5 GHz by Wilson et al. (1970). Observations of other molecular-line transitions have been made towards particular regions of interest. The table indicates that for frequencies up to about 100 GHz most of the lines listed in Volume XVIIB of Transactions of the International Astronomical Union as of astrophysical importance have been involved in some form of survey.

3. LINE SURVEYS OF HII REGIONS

Since longitude and maser line surveys will be discussed elsewhere in this workshop, it is of interest to focus on line observations of southern HII regions. Most of the observations have been based on the HII region surveus at 5 GHz and 408 MHz by Goss and Shaver (1970) and Shaver and Goss (1970) respectively. A later more extensive survey has been made at 5 GHz (Haynes et al. 1978; Haynes et al. 1979) and will no doubt be used for future line observations.

Table II lists HII regions at longitudes between 260° and 358° towards which line absorption or emission has been observed. Generally the source designation is from Goss and Shaver (1970) and the positions and 5 GHz continuum temperatures are from Whiteoak and Gardner (1974). In some cases the positions and temperatures are from Haynes et al. (1979); the latter have been adjusted to the scale used by Whiteoak and Gardner. Where noted, H_2O and NH_3 observations made with other radio telescopes have been added. The HI observations refer only to line absorption; for OH(1.6) the results denote absorption unless emission(E) is noted. The table contains 151 entries. It shows detected transitions only; the individual surveys usually contained many more objects. In most cases observations were made only towards the continuum maxima. A summary of the table is shown in Figure 1. Although the number of detections for the excited states of OH is low, the results have been included in order to demonstrate the lack of information about these transitions.

Details of the CO survey of HII regions being carried out with the Epping radio telescope are probably of particular interest to this workshop. As a first stage of this survey, ^{12}CO observations of the 1-0 transition were made towards a selection (79 positions) of the brightest southern HII regions (Whiteoak et al. 1982). Future mapping will be based on these results. Generally, each line of sight crossed several CO clouds; the brightest features were found to originate in clouds near, and probably associated with, HII regions.

Preliminary comparisons can be made with the results for other molecules observed towards the HII regions. The detected CS, HCN and

HCO^+ emission genrally occurs in the molecular clouds containing embedded HII regions; this is probably a consequence of the higher densities needed to excite these molecules. There are 33 objects with observations of at least three of the molecules H_2CO, CS, HCN, HCO^+ and CO. A comparison between CO and these molecules leads to the following conclusions.

(a) The CO velocities are on average 0.5 km s^{-1} lower than the average velocities for each of the other molecules. The closeness of the velocities suggests that all transitions take place in the same regions of the molecular clouds. Further observations would be required to ascertain whether the small velocity disparity is a real or instrumental effect.

(b) In line with results from other investigations, the CO profile widths are considerably larger than for other molecules. This is presumably caused by high CO optical depths which give rise to saturation effects in the profiles.

(c) In a comparison of profile areas, CO and CS are well correlated. This has been pointed out by Gardner and Whiteoak (1978); the mean ratio of the profile areas (CO:CS) is about 10, implying an abundance ratio exceeding 10^3. There is a weak correlation between CO and HCO^+ which was previously determined by Batchelor _et al._ (1981). There is little evidence of a correlation between CO and either HCN or H_2CO.

The absorption-line spectra of H_2CO (Whiteoak and Gardner 1974) and HI (Radhakrishnan _et al._ 1972; Goss _et al._ 1972; Caswell _et al._ 1975) both show features associated with molecular clouds intercepted by lines of sight towards HII regions. In general these features have well-defined CO counterparts with similar velocities (see Whiteoak _et al._ (1982) for the H_2CO comparison). Thus it can be presumed that any extra CO features must arise in molecular clouds beyond the HII regions. However, little CO was found to be associated with HI and H_2CO features near zero velocity. Further research should show whether this reflects differences in relative abundances, or whether it is merely due to the limited sensitivity of the equipment used in the Epping CO observations.

4. FUTURE OBSERVATIONS

Our information about southern molecular-cloud/HII-region complexes is still lacking in several respects, and I expect that future research by the Division of Radiophysics will include the following:

(a) Surveys of higher transitions of a number of molecules and other isotope species (particularly of CO) will provide temperatures and optical depths. Cloud dimensions, kinematic behaviour, etc. will be investigated mainly by CO mapping with the Epping 4-m telescope.

(b) Surveys involving ions (such as N_2H^+) will add to our knowledge of ion-molecule formation chemistry.

(c) For the investigation of the galactic rotation law in the fourth quadrant, it is necessary to have HII distances determined independently of the kinematic data. Optically determined distances for southern HII regions are generally lacking, partly because the early-

Table I

Line Surveys with the Parkes 64-m or Epping 4-m Telescope

Atom/ Molecule	Freq. (GHz)	Beam-width (' arc)	Type of Survey	References*
HI	1.4	14.5	Continuum sources	1
		14.5†	Continuum sources	2, 3, 4
		15	Longitude grid	5, 6
OH	1.6‡	12	Continuum/OH sources	7, 8, 9, 10
		12	Longitude grid	11, 12, 13, 14
		12	Galactic centre	15, 16
	6.0‡	3.7	Continuum sources	17, 18
	13.4‡	2.2	Continuum/OH sources	19
H_2CS	3.1	6.6	Continuum sources	20
CH	3.3‡	6.4	Continuum sources, dark clouds	21
H_2CO	4.8	4.2	Continuum sources	22
		4.4	Dark clouds	23
		4.4	Galactic centre	24
H_2O	22.2	1.7	Continuum/OH sources	25
NH_3	23.7‡	2.9	Continuum sources	26
SiO	43.1‡	1.6	H_2O/OH masers, variable stars	27
CS	49.0	1.5	Continuum sources	28
HCN	88.6	3.1	Continuum sources, dark clouds	29
HCO^+	89.2	3.2	Continuum sources	30
CO	115.3	2.8	Continuum sources	31
		2.8	Longitude grid	32

**References are as follows.*

1: Kerr and Knapp (1970); 2: Radhakrishnan et al. (1972); 3: Goss et al. (1972); 4: Caswell et al. (1975); 5: Kerr et al. (1976); 6: Kerr et al. (1981); 7: Manchester et al. (1970); 8: Goss et al. (1970); 9: Robinson et. al. (1970); 10: Caswell and Robinson (1974); 11: Caswell and Haynes (1975); 12: Haynes and Caswell (1977); 13: Caswell et al. (1980); 14: Caswell et al. (1981); 15: Robinson and McGee (1970); 16: McGee (1970); 17: Gardner and Whiteoak (1975); 18: Knowles et al. (1976); 19: Balister et al. (1976); 20: Gardner et al. (1980); 21: Whiteoak et al. (1978); 22: Whiteoak and Gardner (1974); 23: Goss et al. (1980); 24: Whiteoak and Gardner (1979); 25: Batchelor et al. (1980); 26: Batchelor et al. (1977); 27: Balister et al. (1977); 28: Gardner and Whiteoak (1978); 29: Whiteoak and Gardner (1978); 30: Batchelor et al. (1981); 31: Whiteoak et al. (1982); 32: Robinson et al. (1982).

†64-m + 18-m interferometer.

‡Several transitions.

Table II

Southern Galactic Continuum Regions (260° < ℓ < 358°) with Detected Line Radiation

Region designation[a] G	Other	Position (1950)[a] R.A. h m s	Dec ° '	Cont. temp.[a] (K)	Detected line radiation
260.8-3.2	Pup A	08 22 27	-42 50.4	0.9	H_2CO
264.3+1.5[e]	RCW 34	08 54 40	-42 54.0	1.1	H_2O[b]
265.1+1.5	RCW 36	08 57 38	-43 33.4	7.3	HI, OH(1.6), CH, H_2CO, H_2O,[b] CS, HCN, HCO^+, CO
267.8-0.9	RCW 38	08 57 42	-47 07.0	2.1	H_2CO
267.9-1.1	RCW 38	08 57 25	-47 19.3	58	HI, OH(1.6, 6), CH, H_2CO, H_2O, CS, HCN, HCO^+, CO
268.0-1.0	RCW 38	08 58 07	-47 20.1	3.0	H_2CO, CO
268.4-0.8		09 00 15	-47 32.6	2.4	H_2CO, NH_3,[c] HCO^+, CO
269.1-1.1	RCW 39	09 01 48	-48 14.6	2.6	OH(1.6), H_2CO, H_2O[b]
270.3+0.8	RCW 41	09 15 05	-47 44.3	0.7	H_2O[b]
274.0-1.1	RCW 42	09 22 47	-51 47.0	9.0	HI, OH(1.6), H_2CO, CS, HCO^+
281.0-1.5		09 57 26	-56 38.0	1.5	H_2CO
282.0-1.2	RCW 46	10 04 55	-56 57.5	8.1	HI, OH(1.6), H_2CO, CO
284.3-0.3	RCW 49	10 22 22	-57 31.8	25.2	HI, OH(1.6), H_2CO, H_2O, CO[d]
285.3-0.0		10 29 38	-57 46.5	5.5	HI, OH(1.6E), H_2CO, H_2O, CO
287.4-0.6	RCW 53	10 41 38	-59 19.3	13.6	HI, OH(1.6), H_2CO, H_2O,[b] CO[d]
287.6-0.6	RCW 53	10 42 50	-59 23.5	12.8	OH(1.6), CO
289.1-0.4		10 54 30	-59 50.4	1.9	H_2CO
290.1-0.8	RCW 54	11 00 52	-60 38.1	1.2	HI
291.0-0.1	RCW 54	11 09 50	-60 22.3	1.6	H_2CO
291.3-0.7	RCW 57	11 09 47	-61 02.6	35.0	HI, OH(1.6,6), H_2CS, CH, H_2CO, H_2O, NH_3, CS, HCN, HCO^+, CO
291.6-0.5	RCW 57	11 12 53	-60 59.2	26.6	HI, OH(1.6), H_2CO, H_2O[b]
291.6-0.4	RCW 57	11 12 53	-60 55.2	15.2	OH(1.6E), H_2CO, H_2O, CO
292.0+1.8	MSH 11-54	11 22 21	-58 59.4	2.4	HI
298.2-0.3		12 07 22	-62 33.1	9.9	HI, H_2CO, H_2O,[b] CO
298.8-0.3		12 12 40	-62 38.4	2.5	H_2CO
298.9-0.4		12 12 46	-62 44.6	8.1	HI, OH(1.6), H_2CO, CO
301.0+1.2	RCW 65	12 32 03	-61 22.9	1.1	OH(1.6E,6), H_2CO, H_2O
301.1-0.2[e]		12 32 31	-62 44.6	0.4	H_2O[b]
301.1+1.1	RCW 66	12 33 13	-61 34.8	1.3	H_2CO
304.6+0.1	SNR	13 02 41	-62 26.5	1.1	HI, H_2CO
305.1+0.2	RCW 74	13 07 07	-62 22.5	2.2	H_2CO
305.2+0.0	RCW 74	13 08 04	-62 29.3	5.7	OH(1.6), H_2CO, CS, HCN, HCO^+, CO
305.3+0.2		13 08 23	-62 17.6	7.0	HI, OH(1.6E), H_2CO, H_2O, CO
305.4+0.2		13 09 21	-62 18.9	11.0	HI, OH(1.6AE), CH, H_2CO, H_2O, NH_3,[c] CS, HCN, HCO^+, CO
305.5+0.4		13 10 46	-62 08.5	1.5	H_2CO

Region designation[a] G	Other	Position (1950)[a] R.A. h m s	Dec ° '	Cont. temp.[a] (K)	Detected line radiation
305.6+0.0		13 11 07	-62 28.9	3.4	H_2CO, CO
307.1+1.2		13 23 02	-61 07.4	1.3	HI
308.8+0.0		13 38 13	-62 00.5	0.9	H_2CO
308.9+0.1		13 39 32	-61 53.7	0.3	OH(1.6E), H_2CO, H_2O[b]
309.8+1.8	13S6A	13 43 44	-60 07.6	3.4	HI
309.9+0.5		13 47 16	-61 26.5	0.7	OH(1.6E,6,13), H_2CO
311.5+0.4		14 00 02	-61 02.4	1.2	H_2CO
311.6+0.3		14 01 17	-61 05.6	1.1	OH(1.6), H_2CO, NH_3,[c] CS, CO
311.9+0.1		14 03 55	-61 13.1	2.1	HI, OH(1.6), H_2CO, H_2O[b]
311.9+0.2		14 03 49	-61 05.1	2.0	HI, H_2CO, CO
316.3-0.0	SNR	14 38 00	-59 49.0	0.5	HI, H_2CO
316.8-0.1		14 41 31	-59 36.9	9.6	HI, OH(1.6AE), H_2CS, CH, H_2CO, H_2O, NH_3,[c] CS, HCN, HCO^+, CO
319.2-0.4		14 59 16	-58 51.7	1.5	H_2CO
319.4-0.0		14 59 23	-58 24.6	1.9	H_2CO
320.2+0.8	RCW 87	15 01 36	-57 19.4	2.8	H_2CO, CO[d]
320.3-0.3		15 06 18	-58 14.8	2.1	OH(1.6E), H_2CO
320.3-0.2		15 06 14	-58 06.4	2.0	H_2CO
321.0-0.5	RCW 91	15 12 08	-58 00.6	2.3	H_2CO
322.2+0.6	RCW 92	15 14 50	-56 28.0	4.3	HI, OH(1.6), H_2CS, CH, H_2CO, H_2O,[b] NH_3,[c] CS, HCN, HCO^+, CO
324.2+0.1		15 29 03	-55 46.4	1.3	OH(1.6E), H_2CO, H_2O, CO
326.2-1.7	MSH 15-56	15 48 26	-56 03.0	3.1	HI
326.5+0.9		15 38 33	-53 49.0	2.5	H_2CO, NH_3,[c] HCO^+, CO
326.7+0.6		15 40 58	-53 57.3	8.7	HI, OH(1.6), CH, H_2CO, H_2O, NH_3,[c] HCN, HCO^+, CO
326.9-0.0		15 45 08	-54 14.5	0.9	OH(1.6), H_2CO
327.3+0.4		15 45 24	-53 40.9	0.8	OH(1.6E), H_2O
327.3-0.5		15 49 13	-54 26.5	16.0	HI, OH(1.6AE,6), H_2CS, CH, H_2CO H_2O, NH_3,[c] CS, HCN, HCO^+, CO
327.8-0.3		15 50 48	-54 00.1	1.2	H_2CO
328.0-0.1		15 50 55	-53 39.3	1.4	OH(1.6), H_2CO
328.2-0.5		15 54 12	-53 52.7	0.6	OH(1.6AE), H_2CO, NH_3,[c] CS, CO
328.3+0.4		15 50 17	-53 02.6	3.1	OH(1.6E), H_2CO, H_2O
328.4+0.2	SNR	15 51 47	-53 08.5	3.3	HI, OH(1.6), H_2CO, CO
328.6-0.5	RCW 99	15 55 48	-53 37.1	3.6	H_2CO, CO
330.7-0.4		16 05 41	-52 07.5	1.1	H_2CO
330.9-0.4		16 06 29	-51 58.6	3.1	HI, OH(1.6AE), H_2CO, H_2O, NH_3,[c] CS, HCN, HCO^+, CO
330.9-0.2		16 06 01	-51 47.4	1.0	OH(1.6AE,6), H_2CO, H_2O, CS, CO

Region designation[a] G	Other	Position (1950)[a] R.A. h m s	Dec. ° '	Cont. temp.[a] (K)	Detected line radiation
331.0-0.1		16 06 22	-51 42.5	1.6	OH(1.6AE), H_2CO, H_2O
331.1-0.5		16 08 24	-51 55.3	1.6	H_2CO
331.3-0.3		16 08 34	-51 39.2	2.3	HI, OH(1.6AE), H_2CO
331.3-0.2		16 07 35	-51 34.7	1.7	OH(1.6E), H_2CO, H_2O,[b] CS, CO
331.4-0.0		16 07 16	-51 23.1	1.8	H_2CO
331.5-0.1		16 08 22	-51 19.5	9.7	HI, OH(1.6AE,6), H_2CO, H_2O, NH_3,[c] CS, HCN, HCO^+, CO
332.2-0.4		16 12 52	-51 09.9	4.5	HI, OH(1.6), H_2CO, CS, CO
332.4-0.4	RCW 103	16 13 45	-50 56.9	1.7	HI, OH(1.6), H_2CO
332.5-0.1		16 13 18	-50 40.3	0.8	H_2CO
332.7-0.6	RCW 106	16 15 59	-50 56.9	4.7	OH(1.6AE), H_2CO, H_2O, CO
332.8-0.6	RCW 106	16 16 25	-50 47.5	3.3	OH(1.6E), H_2CO, H_2O,[b] CS, HCN, CO
333.0-0.4		16 16 52	-50 33.0	6.8	HI, H_2CO, NH_3,[c] CS, HCN, HCO^+, CO
333.0+0.0		16 14 51	-50 12.5	0.9	H_2CO
333.0+0.8		16 11 22	-49 42.2	2.3	H_2CO
333.1-0.4		16 17 15	-50 29.3	8.7	HI, OH(1.6AE,6), H_2CO, H_2O, NH_3,[c] CS, HCN, CO
333.2-0.1		16 15 57	-50 12.3	2.0	OH(1.6AE), H_2CO, H_2O, CO
333.3-0.4		16 17 47	-50 19.2	10.9	OH(1.6), H_2CS, CH, H_2CO, NH_3, CS, HCN, HCO^+, CO
333.3+0.1		16 15 41	-50 02.7	1.0	OH(1.6), H_2CO
333.6-0.2		16 18 26	-49 58.9	27.2	HI, OH(1.6AE,6), CH, H_2CO, H_2O, NH_3,[c] HCN, HCO^+, CO
333.6-0.1		16 17 54	-49 54.1	3.3	H_2CO, CO
335.8-0.2		16 27 22	-48 24.5	1.7	OH(1.6), H_2CO
336.4-0.2		16 30 34	-47 59.3	2.2	OH(1.6), H_2CO
336.4-0.1	RCW 107	16 29 53	-47 57.0	1.8	OH(1.6AE), H_2CO
336.5-1.5	RCW 108	16 36 20	-48 45.6	3.1	HI, OH(1.6), H_2CO, HCN, HCO^+, CO
336.5-0.2		16 30 37	-47 51.1	2.1	H_2CO
336.5+0.0		16 29 30	-47 46.1	1.0	OH(1.6), H_2CO
336.8+0.0		16 30 55	-47 30.6	4.2	OH(1.6AE), CH, H_2CO, CO
336.9-0.1		16 32 08	-47 32.4	2.2	OH(1.6AE), H_2CO, H_2O
337.1-0.2		16 33 02	-47 25.3	5.3	HI, OH(1.6), H_2CO, CO
337.3-0.1		16 33 28	-47 16.8	1.7	OH(1.6), H_2CO, CO
337.4-0.4[e]		16 35 14	-47 19.3	0.7	H_2O,[b] NH_3[c]
337.6-0.0		16 34 32	-46 58.1	1.6	OH(1.6E), H_2CO, H_2O
337.7-0.1		16 34 53	-46 54.7	1.3	OH(1.6AE,6), H_2CO, H_2O
337.8-0.1		16 35 23	-46 52.0	2.0	HI, OH(1.6), H_2CO
337.9-0.5		16 37 28	-47 01.4	5.8	HI, OH(1.6AE), H_2CO, H_2O, NH_3,[c] CO

Region designation[a] G	Other	Position (1950)[a] R.A. h m s	Dec. ° '	Cont. temp.[a] (K)	Detected line radiation
338.0-0.1		16 36 13	-46 45.1	1.8	OH(1.6), H_2CO
338.1-0.2		16 36 57	-46 41.9	1.8	H_2CO
338.1+0.0		16 35 58	-46 36.5	1.9	OH(1.6AE), H_2CO, NH_3,[c] CO
338.4-0.2		16 38 13	-46 29.6	1.6	OH(1.6AE), H_2O
338.4+0.1		16 37 11	-46 18.0	4.7	HI, OH(1.6), H_2CO, CO
338.4+0.2		16 36 30	-46 16.2	4.3	OH(1.6AE), H_2CO, CO
338.9-0.1		16 39 37	-46 00.9	1.1	OH(1.6E), H_2CO
338.9+0.6		16 36 43	-45 34.4	2.5	HI, OH(1.6AE), H_2CO, H_2O
339.6-0.1[e]		16 42 19	-45 33.5	1.0	OH(1.6E), H_2O[b]
340.1-0.2		16 44 31	-45 15.5	0.8	OH(1.6E), H_2CO, H_2O
340.3-0.2		16 45 19	-45 04.2	2.0	OH(1.6), H_2CO
340.8-1.0	RCW 110	16 50 41	-45 12.3	3.6	HI, OH(1.6), H_2CO, NH_3,[c] CO
343.5-0.0		16 55 49	-42 30.3	2.8	HI, H_2CO
344.2-0.6[e]		17 00 40	-42 15.3	1.0	H_2O[b]
345.2+1.0	RCW 116	16 57 10	-40 29.6	3.7	HI, H_2CO, CO
345.3+1.5	RCW 116	16 55 39	-40 09.1	2.9	HI, H_2CO, CO
345.4-0.9	RCW 117	17 06 04	-41 32.1	11.5	HI, OH(1.6AE), H_2CO, H_2O, HCN, CO
345.4+1.4	RCW 116	16 56 12	-40 06.9	5.1	HI, OH(1.6), H_2CO, HCO^+, CO
345.5+0.3		17 00 59	-40 41.9	1.4	OH(1.6E), H_2CO, H_2O, NH_3,[c] CS, CO
345.6+0.0[e]		17 02 46	-40 45.9	2.7	OH(1.6E), H_2O[b]
345.6+0.3		17 01 18	-40 40.6	1.1	H_2CO
347.6+0.2		17 08 11	-39 05.0	2.4	OH(1.6E,6), H_2CO, H_2O[b]
348.2-1.0[e]	RCW 121	17 14 57	-39 16.5	4.8	HI
348.2+0.5	RCW 120	17 08 58	-38 26.0	1.5	H_2CO, H_2O[b]
348.5+0.1		17 11 09	-38 26.6	2.9	HI, OH(1.6), H_2CO, CO
348.7-1.0	RCW 122	17 16 40	-38 54.7	11.6	HI, OH(1.6AE), CH, H_2CO, H_2O, NH_3, CS, HCN, HCO^+, CO
348.7+0.3		17 10 49	-38 07.8	2.3	HI, H_2CO
349.1+0.1		17 13 00	-37 56.7	1.0	OH(1.6E), H_2CO, H_2O
349.2+0.0		17 13 31	-37 58.1	1.3	H_2CO
349.7+0.2		17 14 40	-37 23.2	3.3	HI, H_2CO, CO
350.1+0.1		17 16 06	-37 07.5	2.5	OH(1.6E), H_2CO, CS, HCO^+, CO
351.0+0.7	RCW 127	17 16 28	-36 01.8	5.6	HI, H_2CO, CO
351.1+0.7	RCW 127	17 16 37	-35 55.2	5.2	OH(1.6E), H_2CO, H_2O, CS, HCN, HCO^+, CO
351.2+0.5	RCW 127	17 17 34	-36 00.9	4.0	HI, H_2CO, CO
351.3+0.7	RCW 127	17 17 02	-35 50.1	10.2	HI, H_2CO, CS, HCN, HCO^+, CO

Region designation[a] G	Other	Position (1950)[a] R.A. h m s	Dec. ° '	Cont. temp.[a] (k)	Detected line radiation
351.4+0.7	RCW 127	17 17 18	-35 46.9	15.3	HI, OH(1.6), CH, H_2CO, CS, HCN, HCO^+, CO
351.4+0.7	RCW 127	17 17 35	-35 43.7	4.4	OH(1.6E,6,13) H_2CS, H_2CO, H_2O, NH_3,[c] CS, HCN, HCO^+, CO
351.6-1.3		17 25 56	-36 37.9	8.0	HI, OH(1.6), CH, H_2CO, NH_3, CS HCN, HCO^+, CO
351.6+0.2		17 19 59	-35 51.5	6.4	OH(1.6E), H_2CO, CO
353.1+0.4	RCW 131	17 23 17	-34 33.1	3.3	H_2CO, CO
353.1+0.6	RCW 131	17 22 18	-34 19.9	16.6	HI, OH(1.6), H_2CO, CO
353.2+0.9	RCW 131	17 21 30	-34 08.1	20.9	HI, H_2CO, HCN, CO
353.4-0.4		17 27 12	-34 39.8	3.6	OH(1.6E), H_2CO, H_2O, **CS, HCO^+**, CO
353.5-0.0		17 26 11	-34 21.6	1.3	H_2CO
355.2+0.1	RCW 132	17 30 22	-32 54.1	1.6	H_2CO
357.7-0.1	MSH 17-39	17 37 08	-30 57.2	3.5	HI, H_2CO

Notes to Table: *(a)* Data from Whiteoak and Gardner (1974) unless otherwise indicated. *(b)* Scalise and Braz (1980) or Braz and Scalise (1982). *(c)* Scalise et al. (1981) or Dickinson et al. (1982). *(d)* Gillespie et al. (1977). *(e)* Positions from Haynes et al. (1979).

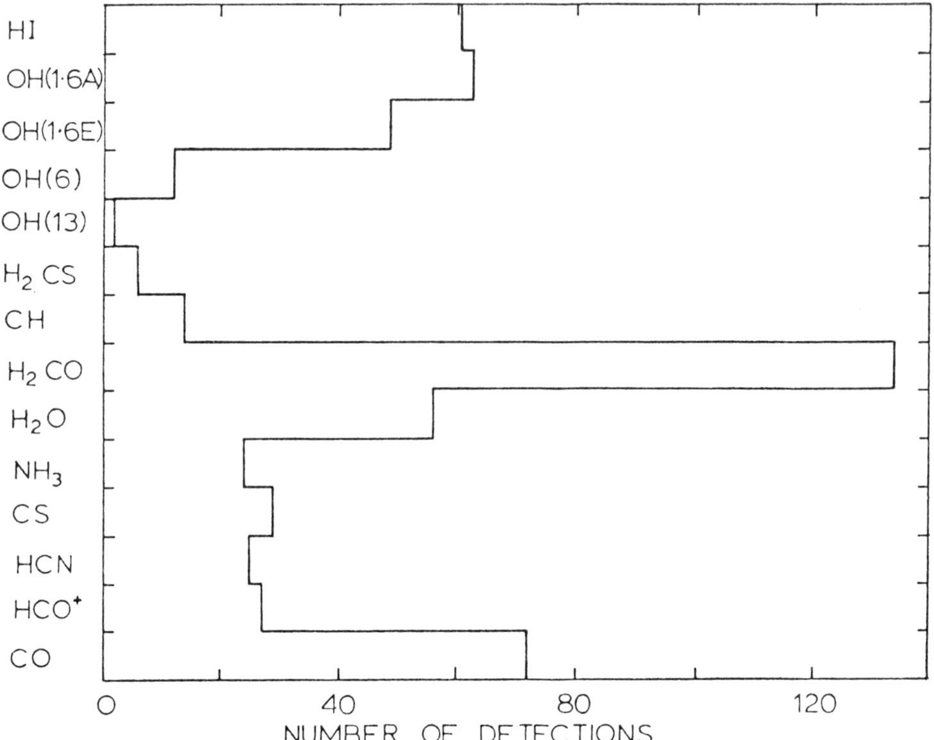

Figure 1. Histogram showing the number of detected lines for HII regions between 160° and 358°.

type exciting stars have not been identified. Proposed objective prism surveys with the 48-inch UK Schmidt telescope should assist in this problem. At the same time, the CO survey of HII regions will be extended to a considerably larger sample of objects.

REFERENCES

Balister, M., Gardner, F.F., and Knowles, S.H.: 1976, Proc. Astron. Soc. Aust. 3. 59.

Balister, M., Batchelor, R.A., Haynes, R.F., KNowles, S.H., McCulloch, M.G., Robinson, B.J., Wellington, K.J., and Yabsley, D.E.: 1977, Mon. Not. R. Astron. Soc. 180, 415.

Batchelor, R.A., Caswell, J.L., Goss, W.N., Haynes, R.F., Knowles, S.H., and Wellington, K.J.: 1980, Aust. J. Phys. 33, 139.

Batchelor, R.A., McCulloch, M.G., and Whiteoak, J.B.: 1981, Mon. Not. R. Astron. Soc. 194, 911.

Batchelor, R.A., Gardner, F.F., Knowles, S.H., and Mebold, U.: 1977, Proc. Astron. Soc. Aust. 3, 152.

Braz, M.A., and Scalise, E., Jr.: 1982, Astron. Astrophys. 107, 272.

Caswell, J.L., and Haynes, R.F.: 1975, Mon. Not. R. Astron. Soc. 173, 649.

Caswell, J.L., and Robinson, B.J.: 1974, Aust. J. Phys. 27, 575.
Caswell, J.L., Haynes, R.F., and Goss, W.M.: 1980, Aust. J. Phys. 33, 639.
Caswell, J.L., Haynes, R.F., Goss, W.M., and Mebold, U.: 1981, Aust. J. Phys. 34, 333.
Caswell, J.L., Murray, J.D., Roger, R.S., Cole, D.J., and Cooke, D.J.: 1975, Astron. Astrophys. 45, 239.
Dickinson, D.F., Gulkis, S., Klein, M.J., Kuiper, T.B.H., Batty, M., Gardner, F.F., Jauncey, D.L., and Whiteoak, J.B.: 1982, Astron. J. (in press).
Gardner, F.F., and Whiteoak, J.B.: 1975, Mon. Not. R. Astron. Soc. 172, 9P.
Gardner, F.F., and Whiteoak, J.B.: 1978, Mon. Not. R. Astron. Soc. 183, 711.
Gardner, F.F., Höglund, B., and Whiteoak, J.B.: 1980, Mon. Not. R. Astron. Soc. 191, 19P.
Gillespie, A.R., Huggins, P.J., Sollner, T.C.L.G., Phillips, T.G., Gardner, F.F., and Knowles, S.H.: 1977, Astron. Astrophys. 60, 221.
Goss, W.M., and Shaver, P.A.: 1970, Aust. J. Phys. Astrophys. Suppl. No. 14, 1.
Goss, W.M., Manchester, R.N., and Robinson, B.J.: 1970, Aust. J. Phys. 23, 559.
Goss, W.M., Radhakrishnan, V., Brooks, J.W., and Murray, J.D.: 1972, Astrophys. J. Suppl. 24, 123.
Goss, W.M., Manchester, R.N., Brooks, J.W., Sinclair, M.W., Manefield, G.A., and Danziger, I.J.: 1980, Mon. Not. R. Astron. Soc. 191, 533.
Haynes, R.F., and Caswell, J.L.: 1977, Mon. Not. R. Astron. Soc. 178, 219.
Haynes, R.F., Caswell, J.L., and Simons, L.W.J.: 1978, Aust. J. Phys. Astrophys. Suppl. No. 45, 1.
Haynes, R.F., Caswell, J.L., and Simons, L.W.J.: 1979, Aust. J. Phys. Astrophys. Suppl. No. 48, 1.
Kerr, F.J., and Knapp, G.R.: 1970, Aust. J. Phys. Astrophys. Suppl. No. 18, 9.
Kerr, F.J., Bowers, P.F., and Henderson, A.P.: 1981, Astron. Astrophys. Suppl. 44, 63.
Kerr, F.J., Harten, R.H., and Ball, D.L.: 1976, Astron. Astrophys. Suppl. 25, 391.
Knowles, S.H., Caswell, J.L., and Goss, W.M.: 1976, Mon. Not. R. Astron. Soc. 175, 537.
Manchester, R.N., Robinson, B.J., and Goss, W.N.: 1970, Aust. J. Phys. 23, 751.
McGee, R.X.: 1970, Aust. J. Phys. 23, 541.
Radhakrishnan, V., Goss, W.M., Murray, J.D., and Brooks, J.W.: 1972, Astrophys. J. Suppl. Ser. 24, 49.
Robinson, B.J., and McGee, R.X.: 1970, Aust. J. Phys. 23, 405.
Robinson, B.J., Goss, W.M., and Manchester, R.N.: 1970, Aust. J. Phys. 23, 363.
Robinson, B.J., McCutcheon, W.H., and Whiteoak, J.B.: 1982, Int. J. Infrared Millimetre Waves 3, 63.
Scalise, E., Jr., and Braz, M.A.: 1980, Astron. Astrophys. 85, 139.

Scalise, E., Jr., Schaal, R.E., Bakor, Y., Vilas Boas, J.W.S., and Myers, P.C.: 1981, Astron. J.: 86, 1939.

Shaver, P.A., and Goss, W.M.: 1970, Aust. J. Phys. Astrophys. Suppl. No. 14, 77.

Whiteoak, J.B., and Gardner, F.F.: 1974, Astron. Astrophys. 37, 389.

Whiteoak, J.B., and Gardner, F.F.: 1978, Mon. Not. R. Astron. Soc. 185, 33P.

Whiteoak, J.B., and Gardner, F.F.: 1979, Mon. Not. R. Astron. Soc. 188, 445.

Whiteoak, J.B., Gardner, F.F., and Sinclair, M.W.: 1978, Mon. Not. R. Astron. Soc. 184, 235.

Whiteoak, J.B., Otrupcek, R.E., and Rennie, C.J.: 1982, Proc. Astron. Soc. Aust. (in press).

Wilson, T.L., Mezger, P.G., Gardner, F.F., and Milne, D.K.: 1970, Astron. Astrophys. 6, 364.

A SOUTHERN SURVEY OF H166α EMISSION FROM THE GALACTIC PLANE

L. Hart, I.N. Azcárate, J.C. Cersosimo and F.R. Colomb
Instituto Argentino de Radioastronomia, Argentina.

ABSTRACT

An extensive survey of distributed low density ionised hydrogen has been completed along the southern part of the galactic plane. The major observations and some preliminary results are presented and discussed. The ionized material along the southern galactic equator exhibits somewhat different longitude and radial distributions than the northern ionized material. Ionized hydrogen in the south is relatively more abundant for R > 6 kpc than in the north. This leads to a total mass of ionized material which is greater in the fourth than in the first quadrant of the Galaxy. The southern radial distribution shows a decline in abundance of ionized material interior to R = 4 kpc which is also characteristic of the northern radial distribution.

I SURVEY

Observations have been made for H166α (1424.734 MHz) radio recombination line emission every 1° in the galactic longitude interval ℓ = 298°-360°-4° at b = 0°. At the frequency of the H166α line the half-power beam-width of the telescope used was 34 arcmin and therefore the surveyed region was not fully sampled. However, experience from previous recombination line surveys of the northern part of the galactic plane at the same frequency and similar angular resolution (Hart and Pedlar, 1976; Lockman, 1976) has shown that a spacing of 1° is sufficient to follow the continuity of large scale features in the ionized gas. Existing continuum and recombination line surveys in the south have concentrated on detecting and observing individual HII regions and complexes of high emission measure and small angular extent. The relatively low observing frequency and large half-power beamwidth used in the present H166α survey make these observations most sensitive to extended regions of low electron, $n_e \sim$ 1-10 cm^{-3}, density.

The sensitivity of the southern H166α survey is very similar to that of the two northern ones. This should allow a detailed comparison

W. B. Burton and F. P. Israel (eds.), Surveys of the Southern Galaxy, 43–47.

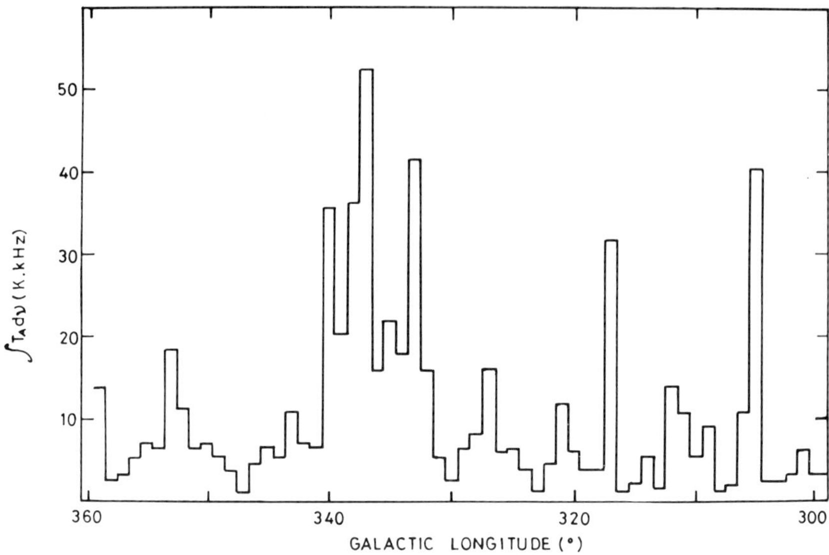

Figure 1. $\int T_A d\nu$ versus longitude for the H166α line.

to be made between them. Measurements of the continuum emission at the recombination frequency were also made at each observed position but these are not presented or discussed further here.

II RESULTS

The power detected in the H166α recombination line ($\int T_A d\nu$, in antenna units) is shown as a function of galactic longitude in Figure 1. The main region of emission occurs between $332^o \leqslant \ell \leqslant 340^o$. Other intense peaks also occur in a few isolated directions. The corresponding northern longitude distribution has a similar main zone of emission in the range $23^o \leqslant \ell \leqslant 31^o$, which is some 3° further away from the Galactic Centre direction than the main southern zone. Some relatively strong sources of emission were found for $\ell < 329^o$ in the southern survey which have no counterpart at $\ell > 31^o$ in the north. However, the northern survey shows an excess of emission over the southern survey in the interval $13^o \leqslant \ell \leqslant 17^o$ by about a factor of two. Apart from this region the two longitude distributions are generally similar for $|\ell| < 20^o$.

The ratio of the total power detected in recombination line emission, integrated over the longitude range $5^o \leqslant |\ell| \leqslant 60^o$, between the southern and northern surveys is 0.9. This value does not necessarily imply that each quadrant contains a similar mass of ionized hydrogen.

The velocity-longitude distribution, shown in Figure 2, is derived from spectra spaced every 1° in longitude in the interval $300^o \leqslant \ell \leqslant 358^o$. The velocity resolution is 15.8 km s^{-1} and contour levels are drawn every 0.04K of antenna temperature starting at 0.02K. Spectra

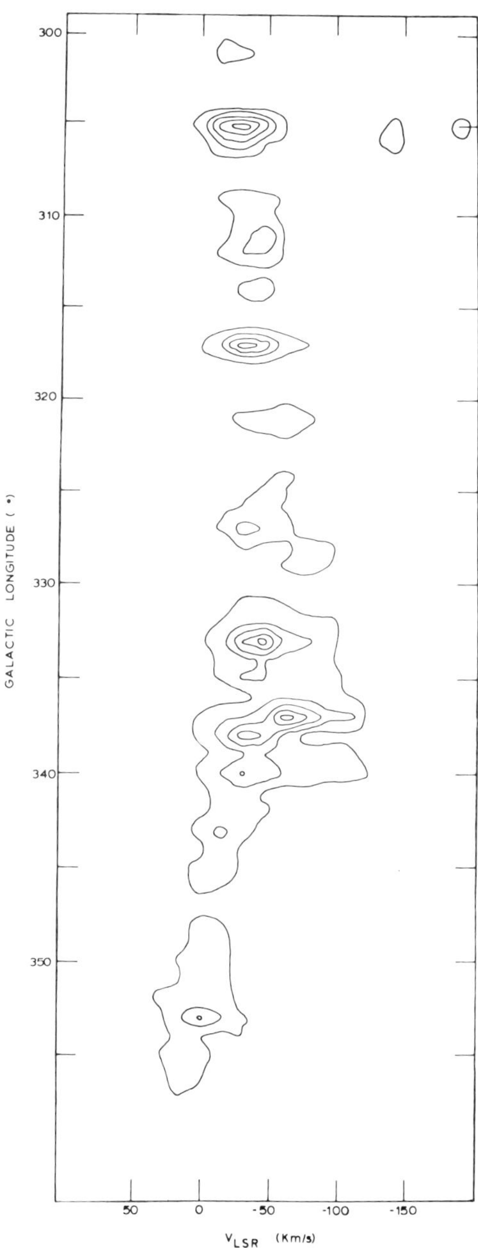

Figure 2. Velocity-longitude distribution at b = 0° for H166α emission.

between $\ell = 333^o$ and $\ell = 340^o$ indicate the presence of more than one velocity component. A similar situation arises in the northern H166α data but at a slightly greater longitude and is due to ionized regions located in two separate arms.

III RADIAL DISTRIBUTION

The radial distribution of the average power in the H166α line per kiloparsec can be derived from the observations by using a rotation curve for the material in the Galaxy. The preliminary radial distribution shown in Figure 3 was derived from spectra in the range $300^o \leqslant \ell \leqslant 350^o$ using Burton's (1971) analytical approximation to the Schmidt (1965) rotation curve. In this model $R_\odot = 10$ kpc. Because of the angular coverage and velocity limits imposed on the data the galactic regions within R = 2 kpc and beyond R = 10 kpc are excluded from the present analysis.

The mean radius of the radial distribution in Figure 3 is 6.5 kpc and the half-peak abundance level is reached at radii of 4 kpc and 9 kpc. For comparison the northern radial distribution has a mean radius of 5.7 kpc and half-peak radii of 4 kpc and 6.5 kpc. The southern distribution thus peaks at a greater radius and is broader and more symmetric than the northern distribution. Put another way, ionized hydrogen is relatively more abundant for R > 6 kpc in the southern half of the Galaxy than in the northern half. Assuming that there is no significant variation in the physical properties of the ionized gas in the first and fourth quadrants and that the emission arises predominantly from extended regions this increased abundance would imply that there is

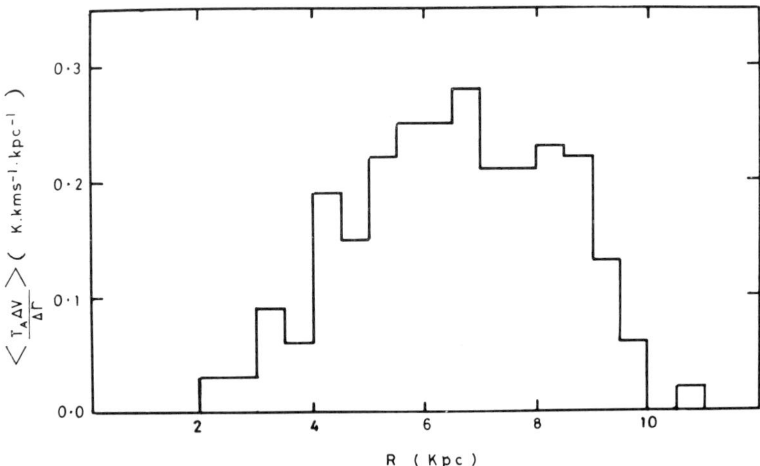

Figure 3. H166α line temperature abundance versus galactocentric radius.

about 60 per cent more ionized mass in the fourth galactic quadrant than in the first.

Both the southern and northern radial distributions show a sharp decline in the abundance of ionized material interior to R = 4 kpc. A similar decline, again at R = 4 kpc, has also been found in the radial abundance of various molecules in both hemispheres. The position of the inner edge of the molecular and ionized gas ring is thus well defined around the whole galaxy. Such a galactic scale phenomenon is presumably intimately connected to the mechanism that is responsible for the generation of spiral arms.

REFERENCES

Burton, W.B.: 1971, Astron.Astrophys. 10, pp.76-96.
Hart, L., and Pedlar, A.: 1976, Mon.Not.R.astr.Soc. 176, pp.547-559.
Lockman, F.J.: 1976, Astrophys.J. 209, pp.429-444.
Schmidt, M.: 1965, Stars and Stellar Systems, 5, pp.513-530.

HI GALACTIC SURVEYS DONE AT THE IAR.

E. Bajaja
Instituto Argentino de Radioastronomía
and Leiden Observatory

1. HISTORY

The Instituto Argentino de Radioastronomía (IAR) was created in Argentine in 1962 by an agreement among the Consejo Nacional de Investigaciones Científicas y Técnicas (CONICET), the Comisión de Investigaciones Científicas de la Provincia de Buenos Aires, and the national universities of Buenos Aires and La Plata. The objective was to provide through this Institute the technical and scientific support to install and operate the radiotelescope, for the observation of the HI 21 cm line, that the Carnegie Institution of Washington (CIW) planned to install in the Southern hemisphere through its Department of Terrestrial Magnetism (DTM).

The site for the observatory was chosen to be at the Parque Pereyna Iraola (longitude = $+58^{\circ}08\overset{'}{.}2$, latitude = $-34^{\circ}52\overset{'}{.}1$) at a distance of 40 km from Buenos Aires and 20 km from La Plata. Construction of the antenna started in November 1963 and was finished in March 1966. The telescope is a 30 m dish with a focal ratio of 0.42 and an equatorial mounting that allows steering from -2 to +2 hours in hour angle and from -10 to -90 degrees in declination. Its surface was adjusted to within 5 mm; its pointing accuracy is about 1 minute of arc.

This antenna is still in operation. A second dish has since been built which is identical to the first one except for the additional facility of being able to move on rails over a distance of 800 m.

The front and IF stages of the receiver were changed as the relevant technology evolved but in 1979 the whole receiver was renewed.

2. PERIOD 1966-1978

The HPBW of the antenna at 21 cm was 30 minutes of arc and its aperture efficiency 52%. The receiver system temperature decreased during this period, because of the improvements made on the receiver,

W. B. Burton and F. P. Israel (eds.), Surveys of the Southern Galaxy, 49–53.

from about 900 K (1966) to 240 K (1978). The backend consisted of 56 10 kHz channel filters spaced 19.84 kHz (4 km s^{-1}) and of 30 100 kHz channel filters spaced 100 kHz. The outputs of the RC integrators were read with a mechanical switch, at a rate of 1.4 seconds of time per channel, digitized and recorded on punched cards.

The observational work at the IAR, up to December 1978, was done with this setup. It consisted essentially in the galactic HI surveys which were the original main objectives for this radiotelescope. Table 1 gives a list of the surveys done between 1967 and 1978 and the publications in which the results are shown and/or discussed. The observations were done, in general, tracking points on a rectangular ℓ,b grid, using the 10 kHz channel filters and two local oscillator frequency settings to make full use of their velocity resolution over the velocity range of 224 km s^{-1}.

An exception is the survey done by Colomb, Pöppel, and Heiles (Ref. No. 11) for which the 56 channels were used with a spacing of 2 km s^{-1}, covering then a velocity range of 112 km s^{-1}, and scanning the sky, with the antenna in transit or slow motion toward the east, at constant declination every 1° between -10° and -90° and for $|b| > 10^{\circ}$. This survey was combined with the one produced from the Northern hemisphere by Heiles and Habing (1974) to produce a whole-sky HI survey outside the galactic plane.

The sky coverage of the surveys listed in Table 1 is given in figure 1.

3. PERIOD 1979-

During 1975 an agreement was reached between IAR-CONICET and NSF-CIW to build a new 21-cm line receiver for the IAR. The receiver was designed and built at DTM with the cooperation of the IAR and installed in 1979. Normal operation started in March 1980. The characteristics of this new receiver were described by Thonnard (1980). The frontend is equipped with a 40 K parametric amplifier and is temperature stabilized at 25° within 0.1 degree. The receiver system temperature is 84 K. A scalar feed at the primary focus diminished substantially the spillover contribution at the cost of a lower resolution (HPBW = 34') but a better aperture efficiency (57%). The larger dynamic range of the first RF amplifier stages, the much smaller spillover contribution, and the use of a narrow filter at the frequency of an airport radar (1360 MHz) have practically eliminated the serious interference present in the old receiver.

The spectrometer consists now of 80 x 75 kHz, 112 x 10 kHz, and 26 x 3 kHz channel filters spaced their widths. A computer controls the observational process and a 112 channel multiplexer makes it possible to read 112 outputs every 6 ms. The output data are stored on floppy disks or on tape. Except for the antenna positioning the observing procedure is done by the computer, including the calibration, following programmed instructions. Three observational modes are possible: load-switch, frequency-switch, and total power.

The main objectives for this new receiver are the observation of HI high velocity clouds (HVC's) and HI in galaxies. The characteristics of the receiver, however, make it suitable for observation also of the continuum and of the recombination line H166α. The OH lines at 18 cm can be also observed with a GASFET amplifier in the frontend.

The HVC survey is being carried out at present as a general survey of the Southern hemisphere below $\delta = -10^{\circ}$. A similar survey is being carried out in the Northern hemisphere from Dwingeloo by Hulsbosch. The observational parameters of both surveys are listed in Table 2, from which it can be seen that they are quite comparable and that they will provide in consequence a rather homogeneous whole sky HVC survey. The IAR survey will require the observation of about 14000 points (grid of $1^{\circ} \times 1^{\circ}$). The first stage has been completed in the form of a $2^{\circ} \times 2^{\circ}$ grid.

Surveying of HI in external galaxies, other than the Magellanic Clouds, has also been undertaken. It is being carried out observing all galaxies catalogued in the Second Reference Catalogue of Bright Galaxies (de Vaucouleurs et al., 1976) having declinations $\delta \leq -40^{\circ}$, radial heliocentric velocities $V_{Hel} \leq 3000$ km s^{-1}, and types $T \geq 3$. An integration time of one hour on and one hour off the objects, using a load-switch mode and the 100 kHz channel filters, fix the upper limit for the detectable HI at about 10^6 $M_{\odot}/kpc^2$/channel.

The Magellanic Clouds are being surveyed in the HI line and in the 21 cm continuum. The 10 kHz channel filters are used for the HI. The observations are done scanning the Clouds, with the antenna fixed in transit, every $0^{\circ}.25$ in declination. The new HI maps obtained with these observations will have a velocity resolution of 2 km s^{-1} and an average r.m.s. noise of 0.06 K (antenna temperature).

In the continuum the receiver has a bandwidth of 40 MHz. The observation of the Clouds is done scanning them as in the HI line but with an angular speed in RA of 10°/min. The expected average r.m.s. noise in the final maps is 50 mJy.

The H166α recombination line is being surveyed at present in galactic HII regions (Hart et al., this Workshop).

For the immediate future we foresee making a complete survey of the Southern hemisphere in the continuum at λ21 cm using the second dish and a new receiver which is being built at present. This work will be done in cooperation with the Max Planck Institut für Radioastronomie of Bonn. A similar survey has been recently published by Reich (1982) for the Northern hemisphere so the Southern one will provide a whole sky continuum survey as 21 cm.

Also in the near future regular observations of the OH lines will be started. On a longer term basis it is planned to use the two dishes as an interferometer of variable spacing, with a correlation receiver.

Table 1: Main galactic HI surveys.

Ref.No.	Publication	Grid (degrees)					
		ℓ_1	ℓ_2	$\Delta\ell$	b_1	b_2	Δb
1.	Goniadzki, D., Jech, A.: 1970, IAU Symp. 38, 157.	230	280	5	-15	-3	1
2.	Garzoli, S.: 1970, Astron. Astrophys. 8, 7.	270	310	1	-3	2	1
3.	Vieira, E.R.: 1971, Astrophys. J. Suppl. 22, 369.	302	310	0.5	2	12	0.5
4.	Garzoli, S.: 1972, CIW Publication No. 629.	270	310	1	-7	2	1
5.	Bajaja, E., Colomb, R.: 1973, CIW Publication No. 632.	220	294	2	-29	-11	2
6.	Pöppel, W.G.L., Vieira, E.R.: 1974, CIW Publication No. 633.	0	12	1	3	17	1
7.	Quiroga, R.: 1974, Astrophys. Space Sci. 27, 232.	281	345	1	-5	5	0.5
8.	Colomb, F.R., Morras, R., Gil, M.: 1976, Astron. Astrophys. Suppl. Ser. 26, 195.	290	314	1	-32	-17	1
9.	Mirabel, I.F.: 1977, Astron. Astrophys. Suppl. 28, 327.	348	360	1	-22	-1	1
10.	Pöppel, W.G.L., Franco, M.: 1978, Astrophys. Space Sci. 53, 91.	348	12	1	3	17	1
11.	Colomb, F.R., Pöppel, W., Heiles, C.: 1980, Astron. Astrophys. Suppl. Ser. 40, 47.	$\delta < -25$, $\Delta\delta = 1$ $\|b\| \geq 10$					
12.	Bajaja, E., Colomb, F.R., Morras, R.: 1980, Astron. Astrophys. Suppl. Ser. 41, 67.	310	325	1	-32	-17	1
13.	Bajaja, E., Morras, R.: 1980, Astron. Astrophys. Suppl. Ser. 41, 121.	220	240	2	-10	2	1
		240	269	1	-10	2	1
14.	Olano, C.A., Pöppel, W., Vieira, E.: 1981, Astron. Astrophys. Suppl. Ser. 46, 41.	320	345	1	18	26	1
		346	350	1	18	20	1

Table 2: Observational parameters for the IAR and Dwingeloo HVC surveys.

Parameter	IAR	Dwingeloo	Units
Velocity resolution	16	16.5	km s^{-1}
Velocity range	-640 to 640	-1050 to 1050	km s^{-1}
System temper.	84	40	K
Integration time	20	15	min.
R.m.s. noise	0.017	0.01	K
HPBW	34	36	arc min.
Grid	1 x 1	1 x 1	degrees
Sky coverage	$\delta < 18$	$\delta \geq 18$	degrees

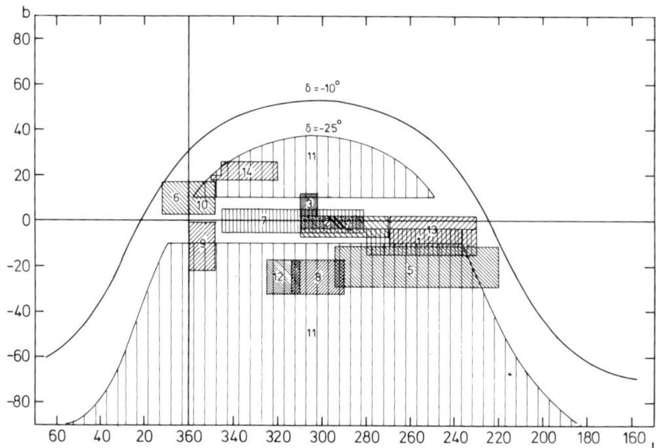

Figure 1: sky areas covered by the IAR HI surveys. The numbers correspond to the references of table 1.

REFERENCES

Heiles, C., Habing, H.: 1974, Astron. Astrophys. Suppl. 14, 557.

Reich, W.: 1982, Astron. Astrophys. Suppl. 48, 219.

Thonnard, N.: 1980, Annual Report of the Director, Department of Terrestrial Magnetism 1979-1980, 581.

Vaucouleurs, G. de, Vaucouleurs, A. de, Corwin, H.G. Jr.: 1976, Second Reference Catalog of Bright Galaxies, Austin: University of Texas Press.

A SURVEY OF HI IN THE SOUTHERN GALACTIC PLANE

Philip A. Riley

Physics Dept., University of Durham, U.K.

ABSTRACT

A 21 cm survey of HI in the Galactic plane, ℓ : 245° - 12°, $|b|<10°$, carried out using the Parkes 64 m radiotelescope, is described. The distribution of atomic hydrogen as a function of Galactocentric radius is derived. A larger mass of gas is found than in most previous studies.

1. INTRODUCTION

A 21 cm survey of neutral hydrogen in the Galactic plane was carried out in 1981, January, using the Parkes 64 m telescope. The primary aim of the survey was the completion of the mapping of the Galaxy in HI so that a comparison can be made between gas column densities and the fluxes of high energy gamma rays measured by the COS B satellite (Mayer-Hasselwander et al. 1982). The area surveyed was ℓ : 245° - 12°, $|b|<10°$, with a sampling interval of 0°.5 in ℓ and 1° in b, this being determined by the time available on the telescope (about 100 hours) and the time required for each observation (about 1 minute).

Apart from the comparison with gamma ray data, the survey is useful for mapping the large scale distribution of HI in the Galaxy. Some results from such a study are described in section 3 of this note.

2. DESCRIPTION OF THE SURVEY

As stated above, the area surveyed was ℓ : 245° - 12°, $|b|<10°$, sampling every 0°.5 in ℓ and every 1° in b. A total bandwidth of 2MHz, divided into 512 channels, was used. For $|b|>2°$ the spectra were centred on 0.0 kms^{-1} giving a velocity coverage -211 kms^{-1} < V_{LSR} < 211 kms^{-1} and a channel width of 0.8 kms^{-1}. For $|b|<2°$, $\ell<326°$ the spectra were centred at +50 kms^{-1} in order to include a

W. B. Burton and F. P. Israel (eds.), Surveys of the Southern Galaxy, 55–58.

significant section of baseline at positive velocities. The integration time at each position was 20 seconds, resulting in an r.m.s. noise of $\Delta T \sim 1K$. The noise diode was calibrated on Hydra A and the standard 21 cm regions S8 and S9 (Williams, 1973). Good consistency between these calibrations was found. S8 was observed regularly in order to monitor any changes in the noise diode, and was used as the calibration standard for the survey as recommended by Weaver and Williams (1973) and Kerr et al. (1976). No significant variations in the noise diode were detected over the 8 days of observation. A fuller description of the survey may be found elsewhere (Strong et al., 1982).

HI column densities calculated from the survey are included in the composite HI map of the Galaxy to be found in this volume.

3. THE DISTRIBUTION OF HI IN THE GALACTIC PLANE

The density of HI in the Galactic plane has been calculated using the standard technique of relating gas velocity to distance from the Galactic centre on the assumption of purely circular velocities. The details of the procedure used are given in Li et al. (1983). The rotation curve used was derived from the data itself and was assumed to be flat beyond the solar circle with $\theta = 250$ kms^{-1}. Some allowance has been made for finite optical depths by the assumption of a uniform spin temperature of 135 K.

Volume densities as a function of galactocentric radius, averaged over the range ℓ : 245° - 345°, are shown in figure 1. Both the volume density at b = 0 and the maximum volume density are shown, the difference between the two giving evidence of the warp in the outer Galactic plane seen in other studies (Baker and Burton 1975; Henderson et al. 1982). The surface density, obtained by integrating the volume density over latitude (see Li et al. 1983), is shown in figure 2. The total mass of HI in the Galaxy implied by this distribution is 5.8 x $10^9 M_\odot$ (about 80% of the mass being at R>10 kpc). A similar study of the region ℓ : 10° - 170° by Li et al. (1983) using the Berkeley HI survey (Weaver and Williams, 1973) found a mass of 5.1 x $10^9 M_\odot$, in good agreement considering that a different region of the Galaxy was studied. The distribution of surface density beyond the solar circle agrees with that found by Henderson et al. (1982). The present results disagree with the northern hemisphere study of Baker and Burton (1975), however. They derive a Galactic HI mass of 2.6 x $10^9 M_\odot$. Most of the discrepancy lies in the very large difference in the surface densities in the outer Galaxy found in the two studies.

4. SURVEY ON MAGNETIC TAPE

A copy of the survey on magnetic tape may be obtained from the

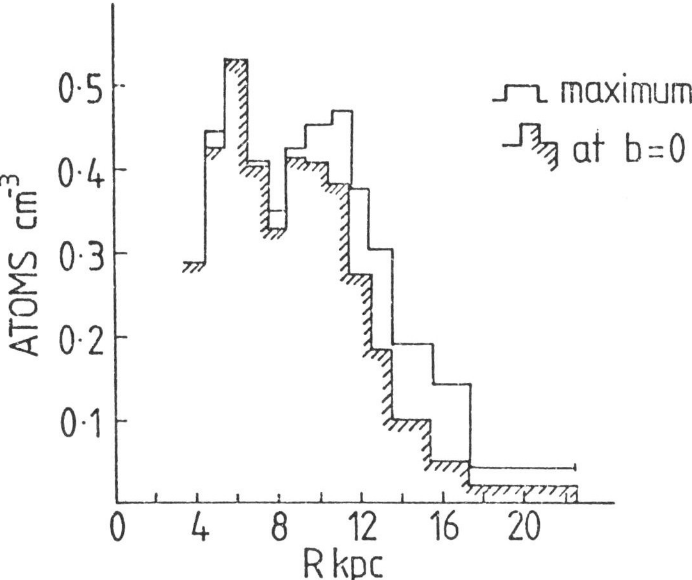

Figure 1. Maximum HI volume density and volume density at b = 0° as a function of Galactocentric radius, averaged over ℓ : 245° - 345°.

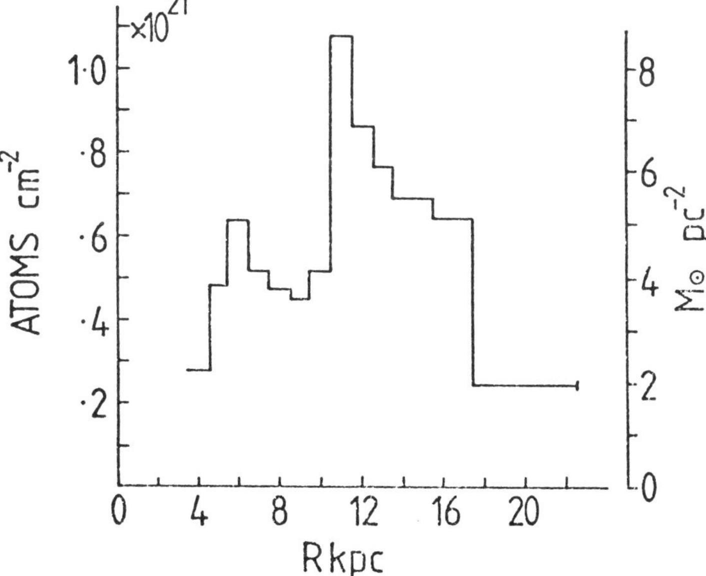

Figure 2. HI surface density as a function of Galactocentric radius, averaged over ℓ : 245° - 345°.

author or from A.W. Strong at the Istituto di Fisica Cosmica, Milan, Italy.

REFERENCES

Baker, P.L., and Burton, W.B., 1975, Astrophys. J., 198, pp.281-297.

Henderson, A.P., Jackson, P.D., and Kerr, F.J., 1982, to be published in Astrophys. J.

Kerr, F.J., Harten, R.H., and Ball, D.L., 1976, Astr. Astrophys. Suppl. Ser. 25, pp. 391-432.

Li, T.P., Riley, P.A., and Wolfendale, A.W., 1983, to be published in M.N.R.A.S.

Mayer-Hasselwander et al., 1982, Astr. Astrophys., 105, pp. 164-175.

Strong, A.W., Riley, P.A., Osborne, J.L. and Murray, J.D., 1982, to be published in M.N.R.A.S.

Weaver, H., and Williams, D.R.W., Astr. Astrophys. Suppl., 8, pp. 1-503.

Williams, D.R.W., 1973, Astr. Astrophys. Suppl., 8, pp. 505-516.

CONTINUUM MAPS AT 843 MHz OF THE SOUTHERN GALAXY AND MAGELLANIC CLOUDS

W.B. McAdam
University of Sydney

The continuum emission from the plane has been observed from the earliest days of radio astronomy: in the decade from Reber's survey to 1958 there were 14 surveys at frequencies from 38 to 1390 MHz and with beamwidths of 25° down to 34'. These surveys blended most of the structure into a narrow ridge along the line of the Milky Way that, in the quadrant towards the centre, coincided with the flat, central layer of H I and helped to define our present system of galactic coordinates (Gum & Pawsey 1960).

Observations along the ridge look through the complete galactic plane and superimpose close, low-brightness regions of large angular extent on distant, luminous sources. Some of the latter are extra-galactic but identification is difficult in the crowded or obscured optical fields. The need has always been for more sensitivity and better resolution.

In the southern hemisphere, completion of the Parkes 64 m dish in 1961 and the Molonglo 1.6 km Cross in 1967 allowed individual sources along the plane to be studied at several frequencies. Parkes surveys were made with beamwidths of 14' to 4' at 1410 MHz (Hill 1968), 2700 MHz (Day, Caswell & Cooke 1972 and related papers) and 5000 MHz (Goss & Shaver 1970). Molonglo 408 MHz surveys were made with the 2.9' beam by Kesteven (1968), Shaver & Goss (1970a) and Green (1974). The matching performances of the two telescopes at 5000 and 408 MHz were used to classify complex structure into nonthermal sources and thermal H II regions on the basis of spectral index (Shaver & Goss 1970b; Caswell *et al.* 1975; Clark & Caswell 1976).

The two telescopes were also used to map sources in the Large & Small Magellanic Clouds - LMC & SMC (McGee *et al.* 1972 and associated papers; Clarke, Little & Mills 1976). Surveys of these neighbouring galaxies are valuable because all sources are at about the same distance and their relative luminosities can be found without the great uncertainty in distance that besets studies in our own galaxy. For this reason, the LMC & SMC should be included as an extension to any program

W. B. Burton and F. P. Israel (eds.), Surveys of the Southern Galaxy, 59–63.

of galactic mapping: there is no evidence that the three galaxies differ in the characteristics of their sources.

By 1978 August, the Molonglo Cross had finished its major projects and observations at 408 MHz stopped. The system was then changed into the Molonglo Observatory Synthesis Telescope (MOST) that observes at 843 MHz in an earth rotation mode using the EW arm alone (Mills 1981). Except for a 43 λ gap in the centre, the complete 1.6 km baseline is filled, with an effective area of 9000 m^2.

There are two mapping modes. (1) Full synthesis: a contiguous set of 64 fan beams covers the central field for 12 h producing a map of size 22 x 22 cosec δ arcmin. (2) Large field: 32 fan beams on each side of the field centre are stepped out over 2 or 3 positions in each 24 s sampling interval to extend the field width to 44 or 66 arcmin. This time sharing reduces the sensitivity, but covers large, bright regions more efficiently and improves the relative precision of contours compared with 4 or 9 adjacent 22 arcmin fields. A complete 12 h synthesis for $\delta < -30^\circ$ produces an elliptic pencil beam 43 x 43 cosec δ arcsec with a first sidelobe -8% and higher lobes of < 1%. A working map is built up in *l,m* aerial coordinates during the observation and is available immediately after it. For $\delta > -30^\circ$ the range of hour angles is limited, but the MOST can observe up to $\delta = +18^\circ$. An observing period may also be shared by many sources giving each a sequence of HA intervals. Maps with less than 12 h cover must be cleaned, as the field has only partial uv coverage.

The sensitivity of the MOST has been established by observations of a high latitude field clear of strong sources (at $l = 320^\circ$; $b = -22^\circ$) on 6 successive nights, producing an average map for a survey of weak sources, and a null map to examine noise and stability. The rms residual level corresponded to a flux density of 0.36 mJy in a full 12 h synthesis. This noise level will be higher in maps with a 44' or 66' time-sharing field, or in maps at more northern declinations where a greater swing in meridian distance is required. However, for a large part of the galactic plane the lowest reliable contour is limited by the dynamic range to better than 1% of the peak flux density. Efforts are directed at improving both the dynamic range and the noise sensitivity. Compared to the 408 MHz maps, the increase to 12 h observing time has improved the sensitivity by almost two orders so that broad regions can be traced to lower brightness levels for all but the steepest spectra.

An example of a MOST map is shown in Fig. 1. This is part of a large field observed in 1982 June during the development of the system when only 32 beams were installed. The strong point source is identified with NGC 5419: it has a flux density of 450 mJy and shows the surrounding negative lobe clearly. The extended source to the east is puzzling. It has a partial shell structure and may be a supernova remnant at high z-distance above the plane. It has a steep spectral index and may be traced 10 arcmin further to the SW at 408 MHz (McAdam &

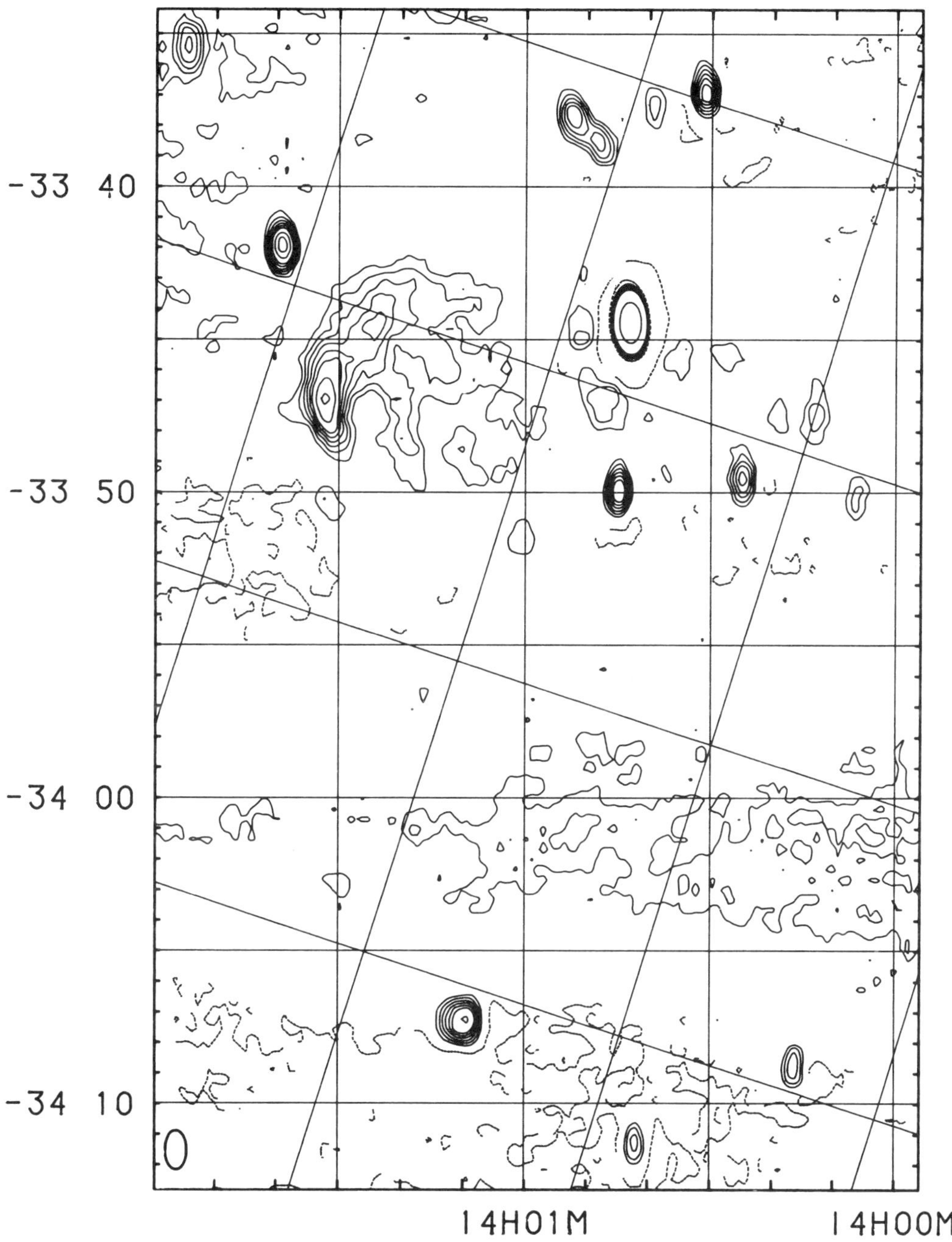

Figure 1. A MOST map of NGC 5419 and the extended source 1401-33. Contour levels are -10 5 10 15 20 25 30 40 50 220 mJy/beam area. The map is part of a 33 x 59 arcmin field with beamwidth (lower left) of 43 x 77 arcsec. The strongest point source coincides with NGC 5419.

Schilizzi 1977). The map shows 7 more point sources and 2 extended sources stronger than 15 mJy. The lowest contour is affected by changes in the zero level that produce broad ridges across the map. These were caused by a radiation coupling between the east and west arms that has now been avoided.

The first synthesis observation was made in 1981 June. For the next year, only 11 or 33 arcmin field widths were possible but exploratory maps were made for many projects. The major program was a search for supernova remnants at positions of likely radio, optical and X-ray candidates in the Magellanic Clouds. Six SNRs were confirmed in the SMC (Mills *et al.* 1982) and preliminary results for a further 29 SNRs in the LMC were reported by Mills 1982 (in "Supernovae Remnants and their X-ray Emission" IAU Symp. No. 101).

Other programs with the MOST include:

Maps:
- a systematic survey of the SMC (now almost completed).
- SNR candidates along the galactic plane.
- compact H II regions.
- planetary nebulae that have optical depth ~ 1 at 843 MHz.
- galactic fields selected from Shaver & Goss (1970a).
- deep fields at pulsar positions.
- Extended extragalactic sources with structure ~ 1 Mpc
- luminosity and structure of bright optical galaxies.

Fan beam observations:
- individual pulse shapes from strong pulsars
- positions and identification of sources
- flux density monitoring of low-frequency variable sources

Many of these programs are long term and will continue with others, not yet started, when the 66 arcmin field is proven. It is hoped to map a 1° width along the galactic ridge from $l = 230$ to 15° and the full galactic centre region.

Improvements to the MOST are likely to continue for some time. The performance limits are being examined and calibration improved. Better preamplifiers have been designed and the required 176 will be installed along the telescope soon.

The MOST offers exceptional sensitivity and resolution at a low frequency. Some spectral information for stronger sources can be obtained with the 50 arcsec beamwidth of the Fleurs synthesis telescope at 1415 MHz but complementary maps at higher frequencies must await the construction of the Australia Telescope.

REFERENCES

Caswell, J.L., Clark, D.H., Crawford, D.F. & Green, A.J.: 1975, Austral. J. Phys. Astrophys. Suppl. No.37.

Clark, D.H. & Caswell, J.L.: 1976, Mon. Not. R. astr. Soc., 174, p.267.

Clarke, J.N., Little, A.G. & Mills, B.Y.: 1976, Austral. J. Phys. Astrophys. Suppl. No.40.

Day, G.A., Caswell, J.L. & Cooke, D.J.: 1972, Austral. J. Phys. Astrophys. Suppl. No.25.

Goss, W.M. & Shaver, P.A.: 1970, Austral. J. Phys. Astrophy. Suppl. No.14, p.1.

Green, A.J.: 1974, Astron. Astrophys. Suppl., 18, p.267.

Gum, C.S. & Pawsey, J.L.: 1960, Mon. Not. R. astr. Soc., 121, p.150.

Hill, E.R.: 1968, Austral. J. Phys., 21, p.735.

Kesteven, M.J.L.: 1968, Austral. J. Phys., 21, p.369.

McAdam, W.B. & Schilizzi, R.T.: 1977, Astron. Astrophys., 55, p.67.

McGee, R.X., Brooks, J.W. & Batchelor, R.A.: 1972, Austral. J. Phys., 25, p.581.

Mills, B.Y.: 1981, Proc. astr. Soc. Aust., 4, p.156.

Mills, B.Y., Little, A.G., Durdin, J.M. & Kesteven, M.J.L.: 1982, Mon. Not. R. astr. Soc., 200, p.1007.

Shaver, P.A. & Goss, W.M.: 1970a, Austral. J. Phys. Astrophys. Suppl. No.14, p.77.

Shaver, P.A. & Goss, W.M.: 1970b ibid, p.133.

THE LARGE-SCALE DISTRIBUTION OF GALACTIC GAMMA-RAY EMISSION

W. Hermsen
Cosmic Ray Working Group, Leiden, The Netherlands

J.B.G.M. Bloemen
Cosmic Ray Working Group and Sterrewacht, Leiden
The Netherlands

ABSTRACT

Gamma-ray skymaps in energy ranges between 70 MeV and 5 GeV, obtained from a complete survey of the Galaxy by the experiment aboard the ESA satellite COS-B, are discussed. Correlations between features of the gamma-ray skymaps and known galactic consituents are evident on various scales: from tens of kpc of the large-scale warp, several kpc of the spiral pattern down to some hundreds of pc of the local molecular cloud complexes. The radial distribution of the gamma-ray emissivity in the Galaxy is compared with the corresponding gas distribution. The variation of the derived emissivity with galacto-centric radius places an upper limit of about a factor 3 on the possible increase in surface density of the interstellar gas between 10 kpc and 5 kpc radius. A possible increase in the cosmic-ray electron density towards the galactic centre is suggested.

1. INTRODUCTION

Gamma radiation in the energy band discussed in this paper (E>70 MeV) is a tracer of the product of the cosmic-ray (CR) density and the interstellar gas density. It originates from the interaction of relativistic CR particles with the interstellar gas (pion decay and bremsstrahlung) and radiation fields (inverse compton). The latter interaction contributes only a small fraction to the observed emission especially at low galactic latitudes (e.g. Kniffen and Fichtel 1981). Therefore, gamma-ray measurements can provide a diagnostic of the CR density in regions where the interstellar gas is well traced at other wavelengths, as well as a diagnostic of the total gas content in cases where the CR density can be assumed to be equal to the local value.

The main advantage of gamma radiation as a tracer is the negligible absorption by the ISM which allows it to reach us from essentially any

W. B. Burton and F. P. Israel (eds.), Surveys of the Southern Galaxy, 65–78.

part of the Galaxy. On the other hand, handicaps are the lack of a distance indicator and the modest angular resolution of e.g. ~ 1° HWHM at energies above 300 MeV (respectively a few degrees at lower energies), together with the low counting statistics of presently available data. Another problem is the quantitatively unknown contribution of compact gamma-ray sources to the total galactic emission (e.g. Bignami et al. 1978, Hermsen 1980, Rothenflug and Caraveo 1980, Riley and Wolfendale 1980, Harding 1981, Harding and Stecker 1981, Salvati and Massaro 1982). Sofar 25 unresolved gamma-ray sources are detected (2nd COS-B catalogue, Swanenburg et al. 1981) of which only two are proven to be compact (the Crab and Vela pulsar). Their latitude distribution is very narrow ($<|b|> \simeq 1.5°$) despite the lower source-detection threshold away from the intense galactic disc emission. Therefore, it is very likely that at least most of the medium latitude emission is diffuse in nature (i.e. originating from the interaction of energetic CR's and local interstellar gas).

The first detailed picture of the galactic disc in the gamma-ray band was obtained from NASA's SAS-2 satellite, which was in orbit between November 1972 and June 1973. Evidence was presented for a correlation with the large-scale structure of the Galaxy (Fichtel et al. 1975), based on a total of ~8000 photons (E>35 MeV). The next gamma-ray satellite in orbit (ESA's COS-B, August 1975 till April 1982) performed the first complete survey (E>50 MeV) of the galactic disc within about 25° from the equator with greatly improved statistical accuracy and significantly better energy measurement. For a description of the instrument and its characteristics see Bignami et al. (1975), Scarsi et al. (1977) or Hermsen (1980).

It is the purpose of this paper to present a detailed picture of the Galaxy in gamma rays using the COS-B data base and to discuss the contribution of the gamma-ray results to our knowledge of the interstellar gas distribution. For references to earlier works on less complete data bases (e.g. SAS-2 data and preliminary COS-B data) see Mayer-Hasselwander et al. (1982).

2. THE GAMMA-RAY SKY

Using data collected during the first four years of COS-B in orbit, Mayer-Hasselwander et al. (1982) published detailed maps of the gamma-ray sky ($|b| \lesssim 20°$) in three energy bands above 70 MeV. These maps, together with a map for the total energy range (70 MeV - 5 GeV), are given in a different presentation on fold-out pages in this volume. Comparison of the skymaps of the photon intensity observed in the three energy ranges (70-150 MeV, 150-300 MeV and 300 MeV-5 GeV) immediately shows the improving angular resolution at higher energies. The three strong gamma-ray sources (Vela at l=263.6, Geminga at l=195.1 and Crab at l=184.6), located in regions of weak galactic emission, provide a good measure of COS-B's instrumental point-spread function in the three energy ranges. It is evident that finer structure is becoming visible with improving angular

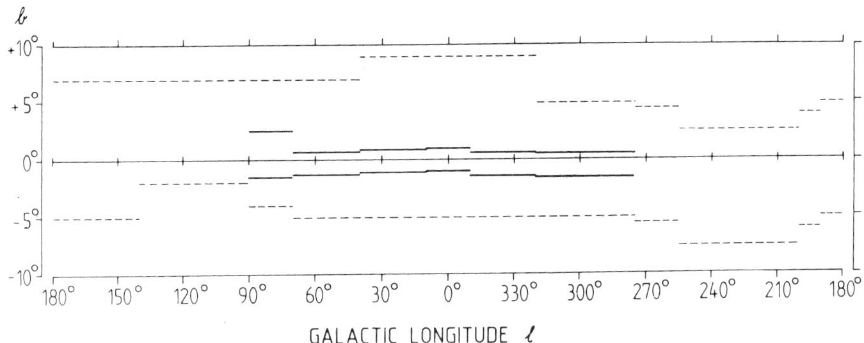

Fig. 1. The intrinsic latitude extent (FWHM) of the narrow component of the galactic gamma radiation derived by an unfolding procedure (solid line) and the latitude extent (FWHM) estimated for the wide component (dashed line).

resolution. The map for the total energy range was derived from "on-axis-rate" maps in the three energy ranges. While photon intensity (ph cm^{-2} s^{-1} sr^{-1}) is the more physical unit, the "on-axis-rate" (on-axis count s^{-1} sr^{-1}; see for definition Mayer-Hasselwander et al. 1982) is more useful for presentation of the spatial properties of the gamma-ray emission observed over the total energy range of the COS-B experiment.

In the gamma-ray maps it is obvious that the major part of the galactic emission is concentrated along the galactic plane. Ignoring for the moment the contribution of unresolved sources, except for the three strong ones mentioned above, the presence of (at least) two components distinctly different in latitude extent has been suggested. This is especially visible in the highest energy range. An analysis of this high-energy data made it possible to seperate the two components and to resolve the intrinsic latitude extent of the disc emission. The result is shown in Figure 1; at all longitudes there appears to be a wide component with a FWHM of $\sim$10° while an intense, much narrower component is superimposed on it towards the inner Galaxy (Mayer-Hasselwander et al. 1982).

An attempt has been made to study the spectral properties of the two components together, following a simple analysis. Namely, the flux ratio F(70-150 MeV)/F(150 MeV-5 GeV) (colour index) of the observed radiation for $|b| < 10°$ has been determined as a function of longitude (Mayer-Hasselwander 1983). Firstly, for all longitudes this colour index is larger than that expected for a pure π° spectrum (signature of the proton-proton collision process). A similar result has been obtained earlier by Paul et al. (1978) and Hartman et al. (1979), interpreted in terms of significant contributions of either bremsstrahlung and/or inverse Compton radiation of CR electrons (Fichtel et al. 1976; Cesarsky et al. 1978; Strong et al. 1978; Lebrun et al. 1982).
Secondly, a systematic variation of the spectral ratio with longitude is indicated; the spectrum at longitudes 50°>l>280° being softer than that of the remainder of the disc (contrary to an earlier, incorrect, result

published in Mayer-Hasselwander et al. 1982). Two possible explanations are: the ratio of relevant CR electrons (>100 MeV) to CR protons (>1 GeV) varies with galacto-centric radius and/or a significant contribution from unresolved compact sources exists in the inner Galaxy with spectra deviating form those resulting from the diffuse processes.

3. THE GALAXY AS SEEN IN GAMMA RAYS

Considering that the gamma-ray sky is devoid of independent distance indicators, one has to resort to correlation with the angular (projected) appearance of known features, located anywhere in the disc. Roughly, one can again distinguish the correlation in distant (>1 kpc) and local ($\lesssim$1 kpc) regions, corresponding to small and higher average latitude values.

Firstly, there is a clear correlation with the 'grand design' of our Galaxy:
- the high intensity towards the inner Galaxy for $|l|<60°$ with an obvious asymmetry between both sides of the galactic centre, the distribution appearing more structured on the southern side,
- the presence of the 'galactic hole' or lack of emission around l=60°, corresponding to a very long line of sight down an interarm region,
- the presence of peaks of emission in Cygnus (l≃75°-80°) and Carina (l≃284°),
- the relative lack of emission in the region l≃210°-260°, a region of the Galaxy particularly poor in gas content,
- the large-scale warp, clearly visible in the gas distribution as the most dominant feature on a galactic scale (see e.g. Kulkarni et al. 1982 and Henderson et al. 1982), is present in the gamma-ray intensity distribution:
a)in the longitude range $100° \lesssim l \lesssim 150°$ the peak of the gamma-ray emission lies at b≃2° (an important fraction of the emission can be attributed to the Perseus arm at an average distance of 2 kpc; the main gas concentration outside the solar circle in this longitude range),
b)in the longitude range 210°<l<260° the relatively weak gamma-ray emission is clearly displaced to negative latitudes,
c)small but significant displacements of the strong ridge of emission are visible in the first quadrant (to possitive latitudes) and especially in the fourth quadrant (to negative latitudes).
Since the warping of the galactic disc is predominantly present outside the solar circle, out to distances of R≃20 kpc (Kulkarni et al. 1982, Henderson et al. 1982) and the main fraction of the gas content of the galaxy is involved in the warping (~80%), significant gamma-ray fluxes could originate at large galacto-centric distances. A first trial to unravel this problem in the second and third quadrants is given by Bloemen et al. (this volume). The work of Schlosser and Feitzinger (1982) suggests that the gamma rays in these quadrants are mainly produced in the local spiral arm; however their analysis and conclusions are in error (see Bloemen et al., this volume).

Secondly, one can consider the correlation between the local matter

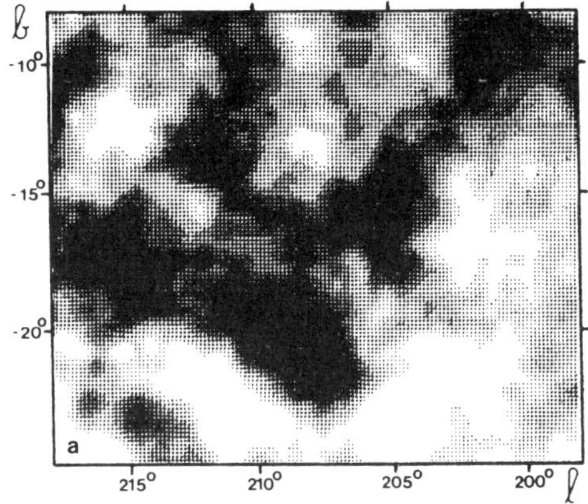

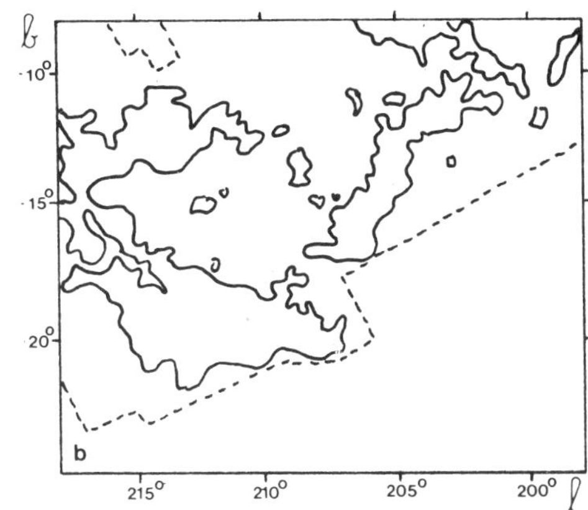

Fig. 2. The Orion region: a. half-tone map showing the gamma-ray intensities (100 MeV<E<5 GeV); b. CO contour plot of 1°K peak antenna temperature (Morris and Thaddeus private communication).

and the medium latitude gamma-ray emission. The known gamma-ray emission from Gould's Belt (Fichtel et al. 1975; Mayer-Hasselwander et al. 1980) is apparent in the data, with in fact now two enhancements resolved; the Orion (Caraveo et al. 1980, 1981) and the ρ Oph cloud complexes. Figures 2 and 3 show the two resolved cloud complexes. The main part of the Orion complex as seen in CO millimeter emission and gamma rays is shown in the first figure. It is evident that a detailed positional correlation is present between figures 2a and 2b, even though the gamma-ray map is seen with appreciably lower resolution. The highest gamma-ray intensities are coincident with the most dense parts of the northern and southern complexes (L1630 and L1641, respectively). The gamma-ray flux of the Orion complex is compatible with that expected for a CR density equal to that observed locally interacting with the gas in the clouds (Caraveo et al. 1980, 1981). Figure 3a shows the gamma-ray map of the ρ Oph region, presenting evidence for resolved structure in gamma rays. This is a result of improved statistics relative to the database of Swanenburg et al. (1981), where this gamma-ray excess was still unresolved. Figure 3b shows the distribution of extinction in this area of the sky after Khavtassi (1960). Also in this case the detailed positional correlation between the gamma-ray distribution and that of the dark clouds at the approximate distance of the ρ Oph cloud complex is striking. The measured gamma-ray flux from the 'upper stream' in the Khavtassi map is higher than from the lower one, consistent with the OH measurements of Wouterloot (1981), who measures the upper stream to be much more massive. A preliminary comparison of the total gamma-ray flux and the rather high mass estimates of Wouterloot (1981), indicates that the gamma-ray flux is about a factor 2 higher than expected, assuming the local CR density. Within the experimental uncertainties this is not inconsistent with the gamma rays originating from the CR-gas interactions. It is important to note that Figures 2b and 3b are just indications of the distribution of molecular gas. Both complexes are imbedded in large amounts of atomic hydrogen (although less structured), that should be taken into account

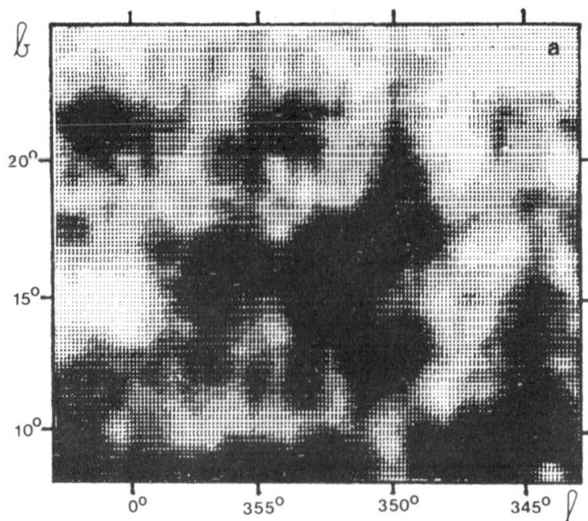

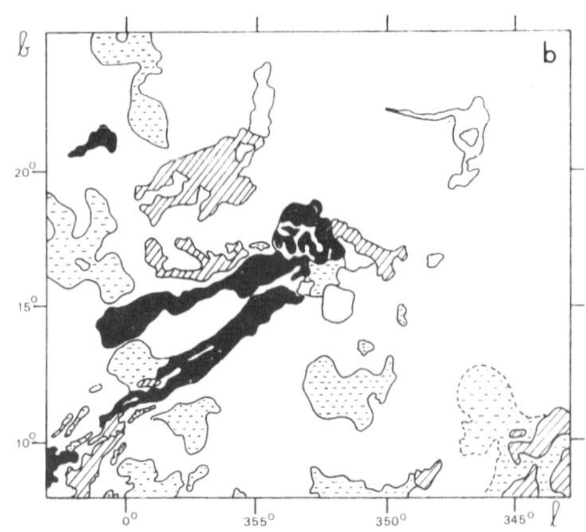

Fig. 3. The ρ Oph region: a. half-tone map showing the gamma-ray intensities (100 MeV $<E<$5 GeV); b. sketch from the atlas of dark clouds by Khavtassi (1960). The dotted, shaded and black regions indicate increasing obscuration. The blank features represent bright diffuse nebulae. For a detailed description of the distances of the dark clouds and the molecular gas connected to them see Wouterloot (1981) and references therein.

when making detailed comparisons with the gamma-ray data. A profound analysis of these cloud complexes by the COS-B Collaboration is in progress. Furthermore, in the large-scale map an altogether new correlation may be seen, which is probably local but independent of Gould's Belt: the region $100° \lesssim l \lesssim 120°$, $10° \lesssim b \lesssim 20°$ (clearly enhanced in gamma rays with respect to the region at equivalent negative latitudes), corresponding to a region of enhanced galactic absorption, as derived from galaxy-count data (Strong and Lebrun 1982) and clearly visible in the stellar reddening map of Lucke (1978). A good candidate for identification of this gamma-ray feature is a section of the disc-shaped structure reported by Dolidze (1980). Because of its similarity to Gould's Belt it is dubbed the 'Dolidze Belt' (Bignami 1981) and has been shown to be visible in the COS-B data base (Bignami 1981, Bignami et al. 1981). Figure 4 shows a sketch of the two local belt systems together with the differences between the average intensities in the range $11° < |b| < 19°$ above and below the plane, where the belts are best visible. The correlation is striking.

4. THE GASEOUS CONTENT OF THE GALAXY AND HIGH-ENERGY GAMMA RAYS

It has been shown that a good correlation exists for the local environment between the gamma radiation observed at medium latitudes and the total gas as estimated from extinction measurements (Lebrun et al. 1982, Strong et al. 1982a; earlier works on this topic Fichtel et al. 1978, Lebrun and Paul 1979, Strong and Wolfendale 1981). The local gamma-ray production rate has been determined from these comparisons. This method, however, cannot be applied to gamma-ray intensities measured at low galactic latitudes. In addition, the observed intensities at these latitudes remain an upper limit for the contribution from gas as long as the magnitude of the discrete-source contribution cannot be estimated unambiguously. Furthermore the gas content of the Galaxy is still poorly known (for discussions on the uncertainties in the derivation of H_2 densities from trace molecules see e.g. Lequeux (1981) and Liszt and Burton (1981). A feasible approach is to compare the observed gamma-ray intensities with those expected from the diffuse emission of the, at 21 cm well mapped, HI alone, under the assumption that the gamma-ray yield per H atom is that measured locally.

HI column densities derived from the surveys of Weaver and Williams (1973), Strong et al. (1982b), Heiles and Cleary (1979) and those given by Heiles and Habing (1974) have been convolved with the COS-B point-spread function in order to ensure a valid comparison. The convolved map is shown in Figure 5b. Figure 5 also gives the gamma-ray intensity map for the total energy range and a map showing the difference between this map and the predicted distribution from HI alone (both maps are preliminary and shown for illustration only). The gamma-ray production rate is taken to be 3.2 x 10^{-25} ph $s^{-1}at^{-1}$ in the energy range 70 MeV-5 GeV (Strong et al. 1982a, Bloemen 1983). In the latter map the three strong sources (Vela, Crab and Geminga) are clearly visible. It is also evident that towards the inner Galaxy ($|l|<50°$), the HI contribution falls short

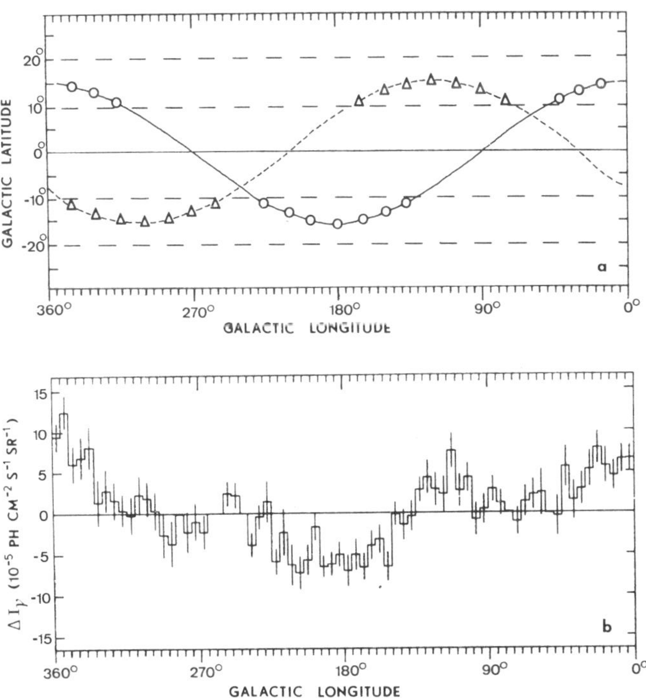

Fig. 4. a. Sketch of the two local belt systems: the Gould Belt (o) and the Dolidze Belt (Δ), see text. b. Difference between average gamma-ray intensities (70 MeV<E<5 GeV) in the regions 11°<b<19° and -19°<b<-11°.

of accounting for the total emission, as first reported by Clark et al. (1968). In addition, the remaining distribution is significantly more narrow than the HI contribution. An evident explanation is that a substantial fraction of the gas in the inner Galaxy is in the form of molecular hydrogen, whose scale height is known to be less than that of HI (Gordon and Burton 1976). However, it should be noted that gamma-ray sources also have a rather small scale height (Swanenburg et al. 1981), and can also account for part of the discrepancy between the HI contribution and the observed profile. In this case, the room for molecular gas is reduced, especially if one considers that the cosmic-ray density may not be uniform and could increase towards the inner Galaxy. Other regions where the HI cannot account for a substantial fraction of the observed emission, indicating the presence of molecular gas and/or sources, encompass the Cygnus complex and Carina arm. Detailed CO observations performed in the first region have revealed the existence of several giant molecular clouds (Cong 1977, Stark and Blitz 1978) whose summed contributions could account for most of this excess (Protheroe et al. 1979, Lebrun et al. 1983).

In most regions in the second and third galactic quadrants only a very small additional (to HI) contribution can be tolerated. For $160° < l < 260°$ around $b=0°$ the predicted HI emission accounts for even more than the observed radiation (except in the direction of the three strong sources and the unresolved source 2CG218-00), leaving no room at all for molecular gas. The problem is solved if the CR density decreases beyond the solar circle, as was first proposed by Dodds et al. (1975). The required small decrease in emissivity is consistent with the results of Bloemen et al. (this volume).

At medium latitudes ($10° \lesssim |b| \lesssim 20°$) Figure 5c gives the gamma-ray prediction of the contribution from the molecular component of the interstellar gas (Strong et al. 1982a and Strong this volume; Bloemen 1983). As a good example of molecular gas already traced at radio and millimetre wavelengths, the resolved structure of the Orion Complex shows up clearly around $l=210°$ (see also Strong, this volume).

5. THE RADIAL DISTRIBUTION OF GAS AND GAMMA RAYS

The three-dimensional distribution of gamma-ray emissivity in the Galaxy can be inferred if one assumes that most of the radiation does not originate in small-scale structure. It is also necessary to impose certain assumptions on the large-scale geometrical structure of the Galaxy before any unfolding procedure, attempting to unravel the spatially varying source function from the observed line-of-sight integrals, can give meaningful results. The aim of the unfolding technique is to convert the two-dimensional distribution of the observed emission into a radial distribution ε (R) of the emissivity versus galacto-centric distance without resorting to physical models of source processes and their distribution in the Galaxy. The parameters of the underlying geometrical pattern of ε (R) are described by Mayer-Hasselwander et al. (1982), who

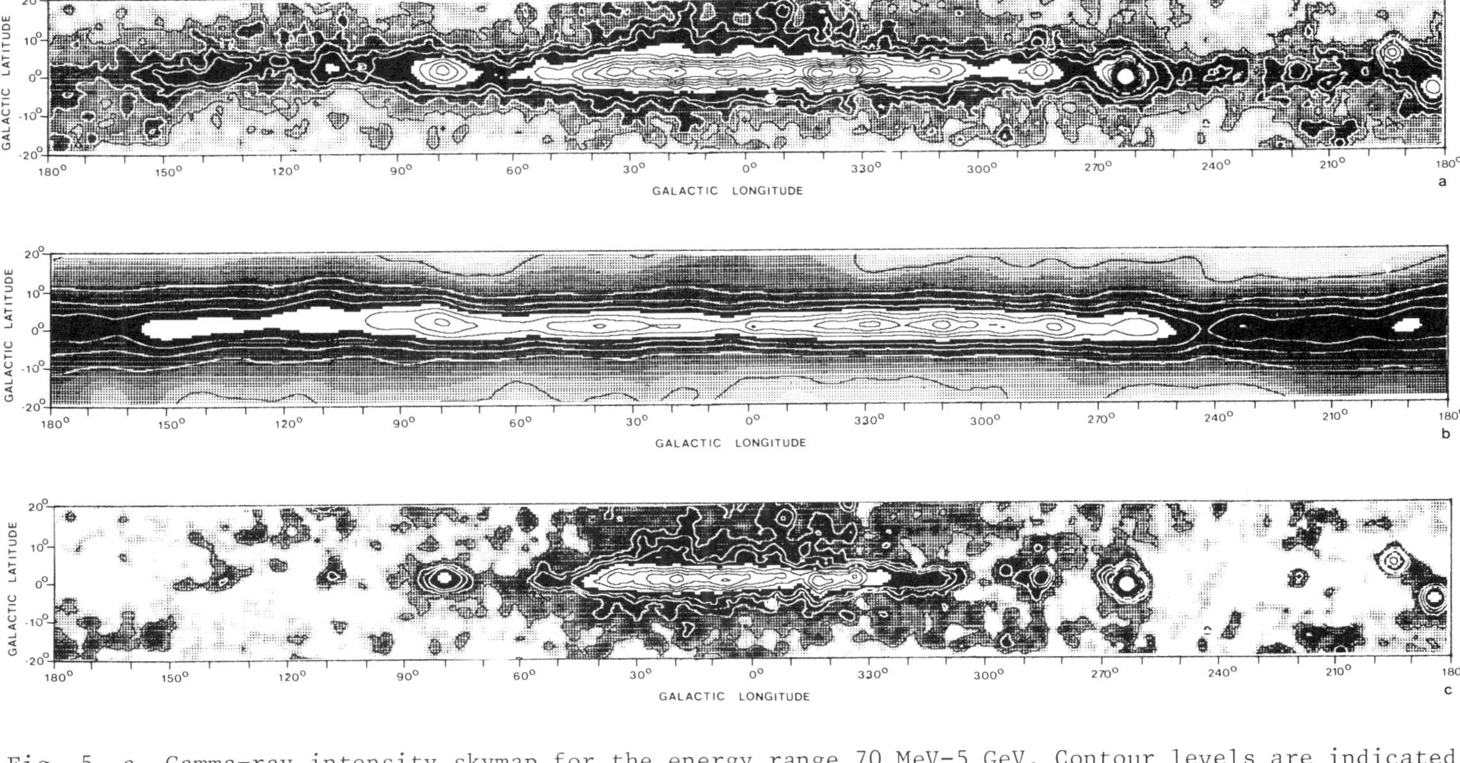

Fig. 5. a. Gamma-ray intensity skymap for the energy range 70 MeV-5 GeV. Contour levels are indicated at (.5, 1, 1.5, 2, 4, 5, 6, 7, 8, 9) x 10^{-4} photon $cm^{-2}s^{-1}sr^{-1}$. b. Skymap of the expected gamma-ray intensity from atomic hydrogen alone (for data origins see text). The HI column densities are convolved with the COS-B pointspread function and with the average observed spectrum of the galactic gamma radiation for the energy range 70 MeV-5 GeV. The local gamma-ray production rate is used. Contour levels are indicated at (.2, .6, 1, 1.4, 2.2, 2.6, 3, 3.4) x 10^{-4} ph $cm^{-2}s^{-1}sr^{-1}$. c. Gamma-ray intensity skymap showing the difference between the skymap presented in a. and the predicted intensities from HI alone given in b.. Contour levels are indicated at (.3, 1, 1.5, 2, 4, 5, 6, 7, 8) x 10^{-4} photon $cm^{-2}s^{-1}sr^{-1}$.

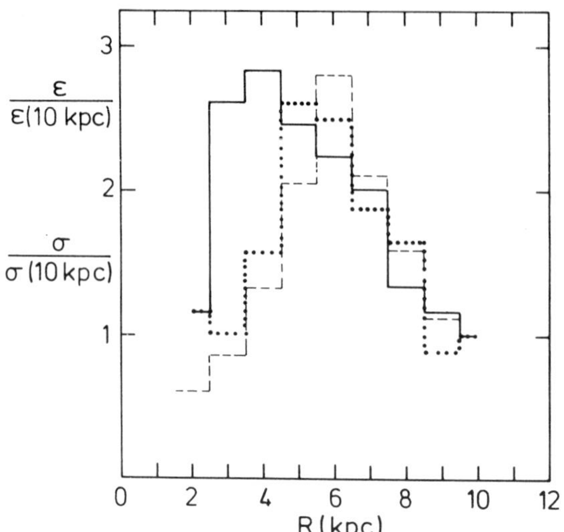

Fig. 6. Comparison of the shapes of the radial distributions of the gamma-ray emissivity (ε ; solid line) and the hydrogen (HI and H_2) gas surface density (σ; dotted line after Gordon and Burton (1976), dashed line after Sanders 1983)). The gamma-ray histogram is an average of the distributions for the first and fourth quadrants, given separately, by Mayer-Hasselwander et al. (1982). Its normalisation value is ε(10 kpc) =2.1 x 10^{-25} photon (>100 MeV) cm^{-3} s^{-1} which agrees with the values for the solar neighbourhood for a gas density of 1 H-atom cm^{-3}.

present the radial distribution of gamma-ray emissivity averaged over 0.5-kpc rings for the first and fourth galactic quadrants seperately. The statistical and systematic uncertainties do not allow us to state that the two distributions are different. Therefore, Figure 6 shows the average in 1-kpc rings, together with similar distributions for the total gas (atomic and molecular hydrogen; for a constant CR density this sum is proportional to the gamma-ray emissivity) as derived from radio and millimetre measurements. The latter results are derived by Gordon and Burton (1976) from CO measurements in the first quadrant between l=10° and l=80° and Sanders (1983) from measurements in the longitude intervals 5°<l<70° and 330°<l<355°. When normalized at R=10 kpc, the two radio plots are in excellent agreement, contrary to earlier results of Solomon and Sanders (1980). The relatively lower values from Sanders for R<5 kpc in Figure 6 are due to a different assumption on the scale height compared to that of Gordon and Burton, namely a scale height decreasing with R in the inner Galaxy respectively a constant one. The deduced absolute value for the H_2 mass surface density at R≃6 kpc , however, is for Sanders still a factor of ~ 1.5 higher than for Gordon and Burton (1976). This difference is due to the use of different conversion factors between the H_2 column density and the integrated CO line intensity. The value used by Sanders (1983) (from Sanders 1981 using Frerking et al. 1981) is 20% above the upper limit set by Lebrun et al. (1983) using COS-B gamma-ray data together with the fully sampled Columbia CO survey and the Berkeley HI surveys in the first galactic quadrant.

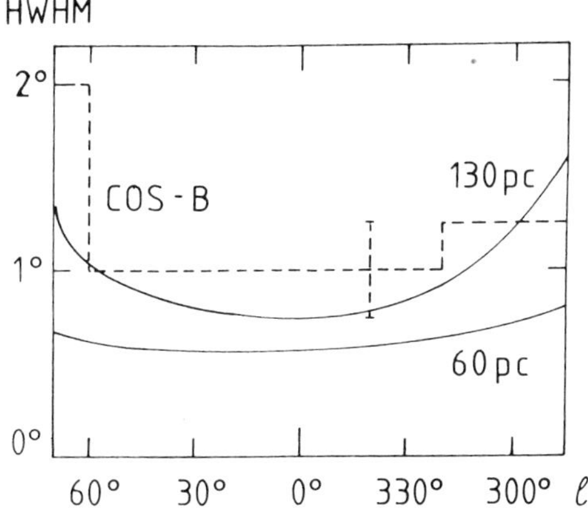

Fig. 7. Latitude extent of the galactic gamma radiation observed (dashed line) and expected (solid lines) for two different scale heights used in the unfolding procedure (Mayer-Hasselwander et al. 1982).

The shape of the gamma-ray distribution in Figure 6 follows nicely the radio-derived ones between 5 and 10 kpc, while the broad peak seems more extended towards the galactic centre in the gamma-ray case. Keeping in mind the appreciable uncertainties, especially for these galactic radii, this broader shape of the gamma-ray distribution, if real, could be explained by a gamma-ray source contribution with a positive gradient towards the galactic centre, or is a hint for an increase in CR density. In addition, the latter possibility together with the indication from the colour index of a softer gamma-ray spectrum in the inner Galaxy than for the rest of the disc emission, could be a hint for a predominantly increase in CR electron density towards the galactic centre.

On the assumption of a uniform CR density, neglecting as unphysical the hypothesis of a negative galactocentric gradient, it is seen that the increase of the total surface density of the gas as traced by COS-B gamma rays cannot be more than a factor of 3 between R=10 kpc and R=5 kpc. However, this value must be regarded as an upper limit, keeping in mind a possible CR density gradient and a significant component due to undetected sources (see Hermsen 1980 and references therein). The ability of the relevant CR's to penetrate dense clouds has been discussed by Skilling and Strong (1976) and Cesarsky and Völk (1978). In addition, the results on the Orion cloud complex (Caraveo et al. 1980, 1981) show that the totality of the gas is involved in the production of gamma rays.

Finally, the question of the scale height of the gamma-ray disc has been addressed. The unfolding was performed with two trial scale heights of 60 pc and 130 pc. Figure 7 compares the resulting model latitude profiles in terms of HWHM with the intrinsic latitude extent of the gamma-ray disc as determined from the COS-B measurements. It is evident that dominant gamma-ray emission with a scale height significantly less than 100 pc is not supported by the data. If the gas in the inner Galaxy were dominated by molecular hydrogen one would expect the scale height of this gaseous disc to be approximately 60 pc (Cohen and Thaddeus 1977).

Therefore, this result can be regarded as independent evidence for a small contribution of molecular gas in the inner Galaxy.

6. CONCLUSIONS

The large-scale distribution of galactic gamma-ray emission as presented in detail by the Caravane Collaboration for the COS-B satellite (Mayer-Hasselwander et al. 1982), shows that the emission originates in astronomical sites of widely different scales ranging from the tens of kpc of the large-scale warp of our Galaxy, to the several kpc of the spiral pattern down to the hundreds of pc of the local environment (the problem of the nature of the unresolved sources lies outside the scope of this paper, see e.g. the review by Bignami and Hermsen 1983). General agreement has been found with known galactic features, in part already seen in the SAS-2 data. Features, newly seen in gamma rays are e.g. the Perseus arm, although a significant fraction of the emission assigned to this arm earlier (Mayer-Hasselwander et al. 1982) originates at distances up to R=18 kpc (see Bloemen et al., this volume), the Carina arm, the resolved structures of the Orion and ρ Oph cloud complexes and the possible correlation with the Dolidze Belt.

The shape of the radial distribution of the gamma-ray emissivity is in full agreement with radio-derived distributions of the surface density of HI plus H_2 between R=5 kpc and R=10 kpc, under the assumption that the gamma radiation is generated by interaction of uniformly distributed CR's with the total gas. However, at smaller galactocentric radii there is an indication for a higher gamma-ray emissivity than expected under the above mentioned assumption. Coupled to the softer gamma-ray spectrum in the galactic centre region this provides an indication for a possible increase in CR <u>electron</u> density towards the galactic centre.

The maximum tolerable <u>increase</u> in the total-mass surface density locally to R=5 kpc is a factor 3, in agreement with the results of Gordon and Burton (1976) and Sanders (1983). However, a positive gradient in CR density towards the inner Galaxy and a significant compact source contribution to the galactic gamma-ray emission will lower this upper limit significantly. The <u>absolute</u> amount of molecular hydrogen tolerable in the molecular ring for the gamma-ray upper limit is slightly smaller than that proposed by Sanders (1983) and in agreement with that of Gordon and Burton (1976) (see also Lebrun et al. 1983).

REFERENCES

Bignami, G.F., Boella, G., Burger, J.J., Keirle, P., Mayer-Hasselwander, H.A., Paul, J.A., Pfefferman, E., Scarsi, L., Swanenburg, B.N., Taylor, B.G., Voges, W.H., Wills, R.D.: 1975, Space Sci. Instr. 1, p.245

Bignami, G.F., Caraveo, P.A., Maraschi, L.: 1978, Astron.Astrophys. 67, p. 149

Bignami, G.F.: 1981, Phil.Trans.R.Soc.Lond. A301, p. 555

Bignami, G.F., Barbareschi, L., Bloemen, J.B.G.M., Buccheri, R., Caraveo,

P.A., Hermsen, W., Kanbach, G., Lebrun, F., Mayer-Hasselwander, H.A., Paul, J.A., Strong, A.W., Wills, R.D.: 1981, 'Proc. 17th Int.Cosm. Ray Conf.' (Paris) 1, p.182

Bignami, G.F., Hermsen, W.: 1983, Ann. Rev. Astron. Astrophys. 21. In press.

Bloemen, J.B.G.M.: 1983, in 'Kinematics, Dynamics and Structure of the Milky Way', ed. W.L.H. Shuter, Dordrecht: Reidel, p. 31.

Caraveo, P.A., Bennett, K., Bignami, G.F., Hermsen, W., Lebrun, F., Masnou, J.L., Mayer-Hasselwander, H.A., Paul, J.A., Sacco, B., Scarsi, L., Strong, A.W., Swanenburg, B.N., Wills, R.D.: 1980, Astron.Astrophys. 91, p.L3

Caraveo, P.A., Barbareschi, L., Bennett, K., Bignami, G.F., Hermsen, W., Kanbach, G., Lebrun, F., Masnou, J.L., Mayer-Hasselwander, H.A., Sacco, B., Strong, A.W., Wills, R.D.: 1981, 'Proc. 17th Int.Cosmic Ray Conf.' (Paris) 1, p.139

Cesarsky, C.J., Paul, J.A., Shukla, P.G.: 1978, Astrophys.Space Sci. 59, p.73

Cesarsky, C.J., Völk, H.J.: 1978, Astron.Astrophys. 70, p.367

Clark, G.W., Garmire, G.P., Kraushaar, W.L.: 1968, Ap.J.(Lett.) 153, p.L203

Cohen, R.S., Thaddeus, P.: 1979, Ap.J.(Lett.) 217, p.L155

Cong,H.: 1977, Ph.D. thesis, Columbia University, NASA Technical Memorandum 79590

Dodds, D., Strong, A.W., Wolfendale, A.W.: 1975, M.N.R.A.S. 171, p.569

Dolidze, M.V.: 1980, Soviet Astronomy Letters 6, p.51

Fichtel, C.E., Hartman, R.C., Kniffen, D.A., Thompson, D.J., Bignami, G.F., Ögelman, H., Özel, M.E., Tümer, T.: 1975, Ap.J. 198, p.163

Fichtel, C.E., Kniffen, D.A., Thompson, D.J., Bignami, G.F., Cheung, C.Y.: 1976, Ap.J. 208, p.211

Fichtel, C.E., Simpson, G.A., Thompson, D.J.:1978, Ap.J. 222, p.833

Frerking, M.A., Langer, W.D., Wilson, R.W.: 1981, preprint Bell Telephone Labs, Holmdel, New Jersey

Gordon, M.A., Burton, W.B.: 1976, Ap.J. 208, p.346

Harding, A.K.:1981, Ap.J. 247, p.639

Harding, A.K., Stecker, F.W.: 1981, Nature 290, p.316

Hartman, R.C., Kniffen, D.A., Thompson, D.J., Fichtel, C.E., Ögelman, H.B., Tümer, T., Özel, M.E.: 1979, Ap.J. 230, p.597

Heiles, C., Cleary, M.N.: 1979, Aust.J.Phys.Astrophys.Supp. 47, p.1

Heiles, C., Habing, H.J.: 1974, Astron.Astrophys.Supp. 14, p.1

Henderson, A.P., Jackson, P.D. Kerr, F.J.: 1982, Ap.J. 263, in press

Hermsen, W.: 1980, Ph.D. thesis, University of Leiden

Khavtassi, J.Sh.: 1960, 'Atlas of Galactic Dark Nebulae', Tiblisi, Abastumani Astrophys.Obs.

Kniffen, D.A., Fichtel, C.E.: 1981, Ap.J. 250, p.389

Kulkarni, S.A., Blitz, L., Heiles, C.: 1982, Ap.J.(Lett.) 259, p.L63

Lebrun, F., Paul, J.A.: 1979, 'Proc. 16th Int.Cosmic Ray Conf.' (Kyoto) 12, p.13

Lebrun, F., Bignami, G.F., Buccheri, R., Caraveo, P.A., Hermsen, W., Kanbach, G., Mayer-Hasselwander, H.A., Paul, J.A., Strong, A.W., Wills, R.D.: 1982, Astron.Astrophys. 107, p.390

Lebrun, F., Bennett, K., Bignami, G.F., Bloemen, J.B.G.M., Buccheri, R., Caraveo, P.A., Gottwald, M., Hermsen, W., Kanbach, G., Mayer-

Hasselwander, H.A., Montmerle, T., Paul, J.A., Sacco, B., Strong, A.W., Wills, R.D.: 1983, submitted to Ap.J.
Lequeux, J.: 1981, Comments on Astrophysics 9, p.117
Liszt, H.S., Burton, W.B.: 1981, Ap.J. 243, p.778
Lucke, P.B.: 1978, Astron.Astroph. 64, p.367
Mayer-Hasselwander, H.A., Bennett, K., Bignami, G.F., Buccheri, R., D'Amico, N., Hermsen, W., Kanbach, G., Lebrun, F., Lichti, G.G., Masnou, J.L., Paul, J.A., Pinkau, K., Scarsi, L., Swanenburg, B.N., Wills, R.D.: 1980, Ann.NY Acad.Sci. 336, p.211
Mayer-Hasselwander, H.A., Bennett, K., Bignami, G.F., Buccheri, R., Caraveo, P.A., Hermsen, W., Kanbach, G., Lebrun, F., Lichti, G.G., Masnou, J.L. Paul, J.A., Pinkau, K., Sacco, B., Scarsi, L., Swanenburg, B.N. Wills, R.D.: 1982, Astron.Astrophys. 105, p.164
Mayer-Hasselwander, H.A.: 1983, in 'Kinematics, Dynamics and Structure of the Milky Way', ed. W.L.H. Shuter, Dordrecht: Reidel, p. 223.
Paul, J.A., Bennett, K., Bignami, G.F., Buccheri, R., Caraveo, P., Hermsen, W., Kanbach, G., Mayer-Hasselwander, H.A., Scarsi, L, Swanenburg, B.N., Wills, R.D.: 1978, Astron.Astrophys. 68, p.L31
Protheroe, R.J., Strong, A.W., Wolfendale, A.W., Kiraly, P.: 1979, Nature 277, p.542
Riley, P.A., Wolfendale, A.W.:1981, Rivista Del Nuova Cimento Vol.4, no.4, p.1
Rothenflug, R., Caraveo, P.A.: 1980, Astron.Astrophys. 81, p.218
Salvati, M., Massaro, E.: 1982, M.N.R.A.S. 198, p.11
Sanders, D.B.: 1981, Ph.D.thesis, SUNY at Stony Brook
Sanders, D.B.: 1983, in 'Kinematics, Dynamics and Structure of the Milky Way', ed. W.L.H. Shuter, Dordrecht: Reidel, p. 115.
Scarsi, L., Bennett, K., Bignami, G.F., Boella, G., Buccheri, R., Hermsen, W., Koch, L., Mayer-Hasselwander, H.A., Paul, J.A., Pfefferman, E., Stiglitz, R., Swanenburg, B.N., Taylor, B.G., Wills, R.D.: 1977, 'Proc. 12th ESLAB Symp.', ESA SP124, p.3
Schlosser, W., Feitzinger, J.V.: 1982, Astron.Astrophys. In press
Skilling, J., Strong, A.W.: 1976, Astron.Astrophys. 53, p.253
Solomon, P.M., Sanders D.B.: 1980, in 'Giant Molecular Clouds in the Galaxy', eds. P.M. Solomon, E.G. Edmunds, Pergamon Press
Stark, A.A., Blitz, L.: 1978, Ap.J.(Lett.) 255, p.L15
Strong, A.W., Wolfendale, A.W., Bennett, K., Wills, R.D.: 1978, M.N.R.A.S. 182, p.751
Strong, A.W., Lebrun, F.: 1982, Astron.Astrophys. 105, p.159
Strong, A.W., Wolfendale, A.W.: 1981, Phil.Trans.R.Soc.Lond. A 301, p.541
Strong, A.W., Bignami, G.F., Bloemen, J.B.G.M., Buccheri, R., Caraveo, P.A., Hermsen, W., Kanbach, G., Lebrun, F., Mayer-Hasselwander, H.A., Paul, J.A., Wills, R.D.: 1982a, Astron.Astrophys. In press
Strong, A.W., Murray, J.D., Riley, P.A., Osborne, J.L.: 1982b, M.N.R.A.S. 201, p.495
Swanenburg, B.N., Bennett, K., Bignami, G.F., Buccheri, R., Caraveo, P.A., Hermsen, W., Kanbach, G., Lichti, G.G., Masnou, J.L., Mayer-Hasselwander, H.A., Paul, J.A., Sacco, B., Scarsi, L., Wills, R.D.: 1981, Ap.J.(Lett.) 243, p.L69
Weaver, H., Williams, R.W.: 1973, Astron.Astrophys.Suppl. 8, p.1
Wouterloot, J.G.A.: 1981, Ph.D. thesis, University of Leiden

GAS CONTENT AND GAMMA RAY EMISSION IN THE FIRST GALACTIC QUADRANT

François Lebrun
Section d'Astrophysique, Centre d'Etudes Nucléaires de Saclay, France
On behalf of the COS-B Caravane Collaboration and the Goddard Institute for Space Studies

ABSTRACT

The Columbia CO survey and the Berkeley HI survey are compared with the COS-B gamma-ray survey. As a first step, the study is limited to the first galactic quadrant and to the high-energy gamma rays (E > 300 MeV). It is found that a simple model, in which uniformly distributed cosmic rays interact with the interstellar gas, can account for the observed gamma-ray emission. Furthermore if the contribution from point sources to the gamma-ray flux is significant, these sources have to have a galactic distribution similar to that of the CO emission.

I. INTRODUCTION

The widely observed 2.6-mm line emitted by the CO molecule (J:1 → 0) is a tracer of the molecular clouds which are believed to contain a significant (Cohen <u>et al</u>. 1980) or dominant (Solomon and Sanders 1980) fraction of the interstellar gas, the remaining part being well depicted by the 21-cm line. The discovery of many gamma-ray sources such as the Crab and Vela pulsars has suggested that the interaction of ambient cosmic rays (CR) with the interstellar gas may not be the only significant source of galactic gamma rays. A detailed comparison of CO and gamma-ray surveys should shed some light on the galactic distribution of interstellar gas and CRs and on the gamma-ray point source contribution.

The COS-B satellite has given a complete picture of the Milky-Way in high-energy gamma rays (Mayer-Hasselwander <u>et al</u>. 1982). The 1.2-m CO-survey telescope at Columbia University has provided a unique opportunity to perform a complete CO survey with a large latitude extent of the first quadrant of the galactic plane (Dame and Thaddeus 1982). The completeness of this CO survey allows for the first time a detailed comparison of the brightness of the 2.6-mm CO and 21-cm HI lines with the gamma-ray intensity. This contribution presents preliminary results of such a comparison performed on the first galactic quadrant and restricted to the high-energy gamma rays (E > 300 MeV).

W. B. Burton and F. P. Israel (eds.), Surveys of the Southern Galaxy, 79–83.

II. DATA AND ANALYSIS

The CO survey of Dame and Thaddeus (1982) has an angular resolution of 1°, adequate for comparison with available gamma-ray data. This wide beam allows a fast and complete sampling over extended regions of the sky. The observations have been made from $\ell = 11.^{\circ}5$ to $97.^{\circ}5$ and from $b = -5.^{\circ}5$ to $10.^{\circ}5$. The spectra have been integrated over all velocities, yielding a quantity hereafter referred to as W_{CO}. The HI data have been taken mainly from the 21-cm line survey of Weaver and Williams (1973) and in smaller part (for $|b| > 10^{\circ}$) from the medium latitude survey of Heiles and Habing (1974). While a spin temperature of 135 K has been adopted in deriving the HI column densities from the Weaver and Williams survey ($|b| < 10^{\circ}$), the 21-cm line has been assumed to be optically thin for $|b| > 10^{\circ}$. The gamma-ray data used are described by Mayer-Hasselwander et al. (1982) and refer to 36 COS-B observation periods of approximately one month duration each. These data cover the entire galactic plane with good statistics. In order to compare the COS-B gamma-ray map with the radio gas-tracer surveys it is necessary to reduce the latter to the same angular resolution by means of a convolution with the point spread function (PSF) of the COS-B experiment. One has to stress the importance of this step and the resulting limitations: for example, gamma rays actually coming from b up to 18° contribute to the gamma-ray intensity measured by COS-B at $b = 9^{\circ}$ in the lowest energy range (70 to 150 MeV) where the HWHM of the PSF is $\sim 3.^{\circ}8$. Noting that the CO survey does not extend beyond $b = 10.^{\circ}5$, it seems then more appropriate as a first step to restrict our study to gamma rays in the high-energy domain where the COS-B angular resolution is the highest (HWHM of the PSF $\sim 1.^{\circ}15$ for $E > 300$ MeV).

Because the derivation of an H_2 column density involves many uncertainties (e.g. Lequeux 1981), a different approach was followed, independent of any interpretation. An attempt has been made to reproduce the observed gamma-ray emission by finding the best values of the parameters A, B, and C in the expression

$$I_p = N_p/\Omega t = A\,N_{HI} + B\,W_{CO} + C \tag{1}$$

where I_p and N_p are the predicted gamma-ray intensity and count, Ωt is the COS-B exposure factor and N_{HI} and W_{CO} refer to the quantities obtained after convolving the HI and CO data with the PSF for $E > 300$ MeV. Values of the parameters A, B, and C were obtained by maximizing the probability of obtaining the observed gamma-ray sky given the predicted one. In other words the likelihood

$$L = \prod_{i,j} \left(N_p^{\,N_o}\, e^{-N_p}/N_o!\right) \tag{2}$$

(where N_o is the observed gamma-ray count in bin (i,j); the product is extended to the whole available sky) has been computed and the (A,B,C) set which maximizes L has been chosen as the best estimate. This procedure is similar to that followed by Lebrun et al. (1982), allowing a straightforward comparison of the results.

Table 1. Best Parameter Values

	This work	Lebrun *et al.* 1982
$A(10^{-25}$ ph at^{-1} cm^{2} s^{-1} sr$^{-1})$	$2.81^{+0.48}_{-0.33}$	2.14 ± 0.27
$B(10^{-4}$ ph sr^{-1} K^{-1} km$^{-1})$	$1.75^{+0.19}_{-0.25}$	
$C(10^{-4}$ ph s^{-1} sr$^{-1})$	10.9 ± 1.3	8.17 ± 0.59

The maximum likelihood estimates of parameters A, B and C are given in Table 1. The predicted gamma-ray emission for these values can be compared with that actually observed to judge the agreement. This can be done by examining Figures 1 and 2. Figure 1a shows clearly that it is possible to reproduce simultaneously the gamma-ray intensity in directions near the galactic center and well away from it without invoking a CR gradient. Figure 1b illustrates that this agreement in the plane is maintained at medium latitudes in all directions. The agreement between the latitude distributions of the observed and predicted gamma-ray emission for various longitude ranges is apparent in Figure 2. It should be noted that the width of these distributions is much larger than the COS-B PSF and that the asymmetries around the plane are also reproduced.

III. DISCUSSION

When performing a multi-linear regression, one has to ensure that the variables are linearly independent. The advantage of the likelihood-ratio method is that the uncertainties derived for each parameter take into account the degree of dependence of the variables. In the present context it appears that W_{CO} and N_{HI} are practically linearly independent. This is illustrated in Figure 1, where the gamma-ray emission expected from HI alone is shown as the dotted line. Clearly this curve is not similar to the one expected from both HI and CO. The good agreement obtained between the gamma-ray intensity and the gas tracers might seem surprising bearing in mind that the interstellar medium is transparent for gamma rays whereas the CO line is typically thick. However because of the differential galactic rotation, clumpiness of the molecular clouds, and the typically narrow lines in the individual cloudlets, shadowing of one cloud by another is unlikely. W_{CO} can then be considered as a tracer of the number of CO clouds along the line of sight. From this point of view the similarity between the distribution of gamma rays and that of the clouds may suggest that, in first approximation, the gamma-ray emission per cloud is uniform. This likely means that the gas mass and the CR density inside the clouds do not vary by large amount from cloud to cloud. Moreover the goodness of the fit indicates that if localized sources contribute significantly to the gamma-ray emission, their global

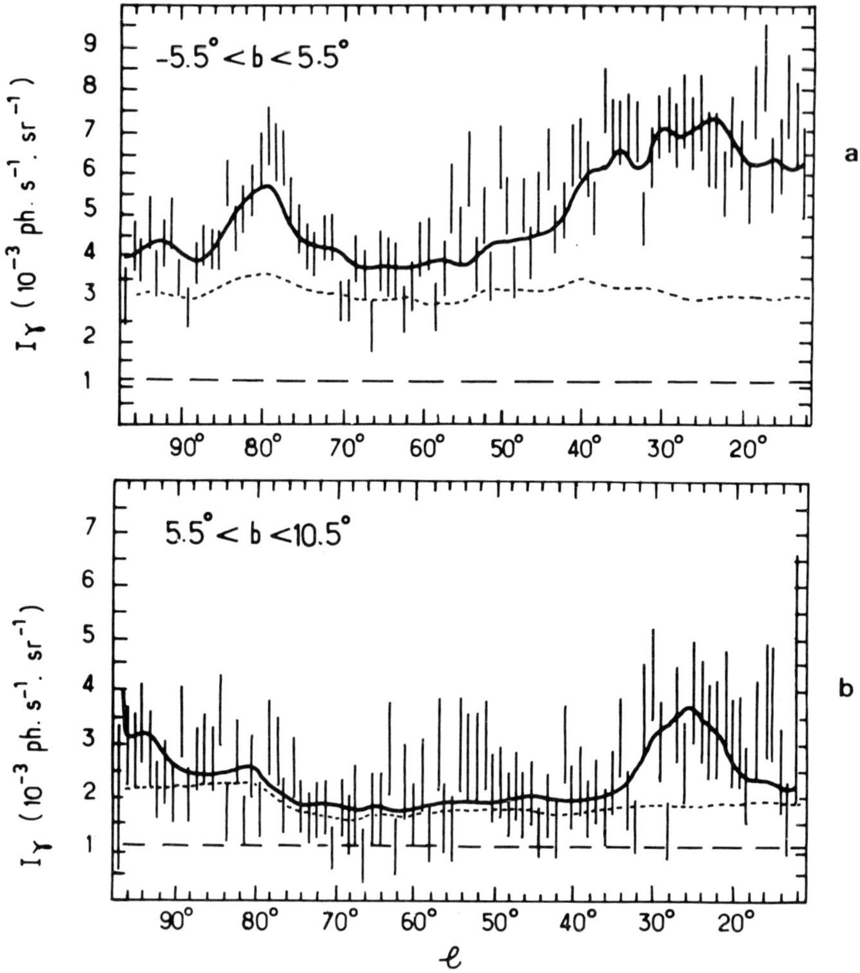

Figure 1. Longitude profiles of the observed gamma-ray emission for E > 300 MeV (error bars) compared with that expected from CO plus HI (solid line) and from HI alone (dotted line). The included background level B is indicated by the interrupted line.

contribution is well represented by one of the right hand terms of equation (1). Because the gamma-ray sources distribution has the same characteristics as the young population of the galactic disk (Swanenburg et al. 1981), their contribution is likely represented by that of the CO emission.

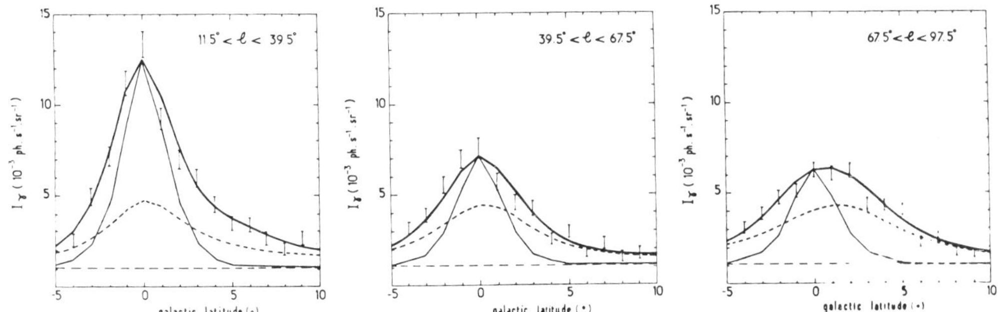

Figure 2. Latitude profiles of the observed gamma-ray emission for E > 300 MeV (error bars) compared with the one expected from CO plus HI (heavy line) and from HI (dotted line). The light line shows the profile expected from emission at B = 0° normalized to the gas tracers expectatioı

REFERENCES

Cohen, R.S., Cong, H., Dame, T.M., and Thaddeus, P.: 1980, Astrophys. J. (Letters) 239, L53.
Dame, T.M., and Thaddeus, P.: 1982, in preparation.
Heiles, C., and Habing, H.J.: 1974, Astron. Astrophys. Suppl. 14, 1.
Lebrun, F. et al.: 1982, Astron. Astrophys. 107, 390.
Lequeux, J.: 1981, Comments on Astrophysics 9, 117.
Mayer-Hasselwander, H.A. et al.: 1982, Astron. Astrophys. 105, 164.
Solomon, P.M., and Sanders, D.B.: 1980, in "Giant Molecular Clouds in the Galaxy", eds. P.M. Solomon and E.G. Edmunds, Pergamon Press, p. 41.
Swanenburg, B.N. et al.: 1981, Astrophys. J. (Letters) 243, L69.
Weaver, H., and Williams, R.W.: 1973, Astron. Astrophys. Suppl. 8, 1.

LIMITS ON THE SURFACE DENSITY OF MOLECULAR HYDROGEN FROM COSMIC GAMMA RAY DATA

P.A. Riley and A.W. Wolfendale,
Physics Department, University of Durham, Durham, U.K.

ABSTRACT

The flux of high energy γ-rays ($E \gtrsim 50$ Mev) from the Galactic plane may be used to place limits on the density of H_2 in the Galaxy. Here the density of H_2 in the molecular ring (R = 6 kpc) is examined. It is shown that the upper limit on the surface density of H_2 is slightly less than estimates based on CO observations and that the most likely value is much less (about 3.5 $M_\odot pc^{-2}$).

INTRODUCTION

Much of the flux of gamma radiation ($E \gtrsim 50$ MeV) from the Galaxy, detected, for example, by the COS-B satellite (Mayer-Hasselwander et al. 1982), is produced by the interaction of cosmic rays with interstellar gas. Insofar as the distribution of cosmic rays is roughly known, it is possible to use γ-ray observations to estimate the amount of H_2 in the Galaxy, this gas component being the subject of much contemporary interest and argument.

The credentials of Gamma Ray Astronomy for the present analyses are best established by two past 'successes'. First, local molecular clouds produce γ-ray fluxes consistent with predictions based on their mass, as deduced from radio observations, and the assumption that cosmic rays pervade the bulk of the clouds, a result which also confirms the assumption that cosmic rays are not excluded from the interiors of dense clouds. (Wolfendale, 1980; Caraveo et al., 1980; Issa and Wolfendale, 1981). Second, the γ-ray emissivity of gas at high latitudes (and, hence 'local' gas), derived using γ-ray fluxes and gas column densities, agrees quite well with what would be expected from our knowledge of the Nuclear Physics of the interaction processes and the local cosmic spectrum (Issa et al., 1981).

W. B. Burton and F. P. Israel (eds.), Surveys of the Southern Galaxy, 85–88.

THE DENSITY OF THE H_2 AT R = 6 kpc

Molecular hydrogen does not play a dominant role locally (apart from its presence in a rather small number of dense clouds) but as one proceeds towards the Galactic Centre the conventional wisdom is that it soon predominates over atomic hydrogen. There is the well known peak in the H_2 surface density in the region of Galactocentric distance, R $\sim$ 5-6 kpc and in the present brief report we choose for our examination the surface density at R = 6 kpc, i.e. σ_{H2} (6 kpc).

The Figure shows values of σ which have been estimated by Solomon and co-workers and Burton et al. from observations on CO. The spread of values which arises comes partly from differences in the CO observations but more particularly from the adoption of different conversions from CO to H_2. The value indicated by an arrow ('Best CO') is probably the best to take for the conventional CO-derivation of σ_{H2} (6 kpc).

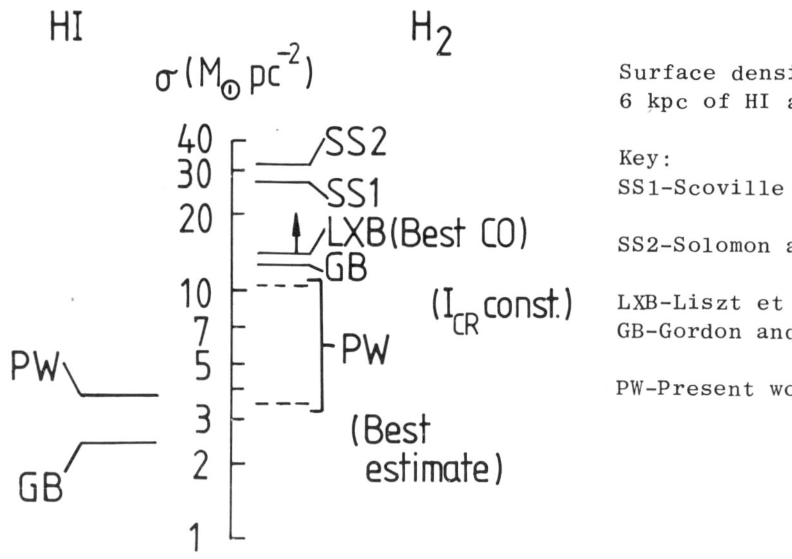

Surface densities at R = 6 kpc of HI and H_2.

Key:
SS1-Scoville and Solomon (1975)
SS2-Solomon and Sanders (1980)
LXB-Liszt et al. (1982)
GB-Gordon and Burton (1976)
PW-Present work

The value of σ_{H2} inferred from the cosmic γ-ray observations can now be examined. In a recent paper (Li et al., 1982) we gave what we regard as the preferred situation for the interrelation of the various parameters; this adopts a cosmic ray gradient of the form I_{CR} α exp-(R/8 kpc) and assumes that 30% of the γ-ray emission in the inner Galaxy is derived from discrete sources. The former is consistent with the distribution of electrons according to the work of Phillipps et al., 1981 (electrons are responsible for much of the γ-flux) and with our estimates of the gradient of cosmic rays in the outer Galaxy - a region where uncertainties due to H_2 density are largely absent. The adoption of 30% for the discrete source contribution comes from our earlier work (Riley and Wolfendale, 1980) and although not definite appears likely. The other important ingredient was our estimate of the distribution of HI (using the Berkeley survey, see the Figure) for σ_{H1} (6 kpc). The derived σ_{H2} (6 kpc) is as indicated; in fact it corresponded to the value derived by us from the Gordon and Burton measurements of CO but with a different calibration (for CO → H_2) which was R-dependent because of the adoption of the so-called metallicity gradient correction. Although there is no doubt that there is a metallicity gradient in the ISM in general, with the ratio O/H (etc.) increasing by about a factor 2 in going from R = 10 kpc to R = 6 kpc (Pagel and Edmunds, 1981), there is considerable argument about its effect on the transformation from CO to H_2 because of the complexity of the interstellar chemistry. In our work we have assumed that $n(H_2) \propto \frac{1}{M} . A(^{12}CO)$

and we have put forward arguments in favour of this from an analysis of the radial dependence of CH and H_2CO in comparison with that of CO (Li et al. 1982) (although we do not regard these arguments as completely satisfactory).

Our best estimate of σ_{H2} (6 kpc) is indicated in the Figure and it is immediately seen that there is a factor of almost 4 discrepancy between the γ-ray estimate of σ_{H2} (6 kpc) and the best estimate from CO.

Although σ_{H2} (6 kpc) = 3.5 $M_\odot pc^{-2}$ is our best estimate, higher values can be allowed if the assumptions are varied. The upper limit occurs when we neglect the contribution from discrete sources completely and assume that the cosmic ray intensity is independent of R for R < 10 kpc; to go further would be quite unphysical. The surface density of H_2 is then σ_{H2} (6 kpc) = 10.1 $M_\odot pc^{-2}$. Even here it is noted that σ is less than inferred directly from the CO data.

There is thus a very real discrepancy between the two methods of determining the surface density of H_2 in the inner ring.

Finally, surface densities of HI are indicated, for completeness. It is disturbing to see that even for this easily measured component there is no concordance. In support of our own estimate we note that it agrees well with the earlier measurements of Van Woerden (1965) and we find essentially the same values at larger R as those determined very recently by Henderson et al. (1982), these also being higher than Gordon and Burton's.

REFERENCES

Caraveo, P. et al.: 1980, Astron. Astrophys. 91, pp.L3-L5.
Gordon, M.A., and Burton, W.B.: 1976, Astrophys. J. 208, pp. 346-353.

Henderson, A.P., Jackson, P.D., and Kerr, F.J.: 1982, Astrophys. J. (in press).

Issa, M.R., Riley, P.A., Strong, A.W., and Wolfendale, A.W.: 1981, J. Phys. G. 7, pp. 973-994.

Issa, M.R., and Wolfendale, A.W.: 1981, Nature, 292, pp. 430-433.

Li, Ti pei, Riley, P.A., and Wolfendale, A.W.: 1982, J. Phys. G. 8, pp. 1141-1154.

Li, Ti pei, Riley, P.A., and Wolfendale, A.W.: 1983, to be published in M.N.R.A.S.

Liszt, H.S., Xiang, D., and Burton, W.B.: 1982, Astrophys. J. (in press).

Mayer-Hasselwander, H.A. et al. (Caravane Collaboration): 1982, Astron. Astrophys. 105, pp. 164-175.

Pagel, B.E.J., and Edmunds, M.G.: 1981, Ann. Rev. Astron. Astrophys. 19, pp. 77-113.

Phillips, S., Kearsey, S., Osborne, J.L., Haslam, C.G.T., and Stoffel, H.: 1981, Astron. & Astrophys. 98, pp. 286-294.

Riley, P.A., and Wolfendale, A.W.: 1981, Nuovo Cimento Vol. 4, No. 4, pp. 1-29.

Scoville, N.Z., and Solomon, P.M.: 1975, Astrophys. J. Lett. 199, pp. L105-L109.

Solomon, P.M., and Sanders, D.B.: 1980, Giant Molecular Clouds in the Galaxy, Ed. Solomon and Edmunds (Pergamon Press), pp. 41-73.

van Woerden, H.: 1956, Trans. I.A.U. 12A, pp. 789.

Wolfendale, A.W.: 1980, I.A.U. Symp. No. 94, pp. 309-319.

ON THE RADIAL DISTRIBUTION OF GAMMA RAYS IN THE OUTER GALAXY

J.B.G.M. Bloemen,
Cosmic Ray Working Group and Sterrewacht, Leiden,
The Netherlands

L. Blitz, University of Maryland, U.S.A. and
Sterrewacht, Leiden, The Netherlands

W. Hermsen,
Cosmic Ray Working Group, Leiden, The Netherlands

ABSTRACT

We describe a new method which has the potential for determining the radial distribution of the diffuse component of galactic gamma rays outside the solar circle. We use the observation that a good correlation exists between gamma-ray intensities and total column densities of the local interstellar gas and that the fractional column density of $H_2 \lesssim 0.1$ HI outside the solar circle. Thus the gamma-ray intensities are shown to be proportional to N(HI). We use the kinematics of the HI to determine the distances from which various fractions of the emission originate in the second and third galactic quadrants. Preliminary results of our analysis show that a significant flux of gamma rays originates from distances as large as 18 kpc from the galactic centre.

1. INTRODUCTION

The gamma-ray experiment aboard the ESA COS-B satellite has mapped the gamma-ray sky in great detail (see e.g. Mayer-Hasselwander *et al.* 1982). The interpretation of this wealth of observational results is however not straightforward. In the first place observational limitations of the gamma-ray data (counting statistics; angular resolution; instrumental background) and of the corresponding data at other wavelengths (the very much needed large-scale surveys of molecules are just becoming available; homogeneous searches for counterparts at different wavelengths in the error boxes of the puzzling gamma-ray sources (Swanenburg *et al.* 1981) are lacking) prevent drawing a direct and unique conclusion about the origin of gamma rays. Second, the intrinsic nature of gamma-ray astronomy itself limits the analysis of the data. These intrinsic

W. B. Burton and F. P. Israel (eds.), Surveys of the Southern Galaxy, 89–98.

problems include the two-component structure of galactic gamma-ray emission (discrete point sources and emission diffuse in nature) with an uncertain relative contribution (Bignami, Caraveo and Maraschi 1978, Protheroe *et al.* 1979, Hermsen 1980, Rothenflug and Caraveo 1980, Riley and Wolfendale 1980, Harding 1981, Harding and Stecker 1981, Salvati and Massaro 1982) and the lack of any direct distance information in gamma-ray astronomy.

The unresolved sources of gamma rays compiled in the 2nd COS-B catalogue (Swanenburg *et al.* 1981) have a very narrow latitude distribution ($\langle b \rangle \cong 1.^{\circ}5$). Therefore they can be ignored in the interpretation of the observed gamma rays at medium latitudes ($10° \lesssim |b| \lesssim 20°$), which are of local ($\lesssim 2$ kpc) origin (Lebrun *et al.* 1982, Strong *et al.* 1982, Bloemen 1983) (See section 2). The gamma-ray emission from local molecular-cloud complexes at known distances can even be studied in detail (see e.g. Caraveo *et al.* 1980, 1981).

Along the galactic plane, especially in the first and fourth quadrants, where the major part of the galactic gamma-ray emission is concentrated, these problems are very serious. The measured gamma-ray intensities are integrals of all contributions of different origin along the line of sight. The radial distribution of the gamma-ray emissivity in the inner Galaxy can be inferred from an unfolding of the observed longitude and latitude profiles (Caraveo and Paul 1979, Mayer-Hasselwander *et al.* 1982). However, certain assumptions on the large-scale geometry of the Galaxy and on the dependence of the emissivity on the distance from the galactic plane have to be made.

In the outer Galaxy (second and third quadrants) this procedure cannot be applied. However, since the typical distance of COS-B gamma-ray sources is larger than 2 kpc (Swanenburg *et al.* 1981) and sources are more easily detected towards the galactic anticentre, all localized sources within a few kpc, in this direction, should be in the 2CG catalogue (in fact only 8 sources). Therefore the majority of the galactic gamma-ray emission in this region is most probably produced by the interaction of energetic cosmic rays with the interstellar gas. As a result, any distance information on the locations of gamma-ray production would yield important clues on the large-scale cosmic-ray distribution in the Galaxy.
It is the purpose of this paper to provide a method for determining the radial distribution of the locations of gamma-ray production in the outer Galaxy. Preliminary results based on an analysis of a small sub-set of the data are presented.

2. GAS-COLUMN DENSITIES AND GAMMA-RAY INTENSITIES

The diffuse component of the galactic gamma-ray emission originates from the interaction of energetic cosmic rays with the interstellar gas and radiation fields. The dominant production process of gamma rays at energies above 70 MeV is the interaction with the gas (i.e. pion decay and bremsstrahlung) and can be investigated best at medium latitudes.

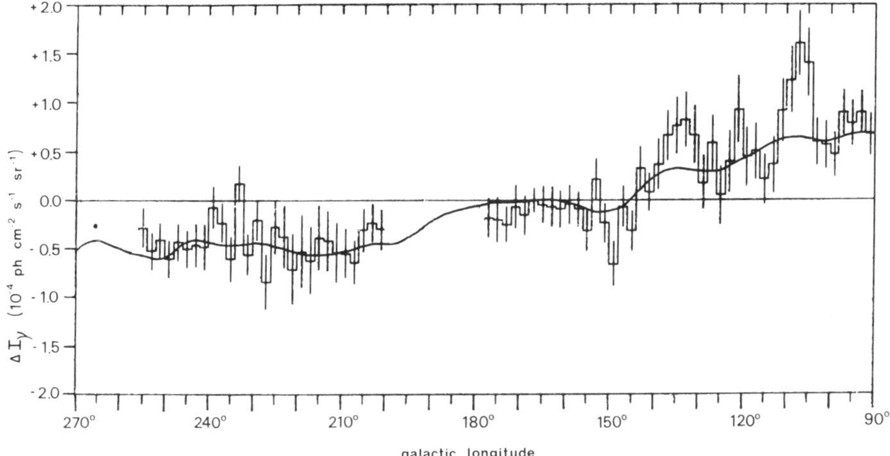

Figure 1. Longitude histogram of the differential gamma-ray intensity ΔI_γ (difference between gamma-ray intensities in the latitude ranges $0°<b<10°$ and $-10°<b<0°$). The solid line shows the corresponding HI estimates, using the gamma-ray production rate as derived in test 2. The clear deviations visible at $l\cong 105°-110°$ and $l\cong 130°-140°$ are indications for the presence of molecular gas and/or point sources not related to the gas. The three strong point sources are excluded.

At those locations no gamma-ray point sources have been observed, despite their greater detectability away from the intense galactic-disc emission. It has been shown that the correlation of the gamma-ray intensities with the total column densities of the local ($\lesssim 2$ kpc) interstellar gas in the latitude range $10°\lesssim|b|\lesssim 20°$ is very good, using the Lick galaxy counts as a tracer of the *total* gas column density (e.g. Lebrun *et al.* 1982 and Strong *et al.* 1982). The contribution of molecular hydrogen is important, because the correlation with atomic hydrogen alone (as derived from the 21 cm surveys) is not as good as with the galaxy counts.

In the outer Galaxy it seems that the molecular hydrogen column densities $N(H_2)$ are much smaller than the atomic hydrogen column densities N(HI). The total mass of atomic hydrogen in the Galaxy is $4.85 \times 10^9 M_\odot$ of which about 80% ($4 \times 10^9 M_\odot$) is situated outside the solar circle (Henderson, Jackson and Kerr 1982), while the amount of molecular hydrogen outside the solar circle is only about $3 \times 10^8 M_\odot$ (Kutner and Mead 1981). Taking into account the latitude extent of the molecular hydrogen we can place an upper limit of $N(H_2)\lesssim 0.2$ N(HI). For a typical line of sight in the outer Galaxy, which excludes some concentrations of molecular gas along the plane (Cohen *et al.* 1980), an upper limit of $N(H_2)\lesssim 0.1$ N(HI) is even more appropriate. Since the distribution of H_2 is also more clumpy (Cohen *et al.* 1980) than the HI distribution, this means, that on a large scale we can ignore molecular hydrogen in the second and third quadrants. However, this basic assumption can be verified in the analysis. The correlation between the gamma-ray intensities and the HI column densities is investigated in the 2nd and 3rd quadrants in the longitude

range 90°<l<270° and latitude range -10°<b<+10°.
Gamma-ray data from the COS-B experiment in the energy range 70-5000 MeV have been used (updated version of the Mayer-Hasselwander *et al.* (1982) database). The strong point sources 2CG184-05 (the Crab pulsar), 2CG195 +04 (Geminga) and 2CG263-02 (the Vela pulsar) have been excluded in the analysis. Five other (unidentified) sources of gamma rays are present in the selected region of the sky, but have not been excluded since they might be related to fine-scale structure in the gas distribution. Their flux is very small compared to the total gamma-ray flux, so that they have a negligible impact on the results presented in this section.
Values of N(HI) are derived from the 21 cm surveys of Weaver and Williams (1973) and Strong *et al.* (1982) for $|b|<10°$ and from Heiles and Habing (1974) and Heiles and Cleary (1979) for $|b|>10°$. HI data outside the selected range have been included for the convolution with the COS-B instrumental point-spread function.
In this investigation two different approaches are followed:

1. We examine whether the *total* gamma-ray intensity I_γ results from the interactions between an uniform cosmic-ray flux and the atomic-hydrogen gas:

$$I_\gamma = \frac{q}{4\pi} \cdot \tilde{N}(HI) + I_B, \qquad (2.1)$$

   where q is the gamma-ray emissivity per hydrogen atom, $\tilde{N}(HI)$ the convolved HI-column density and I_B an isotropic, mainly instrumental background.

2. We examine whether the *differential* gamma-ray intensity ΔI_γ correlates with the differential HI-column density $\Delta\tilde{N}(HI)$ such that:

$$\Delta I_\gamma = \frac{q}{4\pi} . \Delta \tilde{N}(HI) \qquad (2.2)$$

   where ΔI_γ is the difference between gamma-ray intensities at equivalent latitudes above and below b = 0° and $\Delta\tilde{N}(HI)$ is the corresponding quantity for HI column densities. Thus:

$$\Delta I_\gamma = I_\gamma(+b) - I_\gamma(-b) \qquad (2.3a)$$

$$\Delta\tilde{N}(HI) = \tilde{N}(+b) - \tilde{N}(-b) \qquad (2.3b)$$

   It is possible to use this differential technique because both the gamma-ray distribution and the HI distribution are asymmetric with respect to the galactic plane in the second and third quadrants (because of the large-scale warp of the Galaxy). An important advantage is that it eliminates the instrumental background from the analysis of the gamma-ray data.

In both methods, q (in test 1 also I_B) was determined using a least-squares method on bins of 2°x2°. The resulting q values are 10-30% lower than the local gamma-ray production rate as derived by Strong *et al.* (1982) and Bloemen (1983). However, each has systematic uncertainties and a more detailed analysis is in progress. The longitude profiles of the subtracted gamma-ray intensities and the corresponding HI estimates, using the q value derived in test 2, are shown in figure 1. Figure 2 presents a two-dimensional picture of both quantities seperately.

The apparent decrease of the gamma-ray production rate in the outer Galaxy compared to the local value could be due to a decrease in the cosmic-ray density with galacto-centric radius as first mentioned by Dodds *et al.* (1975). The gradient in the cosmic-ray density could be up to about 20% larger than calculated here, since we ignored the molecular hydrogen in our analysis. The importance of the molecular component can be judged by another correlation test of the gamma-ray intensities with the HI column densities as a function of galactic latitude. Since the molecules are strongly concentrated in the physical (warped) plane of the Galaxy, the analysis would yield an apparent decrease of the gamma-ray production rate away from the galactic plane if a significant contribution of molecular hydrogen is present. Preliminary results show that the HI-column densities can reproduce the gamma-ray intensities equally well in each latitude range using only one value of the gamma-ray production rate. Small-scale deviations are certainly present (as expected from concentrations of H_2), but on a large scale the impact of the molecular-hydrogen contribution on the analysis of the gamma-ray data in the second and third quadrants seems *a posteriori* verified to be small.

3. DESCRIPTION OF THE METHOD

As discussed above, gamma radiation is known to be a good tracer of the total gas column density and the gas content of the outer Galaxy can be largely represented by atomic hydrogen. These are the ingredients for a method which determines where the gamma rays are coming from in the second and third quadrants. The basic principle of this method is the analysis of the distribution of HI alone as a tracer of the distribution of the gamma rays. The available HI surveys (mentioned in section 2) together with the knowledge of the kinematics of the Galaxy can be used to construct a 3-dimensional image of the HI distribution. The unambiguous velocity-distance transformation in the outer Galaxy is another reason to restrict this analysis to the 2nd and 3rd quadrants.

The analysis of the HI data starts from the antenna temperatures $T_A(l,b,v)$ of the 21 cm line profiles. $T_A(l,b,v)$ is converted to the brightness temperature $T_b(l,b,v)$ and maps of the HI-column density in small velocity intervals (v_n,v_m) are constucted. Using the conventional procedure (Kerr 1968):

$$N(l,b)_{v_{nm}} = 1.823 \times 10^{18}\ T_s \int_{v_n}^{v_m} \tau(l,b,v)\ dv \text{ at cm}^{-2} \qquad (3.1)$$

$$\tau(l,b,v) = -\ln(1-T_b(l,b,v)/T_s) \qquad (3.2)$$

T_s is the HI spin temperature (assumed to be constant) and τ is the optical depth. Each map is convolved with the point-spread function of

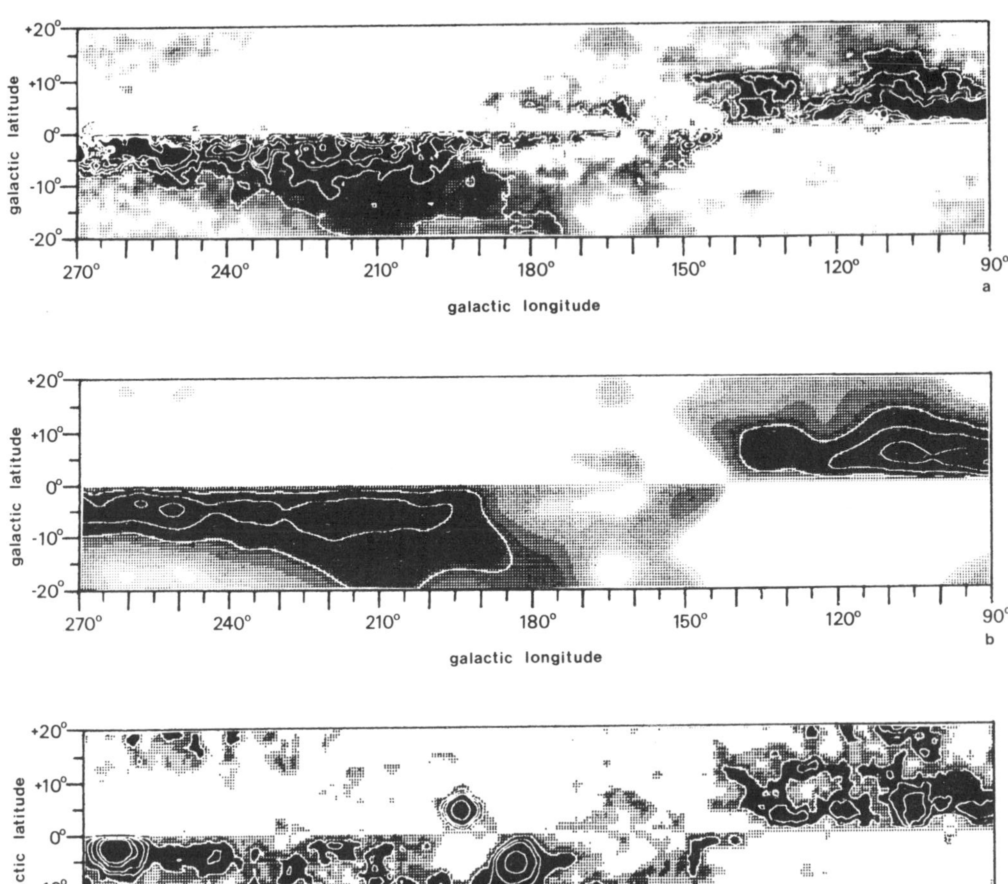

Figure 2. Two-dimensional maps of the differential HI column density and the differential gamma-ray intensity (differences at equivalent latitudes above and below b=0°), i.e. $\Delta N(HI)=N(+b)-N(-b)$ respectively $\Delta I_\gamma=I_\gamma(+b)-I_\gamma(-b)$.

a. the differential HI column density (contour values: (1, 2, 3, 4) x 10^{21} H atom cm^{-2})
b. the same quantity but now convolved with the PSF of the COS-B experiment for the energy range 70-5000 MeV (contour values as in a)
c. the differential gamma-ray intensity in the energy range 70-5000 MeV (contour values: (5, 10, 15, 30, 45) x 10^{-5} ph $cm^{-2}s^{-1}sr^{-1}$). The strong excesses at (185,-6), (195,4) and (264,-3) are related to three strong gamma-ray point sources.

Only positive excesses are indicated.

the gamma-ray experiment in the appropriate energy range. Since data outside the selected region of the sky have to be included in this convolution, parts of at least three of the currently available HI survey are needed (Weaver and Williams 1973, Heiles and Habing 1974 and Strong *et al.* 1982).

Each velocity interval corresponds to a certain distance range, connected to each other by the rotation curve of the outer Galaxy (Blitz, Fich and Stark 1982). Therefore, $N(l,b)_{v_{nm}}$ is a measure of the amount of HI at a certain sky position in a certain galacto-centric distance interval. By adding the different skymaps one can determine the galacto-centric distance within which a certain fraction X of the total HI column density is located:

$$X(l,b) = \{\sum_{v_i=0}^{v_x(l,b)} \tilde{N}(l,b)_{v_i}\}/\{\sum_{v_i=0}^{v_c(l,b)} \tilde{N}(l,b)_{v_i}\} \quad (3.3)$$

where v_c is a cutoff velocity beyond which

$$\tilde{N}(l,b)_{|v|>|v_c|} < 0.01\tilde{N}(l,b)_{|v|<|v_c|}.$$

The same method can be applied for the differential HI-column density (case 2 above) replacing $\tilde{N}(l,b)_{v_i}$ by $\Delta\tilde{N}(l,b)_{v_i}$ in equation 3.3 and using the same cutoff velocity.

Since the correlation of $\tilde{N}(HI)$ with I_γ and $\Delta\tilde{N}(HI)$ with ΔI_γ is very good, as shown in section 2, X(l,b) provides information on the distances from which various fractions of the gamma-ray emission originate.

A complete analysis of the gamma-ray and atomic-hydrogen data is currently underway. However, we present below some preliminary results from a coarse analysis of a few representative longitudes.

4. PRELIMINARY RESULTS

Figures 3 and 4 show latitude-distance plots for X = 50% and X = 75%. Each curve shows the distance within which either 50% or 75% of the HI emission (and thus the gamma-ray emission) is located as a function of latitude. In the analysis the gas is assumed to be optically thin. In figure 3, the plots for the total HI-column density are given (case 1 above) and in figure 4, the plots are shown for the differential HI column density that is, positive minus negative latitudes (case 2). The values of $\Delta\tilde{N}(HI)$ are 20-60% of the corresponding ones of $\tilde{N}(HI)$, values which indicate the magnitude of the asymmetries in the HI and gamma-ray distribution.

We see in both sets of figures a trend to larger distances at lower latitudes. This is not surprising; it means that higher latitude observations are sampling increasingly more local gas. The plots of differential emission show that the corresponding gas is more distant than the total gas at the same latitudes. This can be understood as follows: most of the emission even at low latitudes is coming from

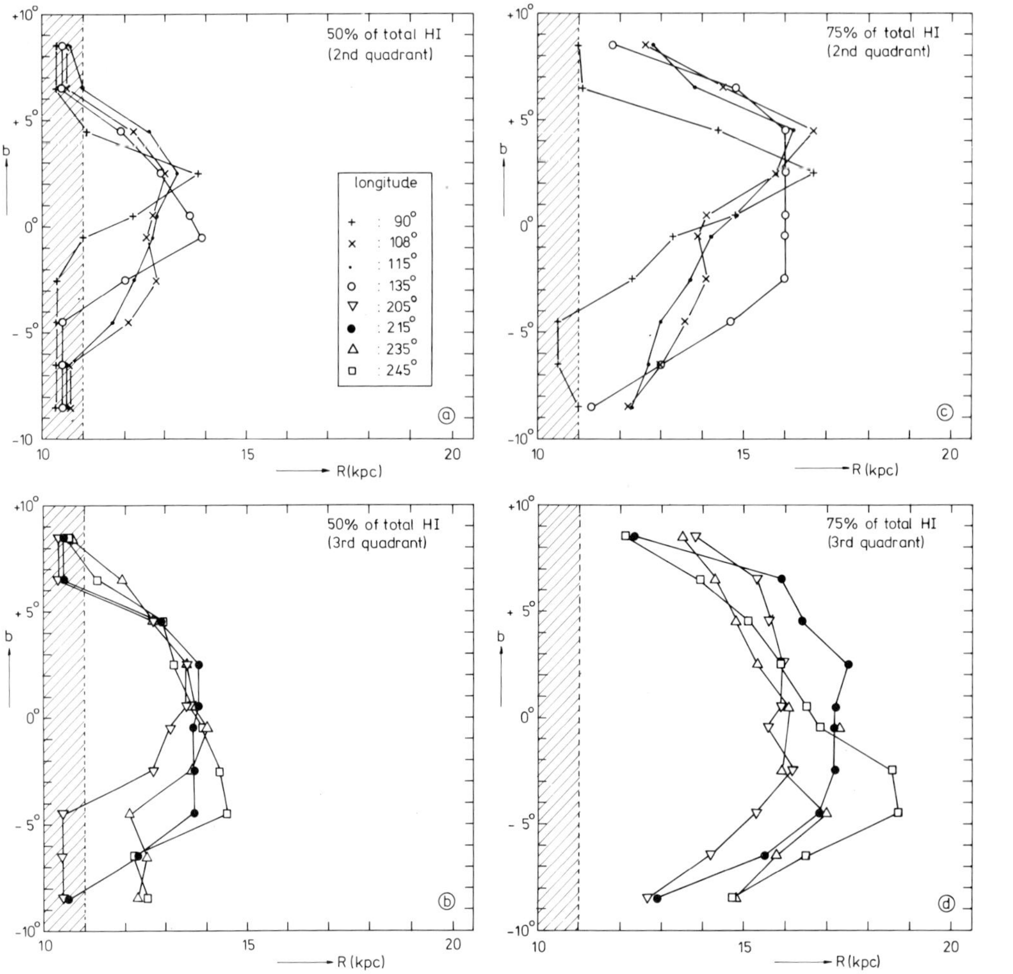

50% of total HI
(2nd quadrant)
longitude
+ : 90°
× : 108°
· : 115°
○ : 135°
▽ : 205°
● : 215°
△ : 235°
□ : 245°
a
75% of total HI
(2nd quadrant)
c
50% of total HI
(3rd quadrant)
b
75% of total HI
(3rd quadrant)
d
b
R(kpc)
+10°
+5°
0°
-5°
-10°
10
15
20

Figure 3. The distance within which either 50% (a and b) or 75% (c and d) of the total HI column density is located as a function of galactic latitude. Within R=10-11 kpc no subdivision has been made.

50% of differential HI
a
75% of differential HI
b
b
R(kpc)
10°
5°
0°
10
15
20

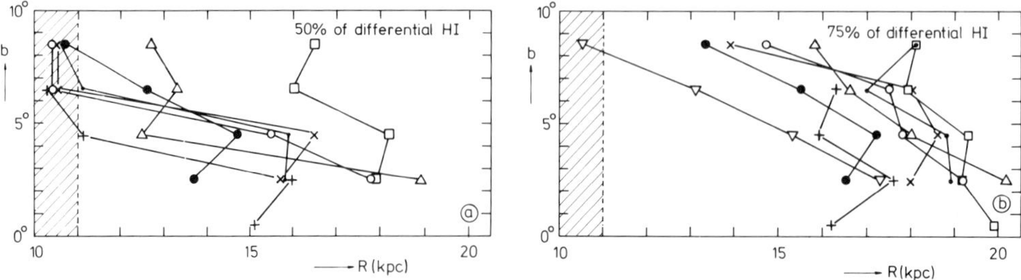

Figure 4. The distance within which either 50% (a) ar 75% (b) of the differential HI column density is located as a function of galactic latitude. The corresponding longitudes are shown in figure 3a. Within R=10-11 kpc no subdivision has been made.

within 13-15 kpc of the galactic centre. However, this relatively nearby gas is more closely confined to b=0° than more distant gas; the more distant gas is subject to the large scale warp (Kulkarni, Blitz and Heiles 1982). The plot of differential emission subtracts out the relatively symmetric nearby gas, leaving the asymmetric component which comes from corresponding more distant gas. Thus $\Delta\tilde{N}$(HI) and the corresponding gamma-ray emission are preferentially sampling the large-scale warp of the Galaxy.

Near $|b|\cong 2°$, $\Delta\tilde{N}$(HI) corresponds to 16 kpc for X = 50% and to 18 kpc for X = 75%. This implies that 25% of the differential emission is coming from distances between 16 and 18 kpc. The gamma-ray production rates derived for these latitudes are not substantially lower than those derived for any outer Galaxy location. This implies that there is a substantial gamma-ray flux up to distances of at least 18 kpc from the centre, and thus the cosmic rays which produce them are also plentiful at these distances.

In a recent study, Schlosser and Feitzinger (1982) claim that the major part of the gamma-ray emission in the second and third quadrants is produced in the local spiral arm (less than 600 pc from the sun). However, their results are biased or wrong in many aspects, e.g.:

- the velocity range used (-10 km $s^{-1} < v < +10$ km s^{-1}) is by definition representative for regions near the solar circle and local gas at higher latitudes
- their 'quantative' analysis is restricted to one gamma-ray, one galaxy-count and one HI isophote, on average in the latitude range $5° < |b| < 10°$, depicting predominantly the local interstellar matter
- their resulting gamma-ray emissivity per H atom is more than three times higher than the local production rate (see e.g. Lebrun *et al.* 1982, Strong *et al.* 1982), but was not noticed because they compared results for different energy ranges; this indicates their severe under-estimates of the column densities due to the limited velocity coverage
- their argument about the asymmetry around b=0° is not applicable locally only but is a large-scale galactic phenomenon out to 20 kpc (Henderson, Jackson and Kerr 1982, Kulkarni, Blitz and Heiles 1982)

We wish to emphasize the preliminary nature of our results, but note also that the method promises to provide important clues to the radial distribution of cosmic rays and gamma rays to large galacto-centric distances.

ACKNOWLEDGEMENTS

L.B. gratefully acknowledges the support of the Organization of the Advancement of Pure Research of the Netherlands (ZWO) and the General Research Board of the University of Maryland. The authors acknowledge the COS-B Collaboration for the permission to use unpublished data to produce figures 1 and 2c.

REFERENCES

Bignami, G.F., Caraveo, P.A., Maraschi, L.: 1978, Astron. Astrophys. 67, p. 149

Blitz, L., Fich, M., and Stark, A.A.: 1982, in preparation
Bloemen, J.B.G.M.: 1983, Proc. 'Kinematics, Dynamics and Structure of the Milky Way', ed. W.L.H. Shuter, Reidel Publishing Company, p. 31.
Caraveo, P.A., and Paul, J.A.: 1979, Astron. Astrophys. 75, p. 340
Caraveo, P.A., Bennett, K., Bignami, G.F., Hermsen, W., Kanbach, G., Lebrun, F., Masnou, J.L., Mayer-Hasselwander, H.A., Paul, J.A., Sacco, B., Scarsi, L., Strong, A.W., Swanenburg, B.N., and Wills, R.D.: 1980, Astron. Astrophys. 91, p. L3
Caraveo, P.A., Barbareschi, L., Bennett, K., Bignami, G.F., Hermsen, W., Kanbach, G., Lebrun, F., Masnou, J.L., Mayer-Hasselwander, H.A., Sacco, B., Strong, A.W., and Wills, R.D.: 1981, Proc. 17th Int. Cosmic Ray Conference (Paris) 1, p. 139
Cohen, R.S., Cong, H., Dame, T.M., and Thaddeus, P.: 1980, Astrophys. J. 239, p. L53
Dodds, D., Strong, A.W., and Wolfendale, A.W.: 1975, Mon. Not. R. astr. Soc. 171, p.569
Harding, A.K.: 1981, Astrophys. J. 247, p. 639
Harding, A.K., and Stecker, F.W.: 1981, Nature 290, p. 316
Heiles, C., and Cleary, M.N.: 1979, Aust. J. Phys. Astrophys. Supp. 47, p. 1
Heiles, C., and Habing, H.J.: 1974, Astron. Astrophys. Supp. 14, p. 1
Henderson, A.P., Jackson, P.D., and Kerr, F.J.: 1982, Astrophys. J. 263 (in press)
Hermsen, W.: 1980, Ph.D.thesis, University of Leiden
Kerr, F.J.: 1968, in Nebulae and Interstellar Matter - Stars and Stellar Systems, Vol. VII, eds. B.M. Middlehurst and L.H. Aller (Chicago: University of Chicago Press) p. 574
Kulkarni, S.R., Blitz, L., and Heiles, C.: 1982, Astrophys. J. 259,p. L63
Kutner, M.L., and Mead, K.: 1981, Astrophys. J. 249, p. L15
Lebrun, F., Bignami, G.F., Buccheri, R., Caraveo, P.A., Hermsen, W., Kanbach, G., Mayer-Hasselwander, H.A., Paul, J.A., Strong, A.W., and Wills, R.D.: 1982, Astron. Astrophys. 107, p. 390
Mayer-Hasselwander, H.A., Bennett, K., Bignami, G.F., Buccheri, R., Caraveo, P.A., Hermsen, W., Kanbach, G., Lebrun, F., Lichti, G.G., Masnou, J.L., Paul, J.A., Pinkau, K., Sacco, B., Scarsi, L., Swanenburg, B.N. and Wills, R.D.: 1982, Astron. Astrophys. 105, p. 164
Protheroe, R.J., Strong, A.W., Wolfendale, A.W., and Kiraly, P.: 1979, Nature 277, p. 542
Riley, P.A., and Wolfendale, A.W.: 1981, Rivista Del Nuova Cimento Vol. 4, no. 4: 1
Rothenflug, R., and Caraveo, P.A.: 1980, Astron. Astrophys. 81, p. 218
Salvati, M., and Massaro, E.: 1982, Mon. Not. R. astr. Soc. 198, p. 11
Schlosser, W., and Feitzinger, J.V.: Astron. Astrophys. (in press)
Strong, A.W., Bignami, G.F., Bloemen, J.B.G.M., Buccheri, R., Caraveo, P.A., Hermsen, W., Kanbach, G., Lebrun, F., Mayer-Hasselwander, H.A., Paul, J.A., and Wills, R.D.: 1982, Astron. Astrophys. (in press)
Strong, A.W., Murray, J.D., Riley, P.A., and Osborne, J.L.: 1982, Mon. Not. R. astr. Soc. 201, p. 495
Swanenburg, B.N., Bennett, K., Bignami, G.F., Buccheri, R., Caraveo, P. A., Hermsen, W., Kanbach, G., Lichti, G.G., Masnou, J.L., Mayer-Hasselwander, H.A., Paul, J.A., Sacco, B., Scarsi, L., and Wills, R.D.: 1981, Astrophys. J. 243, p. L69
Weaver, H., and Williams, R.W.: 1973, Astron. Astrophys. Suppl. 8, p. 1

LOCAL INTERSTELLAR GAS DISTRIBUTION FROM GAMMA-RAY EMISSION

A.W.Strong (on behalf of the Caravane Collaboration)
Istituto di Fisica Cosmica del CNR, Milano, Italy

ABSTRACT.

COS-B data in the latitude range $11^\circ < |b| < 19^\circ$ show the gamma-ray intensity to be closely correlated with the total line-of-sight absorption derived from galaxy counts. It is argued that to a good approximation the gamma-ray intensity is proportional to the total gas column density, and on this basis a map of the angular distribution of local molecular hydrogen is presented. This is compared with a similar map produced from galaxy counts. Several interesting structures are present, including regions of the Southern celestial hemisphere inaccessible to galaxy counts.

1. THE USE OF GAMMA-RAYS AS A LOCAL GAS TRACER.

At present there is no direct method to determine the distribution of total hydrogen column density for the local interstellar medium. The only direct measurements are via UV absorption towards nearby stars (Bohlin et al. 1978); the H_2 column densities are limited to about 100 directions. Indirect tracing of the total hydrogen via interstellar reddening of stars (e.g. Lucke 1978) severely undersamples the interstellar medium at intermediate latitudes, and is more suited to studying the three-dimensional distribution of dust at low latitudes.

Galaxy counts provide uniform sampling of total absorption in the Northern celestial hemisphere, and their use has been studied in detail (Heiles 1976, Burstein and Heiles 1978, Lebrun 1979, Strong and Lebrun 1982, Strong 1982). The main drawback of this method is the effect of various types of fluctuations: variations in gas-to-dust ratio, statistical effects for small numbers of galaxies per square degree and, perhaps most important, the unknown response of the counts to patchy absorption structure.

Surveys in the CO lines should eventually provide complete coverage at these latitudes, but the problems of deriving H_2 column densities from

W. B. Burton and F. P. Israel (eds.), Surveys of the Southern Galaxy, 99–105.

the CO profiles (e.g.Liszt , this Workshop) mean that this method will probably remain an essentially qualitative tracer of molecular hydrogen.

Gamma rays of energies above about 70 MeV produced in cosmic ray - gas interactions (bremsstrahlung and pion-decay) give in principle a linear and fully sampled tracer of the total hydrogen, provided the cosmic ray density can be assumed uniform and the contribution from other components such as discrete sources can be ignored. All gas regardless of scale is completely sampled. One conclusion of the work reported here (see Strong et al. 1982 for a detailed account) is that these assumptions are justified to a good approximation, and that the traditional idea of gamma rays as a diagnostic of cosmic-ray density may perhaps be turned around with advantage to delineate local gas structure.

Lebrun and Paul (1979, 1981) used the SAS-2 gamma-ray database to study correlations with galaxy counts. Lebrun et al. (1982) extended this work to the COS-B data, and showed that the absorption deduced from galaxy counts is much more successful than atomic hydrogen column density as a predictor of gamma-ray intensities at intermediate latitudes. Strong et al. (1982) developed this study to consider the spatial distribution of the gas and its interpretation.

2. COMPARISON OF GAMMA-RAY INTENSITIES WITH ABSORPTION FROM GALAXY COUNTS.

A description of the galactic emission measured by the COS-B satellite is given in Mayer-Hasselwander et al. (1982) and Hermsen (this Workshop) and the reader is referred to these works for details of the gamma-ray observations. The present study is restricted to $11^{\circ} < |b| < 19^{\circ}$ where both galaxy count and gamma-ray data can be used.

To compare gamma-ray intensity I_{γ} and total gas column density we assume

$$I_{\gamma} = (q/4\pi)\, \tilde{N}_{HT} + I_{b} \qquad (2.1)$$

where q is the emissivity per hydrogen atom, N_{HT} is the total gas column density and I_b is a residual background. The tilde denotes convolution with the detector point-spread-function for the energy range considered. To estimate N_{HT} we use the digitized Shane-Wirtanen galaxy count data (Shane and Wirtanen 1967, Seldner et al. 1977). Following Lebrun et al. (1982) we adopt

$$N_{HT} = 2\ 10^{21} \log_{10}(50/N_{g}) \text{ atoms cm}^{-2} \qquad (2.2)$$

where N_g = number of galaxies per square degree (here averaged over 3° by 3° cells). Fits to equation (2.1) gave $q/4\pi$ = (1.4, 0.53, 0.59) 10^{-26} photons s^{-1} sr^{-1} for energy ranges 70-150, 150-300 and 300-5000 MeV respectively. The longitude range 0 - 40° was excluded from this fit since it appeared to lie somewhat above the linear relation of the rest of the data; the choice of region for the fitting is the main source of uncertainty in

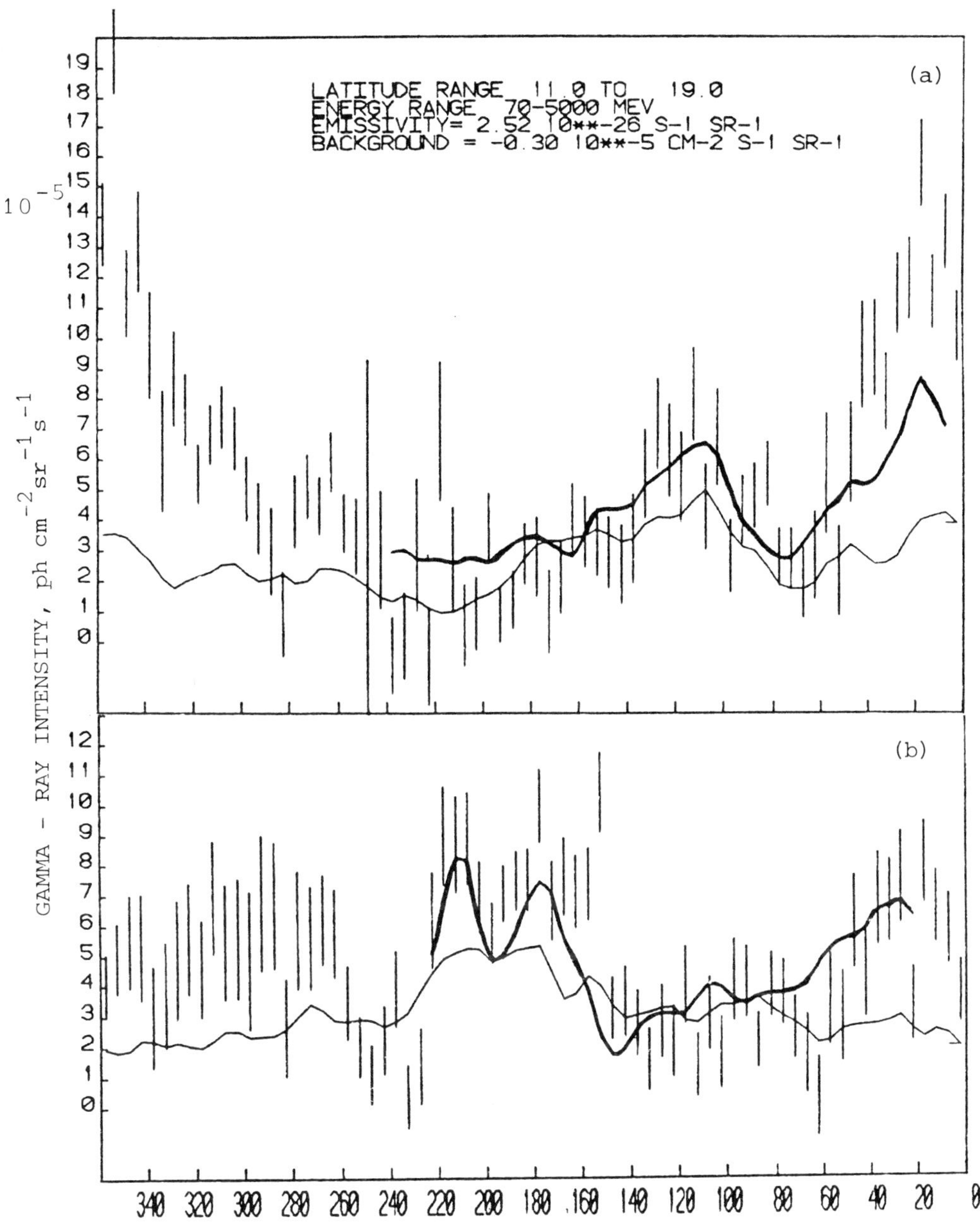

Fig 1. Comparison of gamma rays with galaxy counts and HI (a) $11^{o} < b < 19^{o}$, (b) $-19^{o} < b < -11^{o}$.
Error bars: gamma-ray intensity (70-5000 MeV)
Solid line: prediction for total gas from equations (2.1) and (2.2)
Thin line : prediction for HI using same q values as for total gas
Parameters : $q/4\pi = 2.52\ 10^{-26}\ s^{-1}\ sr^{-1}$, $I_b = -0.3\ 10^{-5}\ cm^{-2}s^{-1}sr^{-1}$.

the q values. At present an uncertainty of 25% should be taken for the above values.

Fig 1 shows the longitude distribution of expected and observed gamma-ray intensity (70-5000 MeV) in 5° by 8° bins. The expected value for total gas is from equations (2.1) and (2.2); the contribution from atomic hydrogen for the same emissivity is also shown. The similarity between the gamma-ray and galaxy-count distributions leads one to the conclusion that they depend on the same quantity, the total gas column density. The effects of variations in cosmic-ray density, presence of point sources, gas - to - dust variations etc. are not large enough to upset the general correlation.

The deviation in $0 < l < 40^{\circ}$, $b > 11^{\circ}$ can be partly explained by the failure of the logarithmic relation (2.2) of absorption to galaxy counts when the absorption is high (see Strong 1982); combined with the uncertainty in the q values there is no reason to consider this region abnormal at present, although the problem deserves further study.

3. LOCAL STRUCTURE IN THE INTERSTELLAR MEDIUM.

Fig 1 illustrates the requirement for a component in addition to atomic hydrogen to reproduce the gamma-ray distribution; the excess is presumably molecular hydrogen. Particular excesses occur in $0 < l < 30^{\circ}$, $b > 11^{\circ}$ in Sco-Oph, $160^{\circ} < l < 180^{\circ}$, $b < 11^{\circ}$ in Tau-Per and $205^{\circ} < l < 215^{\circ}$, $b < 11^{\circ}$ in Orion. The Orion region is discussed in more detail in Caraveo et al.(1980,81);new results on Orion and Oph are given by Hermsen (this Workshop). Another excess appears in $100^{\circ} < l < 120^{\circ}$, $b > 11^{\circ}$, and this corresponds to a feature in the Lucke (1978) interstellar reddening map. Bignami (1981) suggested a relation between this region and a local system dubbed the 'Dolidze Belt'; this would imply a corresponding excess 180° away at negative latitudes, and Fig 1 is consistent with such a feature as part of a much more extended region around $l = 290^{\circ}$.

Perhaps the result of most relevance to this Workshop is the extended region of large H_2 column density for $l > 300^{\circ}$, $b > 11^{\circ}$. This is not covered by the Lick galaxy counts but is revealed by the gamma-ray data. This region should have a high priority for molecular observations.

4. MAPS OF THE LOCAL MOLECULAR GAS FROM GAMMA RAYS.

The correlation between gamma-ray intensity is sufficiently good to suggest using the gamma-ray data to map the molecular gas column density. Fig 2 shows the result of inverting equation (2.1) and subtracting the atomic hydrogen column density derived by 21-cm surveys. The map is the average for three energy ranges. An identical presentation of a map derived from galaxy counts is also shown, allowing critical comparison of the two techniques. Features common to both maps have a good chance of being real gas enhancements; those occuring in only one are more likely

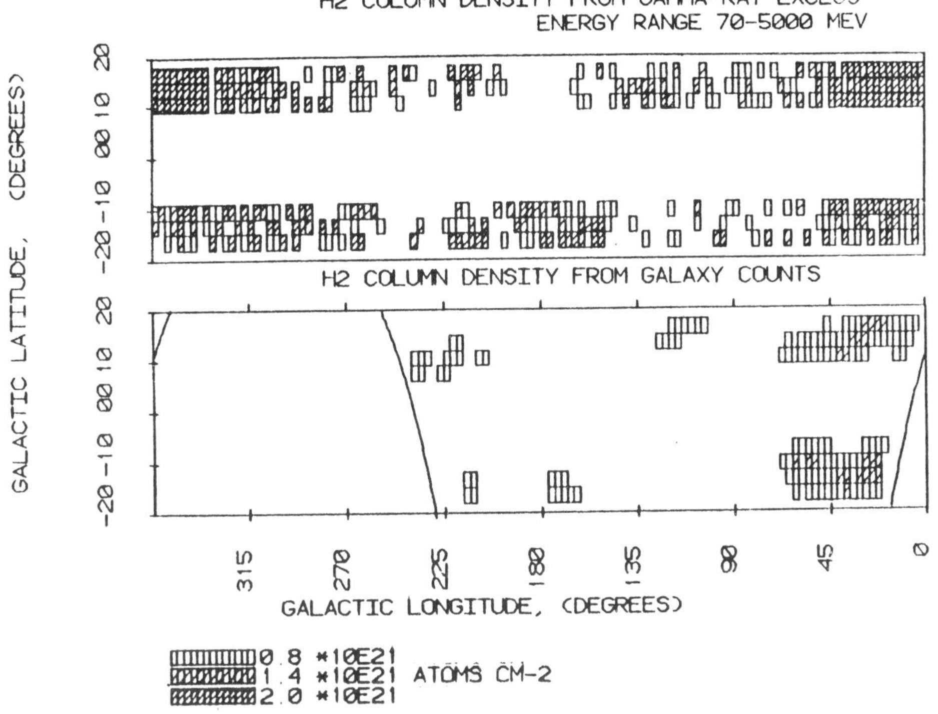

Figure 2. Maps of molecular hydrogen column density at intermediate latitudes, (a) from gamma rays (70-5000 MeV) (b) from galaxy counts. The Southern limit of the Lick counts is indicated by the continuous line. The absolute scale of the map from gamma rays is very sensitive to the adopted emissivity and should be used with caution.

of different origin (statistical, gamma-ray sources etc.). The overall agreement is encouraging and adds credibility to the claim that gamma rays trace mainly the total hydrogen.

5. ACCURACY OF COLUMN DENSITY ESTIMATES FROM GAMMA RAYS.

In Strong et al. (1982) it is concluded that there is evidence in the gamma-ray and galaxy-count data for H_2 column densities of the same order as those for HI in several extended regions at intermediate latitudes. Some authors have preferred to invoke variations in the gas-to-dust ratio rather than additional gas in molecular form when interpreting the galaxy counts (e.g. Burstein and Heiles 1978). It should be emphasized that the column densities given in Fig 2 are very sensitive to the adopted emissivity, particularly because of the subtraction of the HI. As an illustration, if the presently adopted emissivities give $N_{H2}=N_{HI}$ at some point, then if q has been underestimated by 25% the resulting N_{H2} has been overestimated by a factor 1.7. In future it should be possible to improve the gamma-ray estimates, but at present the absolute values should be viewed with caution.

Despite this caveat, the similarity of the maps of Fig 2 shows that the qualitative conclusions are not in doubt. The only alternative would be to assume that the gas-to-dust ratio correlates somehow with cosmic-ray intensity or density of point sources.

Regarding the plausibility of the high N_{H2} values, we can note that several Copernicus UV observations of highly reddened stars in Ophiucus and elsewhere (Bohlin et al. 1978) show column densities for atoms in molecular form comparable to those for atomic gas over distances of more than 100 pc. The high values are therefore not unexpected, but one must remember that there are also many Copernicus stars with negligible H_2 in the same regions, so that conclusions depend critically on selection effects in the sampling of the interstellar medium. Bohlin et al. give 34% as a lower limit on the ratio of atoms in molecular form to those in atomic form.

6. CONCLUSIONS.

Gamma rays are a useful quantitative tracer of total hydrogen at intermediate latitudes on scales greater than a few degrees. Unlike other tracers they provide a linear measure of gas in all phases. It may in future be possible to calibrate other tracers (such as CO) using the gamma-ray data, providing the sampling in such surveys is sufficient. On large scales the error in the gas estimate should eventually be less than the uncertainties involved in estimates based on CO.

The overview of local structures revealed in the gamma rays should provide useful clues as to what to expect and look for at other wavelengths. In particular, in the Southern celestial hemisphere, the longitude range $l > 300^{\circ}$, $b > 11^{\circ}$ is expected to contain much molecular gas.

REFERENCES

Bignami, G.F.: 1981, Phil.Trans.Roy.Soc.Lond.A.301, pp.555-568.

Bohlin, R.C., Savage B.D. and Drake J.F.: 1978, Astrophys.J. 224, pp. 132-142.

Burstein, D and Heiles, C.: 1978, Astrophys.J. 225, pp. 40-55.

Caraveo, P.A., Bennett, K., Bignami, G.F., Hermsen, W., Lebrun, F., Masnou, J-L., Mayer-Hasselwander, H.A., Paul, J.A., Sacco, B., Scarsi,L., Strong, A.W., Swanenburg, B.N., and Wills, R.D.: 1980, Astron.Astrophys. 91, pp. L3-L5.

Caraveo, P.A., Barbareschi, L., Bennett, K., Bignami, G.F., Hermsen, W., Kanbach, G., Lebrun,F., Masnou, J-L., Mayer-Hasselwander, H.A., Sacco, B., Strong A.W., and Wills; R.D.: 1981, Proc. 17th Int. Cosmic Ray Conference (Paris), 1, pp. 139-142.

Heiles, C.: 1976, Astrophys.J. 204, pp. 379-402.

Lebrun, F.: 1979, Astron.Astrophys. 79, pp.- 153-157.

Lebrun, F. and Paul J.A.:1979, Proc. 16th Int. Cosmic Ray Conference (Kyoto) 12, pp. 13-19.

Lebrun, F., Bignami G.F., Buccheri, R., Caraveo, P.A., Hermsen, W., Kanbach, G., Mayer-Hasselwander, H.A., Paul J.A., Strong A.W., and Wills R.D.: 1982, Astron.Astrophys. 107, pp. 390-396.

Lebrun, F.and Paul J.A.: 1981, preprint.

Lucke, P.B.: 1978, Astron. Astrophys. 64, pp.367-377.

Mayer-Hasselwander, H.A., Bennett, K., Bignami, G.F., Buccheri, R., Caraveo, P.A., Hermsen, W., Kanbach, G., Lebrun, F., Lichti, G.G., Masnou, J-L., Paul, J.A., Pinkau, K., Sacco, B., Scarsi, L., Swanenburg, B.N. and Wills, R.D.: 1982, Astron. Astrophys. 105, pp.164-175.

Seldner, M., Siebers, B., Groth, E.J. and Peebles, P.J.E.: 1977, Astron.J. 82, pp. 249-256.

Shane, C.D. and Wirtanen, C.A.: 1967, Pub. Lick. Obs. 22, pp. 1-60.

Strong, A.W.: 1982, Mon. Not. Roy. astr. Soc. 202, (in press).

Strong, A.W. and Lebrun,F.: 1982, Astron. Astrophys. 105, pp. 159-163.

Strong, A.W., Bignami, G.F., Bloemen, J.B.G.M., Buccheri, R., Caraveo,P.A. Hermsen, W., Kanbach, G., Lebrun, F., Mayer-Hasselwander, H.A., Paul, J.A. and Wills, R.D.: 1982, Astron. Astrophys. (in press).

WHAT WE SHOULD EXPECT IN THE SOUTHERN PLANE

Frank N. Bash
University of Texas at Austin

INTRODUCTION

Bash and Peters (1976) proposed that giant molecular clouds (GMC's) could be treated as ballistic particles launched from the Galaxy's spiral arms. Bash (1979) showed that best agreement with observations in the Milky-Way results when the ballistic-particle GMC's are launched from the TASS wave at the post-shock velocity and are abruptly destroyed 40 million years after birth. In the Galaxy, our model assumes the two-arm spiral pattern of Yuan (1969) and it predicts that GMC's will be closely confined to this spiral pattern. Our model does not suggest that all molecular gas is confined to the arms, just that the giant molecular clouds are. The initial post-shock velocity with which GMC's are launched in our model means that their initial velocity is slower than the circular velocity in the tangential direction and with a component directed toward the galactic center so that the Perseus arm molecular clouds at $\ell = 180°$ should have an LSR radial velocity of ~ -15 km s^{-1}.

Since the ballistic particle model's introduction, more and more observers find that CO-emitting giant molecular clouds are confined to spiral arms. Stark (1979) finds the CO clouds to be closely confined to the spiral arms along the minor axis of M31. Cohen, et al. (1980) suggest that CO is confined to spiral arms in the Galaxy. Linke (1981) finds CO to be confined to the optical spiral arms over a large portion of M31 and, especially interesting when compared to the specific predictions of our model, Thaddeus (1982) finds that $\ell = 180°$, the CO emission from the Perseus arm is centered at $v \simeq -15$ km s^{-1}.

In addition to this specific confirmation of our model, Bash (1981) shows that giant HII regions in the Galaxy have observed radial velocities and galactic longitudes consistent with the formation of massive stars in our ballistic molecular clouds. Hilton and Bash (1982) use the observed velocities of young stars near the Sun to confirm that those stars were born in ballistic clouds launched from the spiral arms at the post-shock velocity.

W. B. Burton and F. P. Israel (eds.), Surveys of the Southern Galaxy, 107–112.

CO VELOCITY LONGITUDE DIAGRAMS

In most of our work so far, e.g. Bash (1979), we have made the simple assumption that the lifetime of giant molecular clouds is the same all over the Galaxy. We make that same assumption here for the velocity-longitude diagrams which we display.

Leisawitz and Bash (1982) propose a more sophisticated, physical model in which the lifetimes of GMC's vary systematically over the face of a spiral galaxy and from one Hubble Type to another.

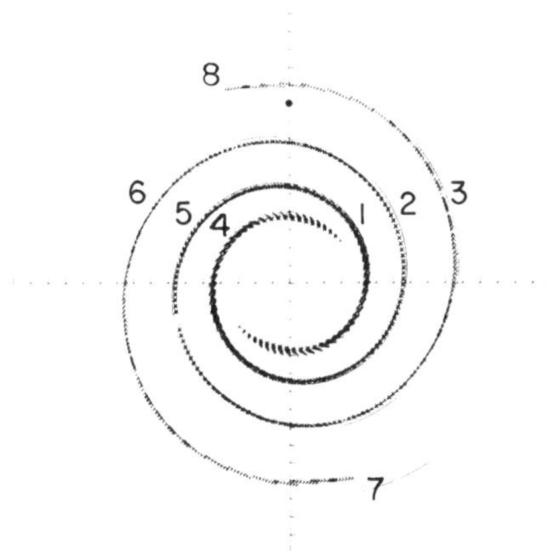

Figure 1 - The Milky Way as seen from the North Galactic Pole. The thin spiral is the density-wave potential minimum and the shaded area shows the location of ballistic particles launched from the TASS wave in the age range 25-40 million yrs. The tick marks are spaced by 1 kpc, the Sun is 10 kpc from the center and the numbers allow a comparison with Figure 2.

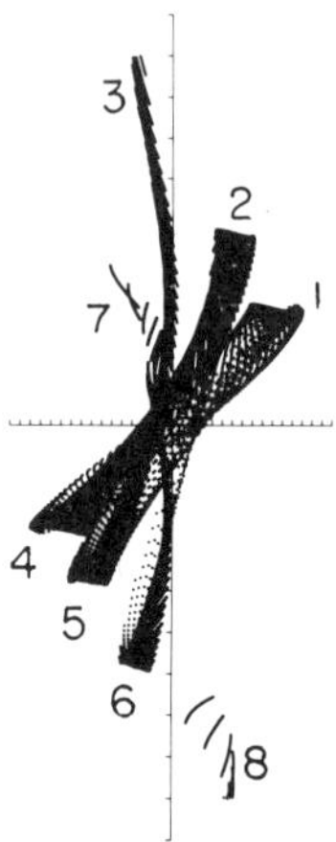

Figure 2 - A velocity-longitude diagram for CO predicted by our ballistic particle model. The abscissa is V_{LSR} and the tick marks are spaced by 10 km s^{-1}. The ordinate is galactic longitude with the tick marks spaced by 10°. The numbers allow one to associate arm features in Fig. 2 with spiral arms in Fig. 1. The GMC's shown here have ages from 0 to 40 million years.

Here, we compute velocity-longitude diagrams for CO by assuming that GMC's are launched from the TASS wave at the post-shock velocity. We use Yuan's (1969) spiral pattern and a hydrodynamic computer program written by Visser to find the location of the TASS wave. The rest of our assumptions are listed in Bash (1981). Ballistic particle GMC's are launched from a set of birthsites spaced by 5° in galactocentric azimuth along the TASS wave. Each particle's orbit is integrated as it moves in the overall, circularly symmetric, galactic gravitational potential perturbed by the potential in the spiral arms and each particle's position and velocity are tabulated each million years.

Figure 1 shows a view of the Galaxy from the North Galactic Pole. In this view the Galaxy rotates clockwise with the Sun's position marked at R_o = 10 kpc. The coordinate grid has tick marks each kpc. The thin spiral pattern is Yuan's (1969) density wave potential minimum and the shaded areas show the locations of ballistic particles launched from the arms with ages from 25-40 million years. We assume that GMC's radiate visible CO spectral lines from 0-40 million years and that star formation occurs 25 million years after the GMC's birth so Figure 1 shows where we believe the Galaxy's giant HII regions are located. A plot of the locations of GMC's over ages of 0-40 million years just fills-in the thin zone between the shaded area and the spiral potential minimum. Therefore we predict that GMC's in our Galaxy lie in an obvious spiral pattern but that they do not move on circular orbits. Our density wave model does not extend inside the Inner Lindblad Resonance so it is not able to describe the "expanding 4 kpc arm" or anything inside it. The numbers on Figure 1 will be used to associate arms on that figure with features in the velocity-longitude diagram.

CO IN THE SOUTHERN MILKY WAY

Figure 2 shows the velocity-longitude diagram for GMC's predicted to result from our model. Galactic longitude is shown along the ordinate with the galactic center in the center of the figure and the tick marks spaced by 10°. The abscissa shows radial velocity with zero in the center, and the tick marks spaced by 10 km s^{-1}. The spiral arm features in Figure 2 are composed of stripes which, in turn, are themselves composed of lines of "X's". Each "X" shows the position of a model GMC each million years and each stripe is the trajectory of a single GMC from birth to age 40 million years. Along some trajectories the model GMC's move rapidly and the "X's" are far apart. Along others the GMC's move slowly and the X's are close together. In some parts of Figure 2 the large density of "X's" is due to several model GMC's having nearly the same values of velocity and longitude. In general we predict that stronger CO signals should be seen in these portions of the diagram where the density of "X's" is larger. The effect of the distance from the Sun to the GMC will be discussed later.

The spiral arm features in Figure 2 are numbered and the numbers correspond with those in Figure 1. Arm 1, the Scutum Arm, is tangent to the line of sight at about ℓ = 30°, v = 130 km s^{-1}. In general arms are seen not to be open loops in Figure 2 but rather broad, curved things. Arm 2, the Saggitarius Arm is tangent at ℓ = 50°, v = 80 km s^{-1} and Arm 3, the Perseus Arm is never tangent to the line-of-sight. In Figure 2 no point is plotted between $\ell = 90°$ and $\ell = 270°$. Arms 1, 2 and 3 lie on the northern side of the Galaxy and arms 4, 5, and 6 are the respective extensions of those arms into the southern side. Arms 7 and 8 are the "tails" of the Perseus Arm across the line through the Sun and galactic center. The tail is cut-off entirely arbitrarily in the model.

If we concentrate on the two quadrants $v > 0$, $90°>\ell>0°$ (northern) and $v < 0$, $360°>\ell>270°$ (southern) we note that only two prominent arm features enter the northern quadrant whereas three are seen in the southern one. Arm 4 lies inside the Scutum arm and is tangent to the line-of-sight at $\ell = 335°$, $v = -140$ km s^{-1}, the southern Scutum arm (Arm 5) at $\ell = 322°$, $v = -90$ km s^{-1} and the southern Sagittarius arm (Arm 6) at $\ell = 300°$, $v = -40$ km s^{-1}. It is interesting to see that Arm 6 is predicted to be rather clearly separated from Arms 4 and 5 in the velocity-longitude diagram.

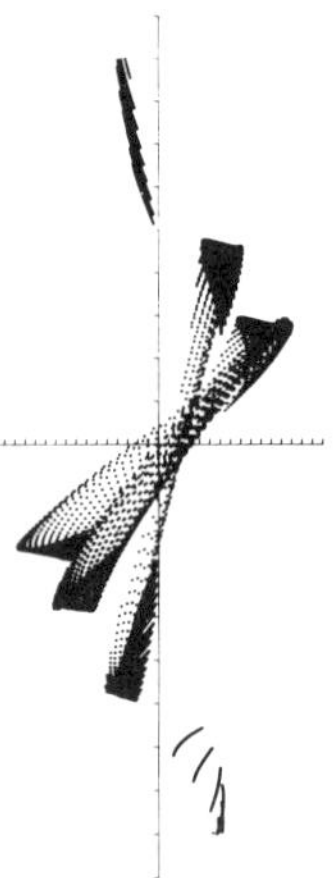

Figure 3 - Same as Figure 2 except that it excludes GMC's on the far side of the Galaxy from the Sun, that is those below the horizontal line passing through the galactic center in Figure 1.

Figure 3 shows an attempt to include, in our predicted velocity-longitude diagram, the effect of the distance from the Sun to the GMC's. Figure 3 is exactly the same as Fig. 2 except that only model GMC's on the Sun's side of the galactic center are shown. That is, those GMC's below the horizontal line through the galactic center in Figure 1 have been eliminated. Now whole features have disappeared like that labeled Arm 7 in Figure 2 and the portion of Arm 3 below $\ell = 50°$. The density of "X's", predicting the observed CO signal strength, is still rather large at the terminal ends of the arms but is surprisingly low along the main body of the northern Sagittarius arm feature (Arm 2). These GMC's are rather close to the Sun, especially near $\ell = 0°$, but they cover large angles in one time-step giving a lower density of X's in Figure 3 and lower predicted CO strength in that region of the diagram.

Although the Galaxy's arms aren't smooth and regular and some CO is present between the arms, we predict the three arm features (Arms 4, 5,

6) shown in Fig. 3, with the terminal ends showing the strongest CO signal, will be found in velocity longitude diagrams from the southern surveys.

Robinson, McCutheon and Whiteoak (1982) have published preliminary results of a southern galactic plane CO survey. The coverage available at the time and the quality of their published velocity-longitude diagram does not allow verification of the above predictions but the authors do comment that the feature which is seen in northern surveys at $v \sim +10$ km s^{-1}, $15° < \ell < 40°$ is not present in their survey in the region $v \simeq -10$ km s^{-1} and $300° < \ell < 340°$. In terms of the global spiral structure, Figures 1, 2 and 3 show Arm 3 which arises from the Perseus Arm but we predict no analogous feature on the southern diagram except for a narrow, curved feature in Figure 2 arising from the Sagittarius arm on the other side of the Galaxy. If the Perseus Arm contained enough GMC's to produce observable CO and if it extended farther than the cut-off in Figure 1 then we should see a feature at $v \simeq -10$ km s^{-1} and $300° < \ell < 340°$.

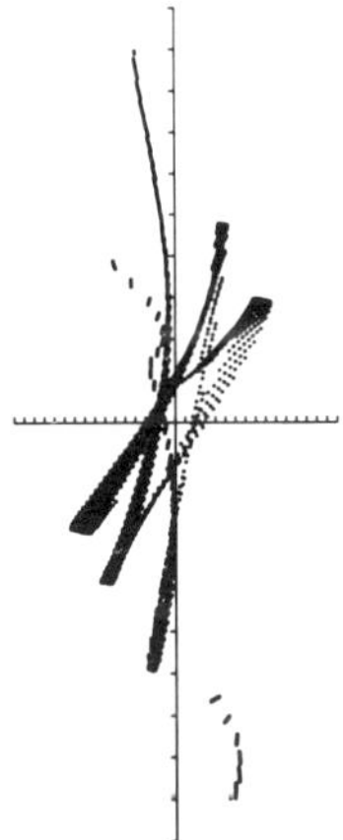

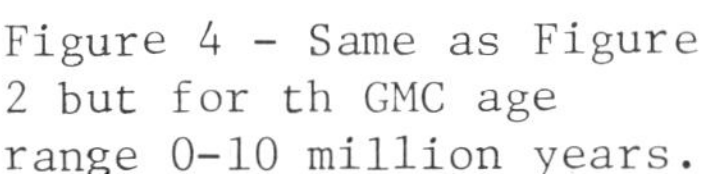

Figure 4 - Same as Figure 2 but for th GMC age range 0-10 million years.

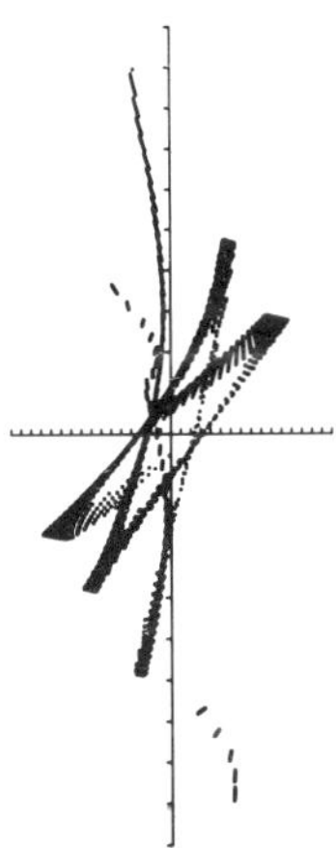

Figure 5 - Same as Figure 2 but for the GMC age range 11-20 million years.

Finally, Figures 4, 5, 6, and 7 show the effect of their age on the expected velocities and longitudes of GMC's. In these successive figures, the 0-40 million year age span for GMC's shown in Figure 2 is broken into 4, 10 million year age ranges (11 million years for Figure 4). Many detailed changes can be seen in the shape, width and curvature of the spiral arm features as the age range changes but the most obvious and observable change is the increase of the value of the terminal velocity of each arm with age. Given a galactic rotation curve, the observed CO terminal velocity is a function of the maximum age of the GMC's. In the context of the current model, a maximum GMC age of 40 million years is the best fit to northern hemisphere CO surveys. Again, if this model is correct, the same maximum age should fit the southern data best.

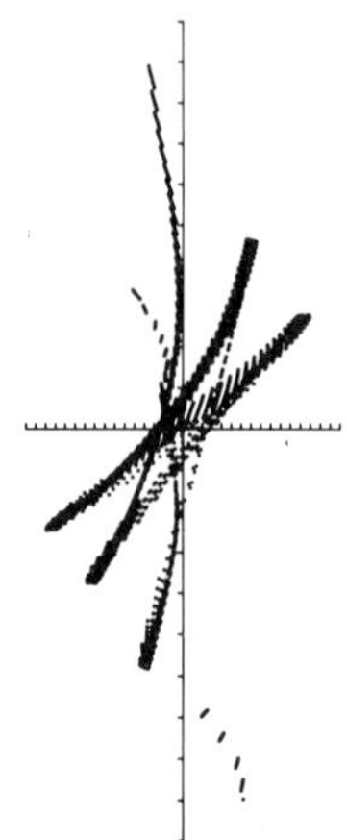

Figure 6 - Same as Figure 2 but for the GMC age range 21-30 million years.

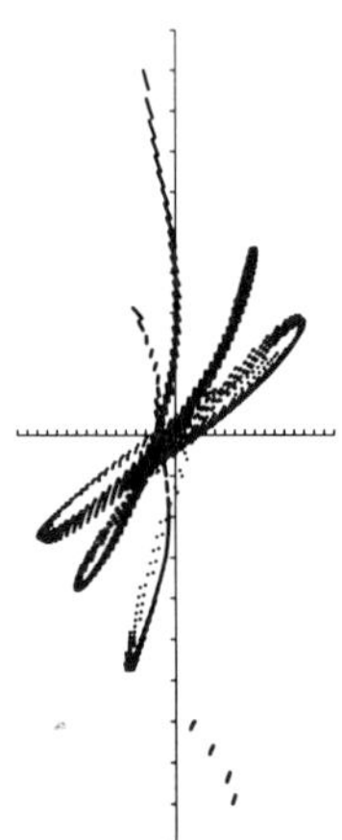

Figure 7 - Same as Figure 2 but for the GMC age range 31-40 million years.

SUMMARY

We have used our ballistic particle model to make several predictions about what should be found in CO surveys of the southern galactic plane.

REFERENCES

Bash, F. N. 1981, Astrophys. J., 250, 551.
Bash, F. N. 1979, Astrophys. J., 233, 524.
Bash, F. N., and Peters, W. L. 1976, Astrophys. J., 205, 786.
Cohen, R. S., Cong, H., Dame, T. M., and Thaddeus, P. 1980, Astrophys. J., 239, L53.
Hilton, J. and Bash, F. N., 1982, Astrophys. J., 255, 217.
Leisawitz, D., and Bash, F. N. 1982, Astrophys. J., to appear in Aug. 1 issue.
Linke, R. A. 1981, Extragalactic Molecules (Green Bank: National Radio Astronomy Observatory).
Robinson, B. J., McCutcheon, W. H., and Whiteoak, J. B. 1982, International Journal of Infrared and Millimeter Waves, 3, 63.
Stark, A. A. 1979, Galactic Kinematics of Molecular Clouds (Ann Arbor: University Microfilms International).
Thaddeus, P. 1982, private communication.
Yuan, C. 1969, Astrophys. J., 158, 871.

RECENT 21CM SURVEYS OF THE SOUTHERN MILKY WAY AND THE DISTRIBUTION OF HI BEYOND THE SOLAR CIRCLE

F. J. Kerr
University of Maryland

Abstract

Two recent 21cm surveys with the Parkes 18m and 65m telescopes are discussed, together with a new compilation of high-resolution 21cm data right around the galactic plane. The most clearcut results are for the outer Galaxy, where the Parkes 18m and Hat Creek 26m surveys have been used to study the distribution of HI based on a circular rotation model. The surface density distribution is discussed, and also the warping and the thickness of the HI layer. The total mass of HI in the Galaxy is found to be 4.8 x 10^9 $M_\odot$.

Time will not permit an overall review of the large-scale structure of atomic hydrogen. Also, the results of the other recent study (by Kulkarni, Blitz, and Heiles) are being discussed separately. This paper will therefore concentrate on the recent work of the Maryland group.

1. Outer region

In two recent large-scale studies, special attention has been directed to the outer regions of the Galaxy, because the problem of the distance ambiguity is avoided there. One of these studies is based on comparisons between southern and northern data. The southern results come from a completely sampled Southern Milky Way neutral hydrogen survey, which was carried out at Parkes several years ago by F. J. Kerr, P. F. Bowers, and M. Kerr, using the 18-meter telescope and covering a range $\ell = 240°$ to $350°$ in galactic longitude, and $b = -10°$, to $+10°$ in galactic latitude. The velocity resolution was 2.1 km s^{-1} , and the spatial resolution 48 arcmin. These data have been combined with the analogous northern survey by Weaver and Williams (1974) with the Hat Creek 26-meter telescope in the outer region study of Henderson, Jackson and Kerr (1982). Optical depth effects

W. B. Burton and F. P. Israel (eds.), Surveys of the Southern Galaxy, 113–116.

were taken into account in this study. In this analysis, a flat rotation curve was assumed outside the solar circle, because the observational evidence reported for a rising curve (Blitz 1979) is limited so far to a rather small range of galactocentric angles, and other galaxies seem to show flat curves.

As mentioned above, the related study by Kulkarni, Blitz, and Heiles (1982), which uses a new analysis of the data of Weaver and Williams, is being described in another paper in this workshop.

1.1 Density pattern

Henderson et al. (1982) examined the overall density pattern for the HI in the outer Galaxy by integrating over the whole latitude range and then plotting out the density projected on to the galactic plane. They show that the pattern is dominated by spiral features on both sides of the diagram, including the well-known Perseus arm, Carina arm, and the "Outer arm". A strong feature near R = 11 kpc and φ (galactocentric angle) = 340° extends over a wide latitude range, and is probably a relatively local feature with a non-circular velocity, receding from the Sun.

The bilateral symmetry which would be expected from a two-armed "grand design" is clearly not present. In addition, the HI extends further out in galactocentric radius in the third and fourth quadrants of longitude (the "southern" side) than in the first and second quadrants. This interpretation depends on the correctness of the assumption of circular motion, which determines the conversion from velocity to distance. Another striking feature of the pattern is the unbalanced appearance near the blank regions around ℓ = 0° and 180°. There would be a more symmetrical appearance if the local standard of rest were postulated to move outwards by 7-10 km s^{-1} with respect to the center of mass of the Galaxy (Shuter 1981; Henderson et al. 1982). Alternatively the motions might be elliptical, or indeed the structure could be quite asymmetrical.

The radial distribution of the mean projected density of the HI can be obtained by summing over φ for various ranges of R. This has been done for the two intervals φ = 30° to 150° and φ = 210° to 330°. These summations show again the large departure from bilateral symmetry, when distance estimates are based on circular motion. The total mass of HI outside R = 11 kpc is found to be 1.12 x $10^9 M_\odot$ and 1.49 x $10^9 M_\odot$ for the two quoted intervals of R. After allowing for the mass in the inner region, and in the gaps near ℓ = 0° and 180°, Henderson et al. (1982) derived a figure of 4.8 x $10^9 M_\odot$ for the total HI mass in the Galaxy, of which 81% lies outside R = 11 kpc. These mass estimates are all very dependent on the assumptions of circular motion and a flat rotation curve.

1.2 Warp

Discussions of the outer warp of the Galaxy have always depended on earlier and lower-sensitivity surveys. The shape of the warp has been more precisely delineated in the recent studies. Again we have a striking departure from symmetry. For the north, the warping increases progressively and rapidly with R, whereas in the south the centroid of the HI distribution reaches a maximum negative value of -850 pc, near R = 17 kpc and $\varphi = 260^{\circ}$, and then returns to the plane at larger R. The axis of the warp runs from 260° to 80° in φ; this direction is inclined about 10° to the Sun-center line, and points approximately in the direction of the Large Magellanic Cloud. There is an interesting "scalloping" effect in the outermost part of the warp at $R > 24$ kpc, which appears on both sides of the map.

1.3 Layer thickness

The third quantity examined was the neutral hydrogen layer thickness. The recent studies refine the long-known result that the thickness increases rapidly with R in the outer regions, approaching a value of 2 kpc near the outer edge of the Galaxy. Henderson et al. find that the apparent thickness is less near the Sun than in the far part of the Galaxy beyond the central region, but this result, which occurs on both sides of the direction to the center, is probably an artefact of the observations. The calculations for the regions of the Galaxy at the greatest distance are affected by HI emission from nearby regions, either as "stray radiation", or through local material at noncircular velocities being attributed to greater distances.

1.4 Terminal velocities

The terminal velocities have also been re-examined, through the line profiles obtained for the outer regions on the two sides of the Galaxy. The newer, more sensitive data show that the southern terminal velocities are about 10-15 km s^{-1} higher than those for the northern side (Jackson and Kerr 1981). This result may be influenced by differences in the warp on the two sides, but a straightforward interpretation suggests a value of 250 km s^{-1} for the circular velocity at the Sun, Θ_0, rather than the smaller value suggested by some previous workers, and also indicates that the hydrogen density probably falls off more slowly than the exponential shape previously assumed.

2. Inner region

Interpretation of HI observational data in the inner part of the Galaxy is more difficult because of the distance ambiguity there. D. L. Ball at Maryland has been looking at the data from our two recent southern HI surveys (Kerr, Bowers and Henderson 1981; Kerr, Bowers and Kerr 1982). He has been carrying out a modeling study, based on a range of density-wave parameters; this study has used both the linear and nonlinear theories, and also introduces the z-dimension so that comparisons can be made with the full latitude range of the observational data. An important conclusion from this work is that the Galaxy is more irregular than used to be thought. Each observed feature must be fitted separately to an appropriate model, with different parameters, rather than to a single grand design.

3. Galactic equator diagram

We have put together a new presentation of high resolution 21-cm data right around the galactic equator, in the form of a detailed array in the longitude-velocity plane. The data come from three Maryland surveys, (i) $\ell = 241^\circ - 349^\circ$, by Kerr, Bowers and Henderson (1981) with the Parkes 64-meter telescope, (ii) $\ell = 349^\circ - 11^\circ$, by Sinha (1979) with the Green Bank 43-meter telescope, and (iii) $\ell = 11^\circ - 231^\circ$, by Westerhout (1973, 1982) with the Green Bank 92-meter telescope. A color version of this overall ℓ-v plot will be published shortly and the compilation is also available on magnetic tape, with profiles every 3 arcmin. In addition to the main spiral features, a very large amount of detail can be seen in this combined plot of the hydrogen distribution, including a substantial number of "holes" in the hydrogen in limited longitude and velocity ranges.

References

Blitz, L.: 1979, Astrophys. J. Letters, 231: pp. L115-L119.
Henderson, A. P., Jackson, P. D., and Kerr, F. J.: 1982, Astrophys. J., in press.
Jackson, P. D.: 1976, Astron. Astrophys. Suppl. 25, pp. 449-452.
Jackson, P. D., and Kerr, F. J.: 1981, Bull. Amer. Astron. Soc., 13, p. 538.
Kerr, F. J., Bowers, P. F., and Henderson, A. P.: 1981, Astron. Astrophys. Suppl. 44, pp. 63-75.
Kerr, F. J., Bowers, P. F., and Kerr, M.: 1982. In preparation.
Kulkarni, S., Blitz, L., and Heiles, C.: 1982, Astrophys. J. Letters,
Shuter, W. L. H.: 1981, Mon. Not. Roy Astron. Soc., 199, pp. 109-113.
Sinha, R. P.: 1979, Astron. Astrophys. Suppl., 37, pp. 403-463.
Weaver, H. F., and Williams, D. R. W.: 1974, Astron. Astrophys. Suppl., 17, pp. 1-249.
Westerhout, G.: 1973, Second Maryland-Green Bank 21-cm Line Survey. University of Maryland.
Westerhout, G., and Wendlandt, H. U.: 1982, Astron. Astrophys. Suppl., 49, pp. 143-245.

MILKY WAY SPIRAL STRUCTURE: A NEW LOOK

Leo Blitz
University of Maryland

ABSTRACT

The spiral structure of the outer Milky Way based on two recent HI studies is shown to imply that beyond R = 10 kpc, the Galaxy is a regular, four-armed spiral galaxy. The spiral pattern is extrapolated inward and compared to the results of northern and southern hemisphere CO surveys. The positions of the maxima of the CO emissivity can be reproduced which suggests that at least part of the emissivity of the "molecular ring" is related to spiral structure. Comparison of survey results with predicted ℓ-υ diagrams indicates that although a number of features seen in the CO surveys can be explained, the inner Galaxy is more complex than an extrapolation of the relatively simple outer structure would suggest.

The mapping of the spiral structure of the Milky Way has been a notoriously difficult task. Among the many problems have been the confusion caused by velocity streaming (Burton 1971, 1972), the distance ambiguity in the inner Galaxy, and the lack of objective criteria for identifying spiral arms. The last difficulty has introduced an unwelcome degree of subjectivity to spiral structure studies, and there is as yet no consensus on what the global appearance of the Milky Way is when viewed normal to the disk. For example, one would like to know to what degree the spiral arms are regular and symmetric, what is the length scale of the spiral features, and to what degree the structure is dominated by "spurs:" features that can be identified only over a small fraction of a radian.

Because there is no distance ambiguity for material beyond the solar circle, analysis of the HI in the outer Galaxy has the potential for providing a clearer picture of the spiral arm pattern than can be obtained in the inner Galaxy. Two independent, complementary analyses of the HI beyond the solar circle by Kulkarni, Blitz and Heilies (1982 hereafter KBH) and Henderson, Jackson and

W. B. Burton and F. P. Israel (eds.), Surveys of the Southern Galaxy, 117–125.

Kerr (1982 hereafter HJK) are used here to examine the spiral structure in the outer Galaxy. The observed structure in the outer Galaxy is then used to predict the appearance of longitude-velocity plots of the gas in the inner Galaxy, assuming that the spiral pattern is unchanged inwards to a distance of 4 kpc from the center.

SPIRAL ARMS IN THE OUTER GALAXY

For the purpose of defining the spiral arms, the longitude-velocity plots of HI surface density from KBH are used. These are preferable to those of HJK because no distortions are introduced in going from observed ℓ-v space to the transformed R,θ space of HJK. The KBH plots are reproduced in Figure 1. The shaded contours show a high degree of coherence and connectedness. In the second and third quadrants, features marked A and B are well separated except near $\ell = 90^o$. Feature B appears to continue across the anticenter; the continuity is determined on the basis of velocity. The surface density of feature C is underestimated, especially in the second quadrant because the analysis did not consider gas at $|b| > 10^o$. In any event, feature C also appears to be continuous across the anticenter.

KBH have called feature A the Cygnus arm; features B and C correspond to the Perseus and Orion arms respectively. All of the features are at increasing galactic distance with increasing longitude, thus the features are trailing spiral arms. The linear extents of the arms in Fig. 1 are 25 kpc, 20 kpc and 4 kpc for the Cygnus, Perseus and Orion arms respectively. Thus the outer Galaxy possesses well defined, large scale, coherent spiral arms. Assuming that the arms are logarithmic spirals, the mean pitch angles for the Cygnus and Perseus arms are 22° and 27°, considerably higher than the pitch angles ususally derived for the features in the inner Galaxy (e.g. Lin, Yuan and Shu 1969). The pitch angles are, however, dependent on the assumed rotation curve. For a flat rotation curve (which is what was used in the HJK analysis), the pitch angles for the two arms are 16° and 20° respectively.

If the Perseus and Orion arms are plotted in galactocentric coordinates, the pattern in Figure 2 is obtained. The angle that the arms make with the galactic center at a given radius is close to 90°. Thus, one would expect that if the spiral pattern of the Milky Way were reasonably symmetric, the Galaxy should have 4 arms beyond the solar circle, and the positions of the additional arms would be determined from the locations of the Perseus and Cygnus arms (the Orion arm appears to be a spur). These arms are shown as fine lines in Figure 2, assuming a pitch angle of 22°; the dotted lines in Figure 2 are extrapolations of the four arms to a galactocentric radius of 4 kpc.

One of the major differences between the KBH and HJK analysis is that HJK have included data from the southern hemisphere; thus their

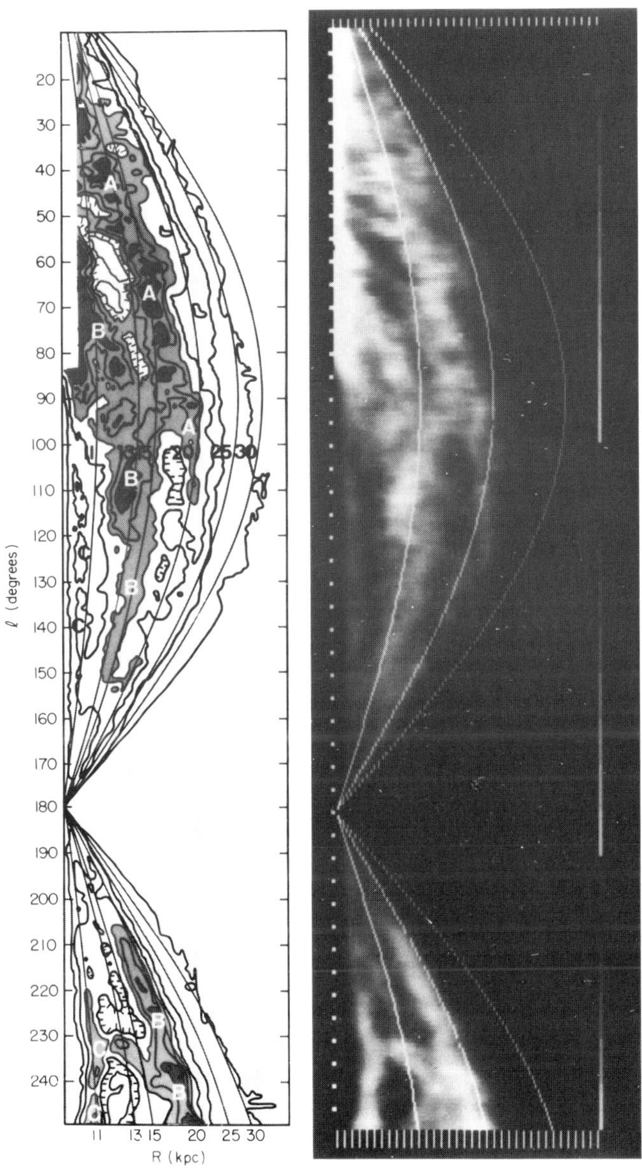

Figure 1 - Longitude-velocity plot from KBH of HI surface density beyond the solar circle. The left hand panel is a contour map with contour intervals are 1 $M_\odot$ pc^{-2}, and the right hand panel is a gray scale plot with density intervals of 0.5 $M_\odot$ pc^{-2}. Lines of constant galactocentric radius are shown which correspond to R = 13, 20 and 30 kpc on the gray scale plot. The features described in the text as the Cygnus, Perseus and Orion arms are marked A, B and C respectively.

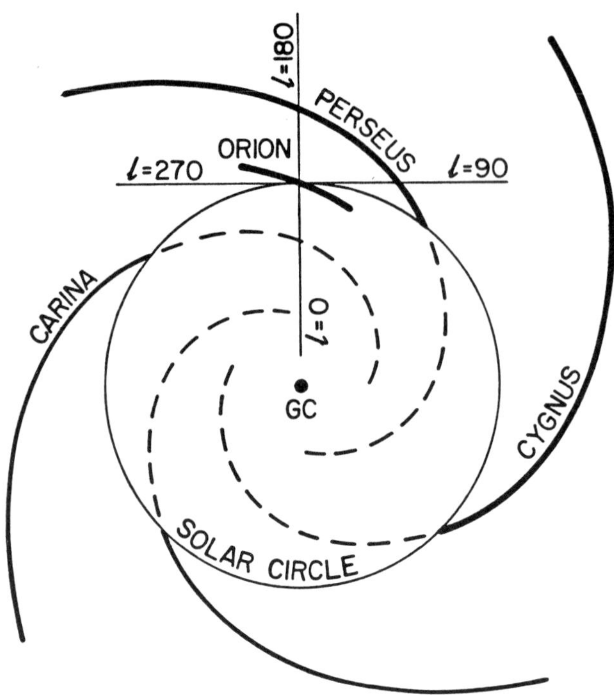

Figure 2 - A face on view of the spiral arm ridge lines from Figure 1. The heavy dark lines are from the data. The fine dark lines are drawn assuming the outer galaxy has 4-fold symmetry. The dashed lines are an extrapolation of the outer arms inward to a distance of 4 kpc from the center.

analysis is particularly useful to see if there are features which correspond to the two additional arms implied by a symmetric Galaxy. Figure 3 shows the major arms in Figure 2 corrected for a flat rotation curve superimposed on the HI surface density plot of HJK. There is good agreement between HJK and KBH on the locations of the Perseus and Cygnus arms. In fact, Figure 3 shows that the Perseus arm extends only about 10° in longitude beyond what is shown in Figure 1. Note particularly, the good agreement between the position of the expected arm near ℓ = 260° and the position of the major HI feature which corresponds to the Carina arm. The arm has an extent comparable to that of the Perseus and Cygnus arms, but appears to have a smaller pitch angle. The fourth arm cannot be followed for most of its length because it passes behind the galactic center. Nevertheless a strong HI feature is seen at the position where the arm is expected to cross the solar circle. Sills (1982) has found an HI feature which may correspond to the extension of the arm to low longitudes.

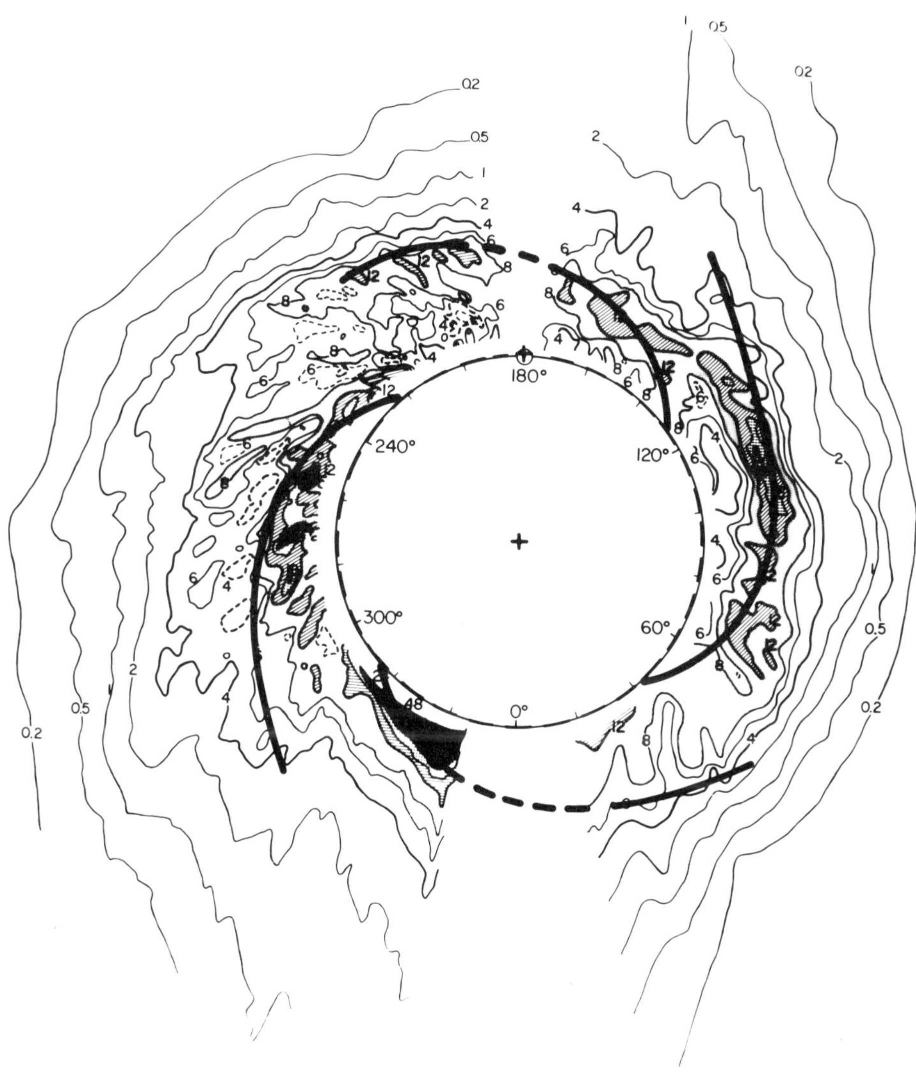

Figure 3 - A superposition of the arms from Figure 2 onto the HI surface density plot of HJK. The pitch angles of the arms are corrected for a flat rotation curve which was assumed in the HJK analysis.

It seems reasonable to interpret Figure 3 as follows: beyond the solar circle, the Milky Way is a relatively symmetric four-armed Galaxy with a structure dominated by regular, large scale spiral arms. The pitch angles of the arms seem to vary somewhat, but such a variation could be at least partially due to asymmetries in the large scale velocity field of the Milky Way. The mean pitch angle of the arms is ~22° assuming that the arms are logarithmic spirals and the rotation curve is similar to that of Blitz, Fich and Stark (1980). There seem to be several HI spurs present: the Orion arm, for example, does not connect to any large scale feature in the HJK map. The connection in velocity between the Perseus and Cygnus arms near $\ell = 90^{o}$ may represent another spur. The four outer arms cross the solar circle at longitudes of 23°, 66°, 288° and 338° with uncertainties of ~ $\pm 5^{o}$.

The crossing points are particularly important for understanding how the large scale spiral pattern of the outer Galaxy relates to that of the inner Galaxy. Any spiral pattern deduced from observations of the inner Galaxy must be continuous across the solar circle and the crossing points are the boundary conditions which must be satisfied. Kennicutt's (1982) study of the structure of "grand design" spirals shows that pitch angle variations along the spiral arms and from arm-to-arm are common, but a sudden jump in pitch angle across the solar circle is a serious flaw in any model of spiral structure.

EXTRAPOLATION TO THE INNER GALAXY

The inner Galaxy has been well surveyed at small longitudes (Cohen, Cong, Dame and Thaddeus; 1981 - hereafter CCDT), and the papers presented by Robinson and Manchester at this meeting provide the first reliable survey results at large longitudes. CCDT have shown that the inner Galaxy CO is well correlated with the well known HI enhancements, but the CO shows greater contrast. The CO data should therefore provide a good test as to how well the outer Galaxy spiral pattern can be extrapolated inward. It is important to keep in mind Burton's (1971, 1972) admonitions about interpreting structure in velocity space as density structure. The tests below are, in principle, investigating whether the spiral arms identified in _velocity_ space are continuous across the solar circle.

i) Radial Gas Distribution

If CO is concentrated in spiral arms, one expects to see enhancements in the CO at longitudes corresponding to lines of sight which are tangent to the arms. The reason for this is that small changes in longitude correspond to large changes in length measured along spiral arms, which encompass a large number of molecular clouds. Thus, when integrated over velocity, the CO emissivity of Figure 2 shows that in the northern hemisphere (low longitudes) there is one line of sight near $\ell = 30^{o}$ which is tangent to the spiral arms, but in the southern hemisphere thre are two, one near $\ell = 325^o$ and one near $\ell = 340^o$. Of the two expected maxima in the south, the one near $\ell = 325^o$ should be larger because a given longitude change near the targent point corresponds to a larger path length change along the arms.

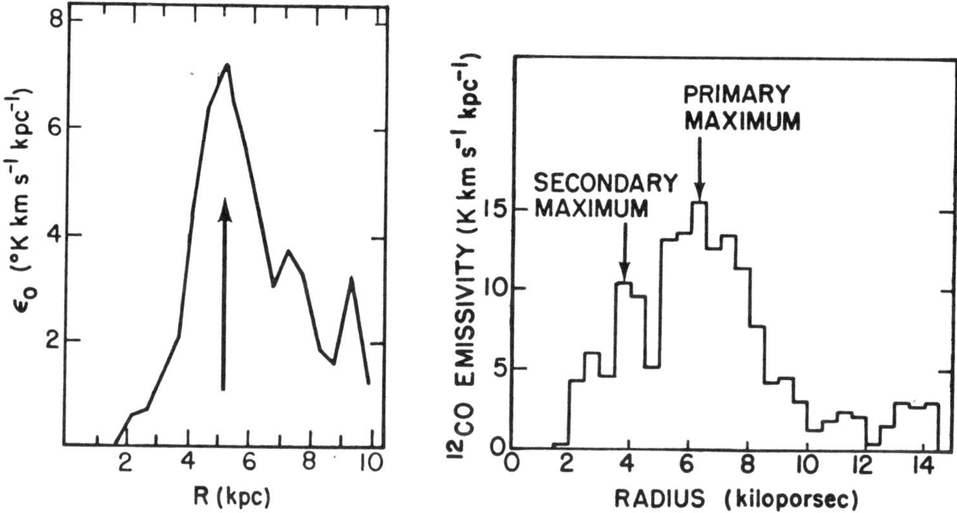

Figure 4 - The positions of the CO maxima expected from extrapolating the outer Galaxy spiral structure inward are shown by arrows. The graphs are CO emissivity reproduced from Cohen, Tomasevich and Thaddeus (1979) in the North and from Robinson, McCutcheon and Whiteoak (1981) in the South.

Figure 4a shows the velocity integrated CO emissivity for the low longitude gas from Cohen, Tomasevich and Thaddeus (1979) and the position of the expected CO maximum from an extrapolation of the outer Galaxy pattern. (Other, less completely sampled CO surveys show qualitatively similar results). In Figure 4b, similar results are shown for the high longitude gas, which shows two maxima at longitudes (which have been converted to distance in Fig. 4 using the inner Galaxy rotation curve) close to those expected from the four-armed outer Galaxy model. There is thus good agreement between the expectations of the extrapolation and the locations of the CO maxima.

ii) Longitude - Velocity Diagrams

Assuming that all of the CO is in the spiral arms, it is possible to construct a longitude-velocity plot of the midpoints of the spiral arm pattern shown in Figure 2 using the rotation curve of Burton and Gordon (1978). The resulting longitude-velocity plots are shown in Figure 5 and are calculated for distances larger than 4 kpc from the center.

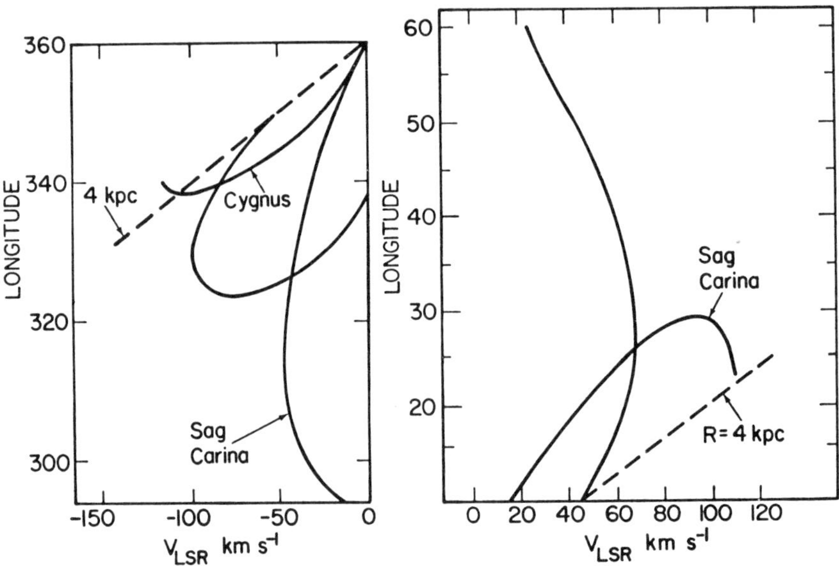

Figure 5 - Expected longitude-velocity plots of CO ridge lines deduced from the spiral pattern given in Figure 2 assuming that CO occurs only in the major arms. The dashed line marked 4 kpc represents the inner boundary of the molecular ring. The left and right hand panels represent southern and northern hemisphere observations respectively.

Comparison of Figure 5 with the longitude-velocity plots of CCDT and Robinson and Manchester show good agreement with some features, poor agreement with others. In the CCDT maps for example, the feature identified as the Scutum arm is well matched by the feature in 5b which appears between longitudes of 10° and 30°. Other CO features in the region of $\ell = 30^{\circ}$ to $\ell = 45^{\circ}$ in the velocity range of $v = 30 - 50$ km s^{-1} (identified with the Sagittarius arm), are however not seen in Figure 5.

To what degree the lack of agreement is due to spurs, and to what degree it is due to the inapplicability of extrapolating the outer Galaxy structure inside the solar circle is not clear. Galaxies sometimes show different structures in their inner and outer regions, so it is not surprising that extrapolation from one part may not apply to another.

We conclude the following: Beyond 10 kpc from the center, the Milky Way possesses a relatively symmetric 4 armed spiral pattern with long coherent arms and a few spurs. Extrapolation of this pattern inward predicts the locations of maxima in the radial CO emissivity, but does not reproduce all of the features of the CO emission on the ℓ-v diagrams. The inner Galaxy therefore has a more complex, possible more chaotic structure than that found in the outer Galaxy.

REFERENCES

Blitz, L., Fich, M. and Stark, A. A. 1980 in Interstellar Molecules, B. H. Andrew, ed. Reidel: Dordrecht p. 213.

Burton, W. B. 1971, Astron. Ap. 10, 76.

Burton, W. B. 1972, Astron. Ap. 19, 51.

Burton, W. B. and Gordon, M. A. 1978, Astron. Ap. 63, 7.

Cohen, R. S., Tomasevich, G. R. and Thaddeus, P 1979 in the Large Scale Characteristies of the Galaxy, W. B. Burton, ed. Reidel: Dordrecht p. 53.

Cohen, R. S., Cong, H. I., Dame, T. M. and Thaddeus, P. 1980, Ap. J. (Letters) 239, L53.

Henderson, A. P., Jackson, P. D. and Kerr, F. J. 1982, Ap. J., 263, 116.

Kennicutt, R. and Hodge, P. 1982, Ap. J. 253, 101.

Kulkarni, S. R., Blitz, L. and Heiles, C. 1982, Ap. J. (Letters), 259, L63.

Lin, C. C., Yuan, C. and Shu, F. H. 1969, Ap. J. 155, 721.

Robinson, B. J., McCutcheon, W. H. and Whiteoak, J. B. 1982, Int. J. of Infrared and Millimeter Waves, in press.

Sills, R. M. 1982, Ph.D. Dissertation, University of California, Berkeley.

THE MASSACHUSETTS - STONY BROOK CO SURVEY OF THE GALACTIC PLANE

D. B. Sanders
Five College Radio Astronomy Observatory
University of Massachusetts

1. INTRODUCTION

Over the past four years the 14-meter millimeter wave telescope of the Five College Radio Astronomy Observatory has been used to survey the galactic plane in the CO emission line. These observations have revealed the extent of the molecular disk in both R and z from the galactic center out to 16 kpc, the outer edge of detectable CO emission, and have allowed a determination of the characteristics of discrete molecular clouds. The final galactic survey, a two year project to be complete by mid-1983, will provide a map of the entire first quadrant at 3 arc minute spacing, increasing our previous sample by nearly an order of magnitude and allowing an unprecedented view of individual molecular clouds throughout the disk on all size scales between a few parsecs and many kiloparsecs.

The importance of CO millimeter line surveys has been evident since the first surveys (Scoville and Solomon 1975; Burton et. al. 1975; Cohen and Thaddeus 1977) revealed large concentrations of molecular gas in the inner galaxy - in the galactic center ($R \leqslant 1$ kpc) and in a 'ring' at $R = 4$-8 kpc ($R_{\odot} = 10$ kpc) - a distribution strikingly different from the flat radial distribution of HI. Scoville and Solomon 1975, Gordon and Burton 1976, Solomon and Sanders 1980, and Liszt, Xiang and Burton 1981, have interpreted this emission, supplemented by ^{13}CO data, as indicating that the mass of H_2 in the ring exceeds the HI mass by a factor of three to five. Strip surveys of ^{12}CO (Solomon, Sanders and Scoville 1979) and ^{13}CO (Liszt, Xiang and Burton 1981) emission at 3 arc minute spacing using the 1 arc minute beam of the 36-foot NRAO antenna and full grid maps at 3 arc minute spacing over regions one degree square using the 1.6 arc minute beam of the BTL 7-meter antenna (Stark 1979), reveal that much of the emission comes from giant clouds with diameters larger than 20 pc and masses greater than 10^5 $M_{\odot}$.

This paper presents recent results from the Massachusetts-Stony Brook galactic plane surveys including: 1) a comparison of the northern and southern CO distributions provided by a ^{12}CO and ^{13}CO

W. B. Burton and F. P. Israel (eds.), Surveys of the Southern Galaxy, 127–136.

survey along the galactic equator in the fourth quadrant using the NRAO 36-foot telescope and the 4.6-meter Univ. of Texas MWO telescope; 2) an ℓ-b survey of the entire first and second quadrants out to b $\simeq$ 2.5°; 3) a measure of the spatial contrast in CO emission in the inner galaxy on scales of a few kpc^2 from our most recent 3 arc minute survey; and 4) a determination of the mass spectrum of molecular clouds.

2. THE RADIAL DISTRIBUTION OF CO AND H_2

The ringlike distribution of CO emission in the inner galaxy is illustrated in Figure 1. The northern CO distribution shows the well known peak at R ~5.5 kpc where the midplane emissivity, J_o, is five times that at R = 3 and 10 kpc. The southern distribution exhibits a similar ring structure, but with a more 'fanned out' appearance and a less prominent peak near R ~6 kpc; however, the integrated emissivity between 3 and 9 kpc in the north and south is identical to within ±10 % indicating equal amounts of molecular gas in both hemispheres. The shape of our southern CO distribution is similar to that of Robinson, McCutcheon and Whiteoak (1982) who have obtained a much more extensive CO survey of the galactic equator using the 4 meter Australian millimeter telescope.

In both the north and south the radial distribution of ^{13}CO emission mimics the ^{12}CO - the ratio, $J(^{12}CO) / J(^{13}CO) \simeq 6$ at all radii between 3 and 9 kpc. The similarity of the distributions for the two isotopes whose optical depths differ by a factor of ~75 can be understood if the CO emissivity is determined primarily by the number of clouds whose mean properties (e.g. temperature, density) are similar (Solomon, Scoville and Sanders 1979).

The H_2 column density and volume density are related to the CO integrated intensity and emissivity by the expressions,

$$N(H_2) = 3.6 \times 10^{20} \int T(^{12}CO)\, dv \qquad (cm^{-2}) \qquad (1)$$

$$n(H_2) = 0.12\, J(^{12}CO) \qquad (cm^{-3}) \qquad (2)$$

where the numerical constant in equation 1 has been determined from measurements of CO integrated intensity and visual extinction in nearby dark clouds and from virial theorem mass estimates for giant molecular clouds in the galactic ring (Sanders, Solomon and Scoville 1983). Our conversion factor (3.6×10^{20}) is intermediate compared to other published values; Blitz and Thaddeus (1980) use 7×10^{20} while Dickman's (1978) LTE value for small diffuse clouds favored by Blitz and Shu (1980) is 2×10^{20}. Liszt (1982) has summarized all published conversion factors and finds a mean of 5×10^{20}.

The total CO content and mass of H_2 in the galactic disk has been determined from extensive observations in ℓ and b (Sanders, Solomon and

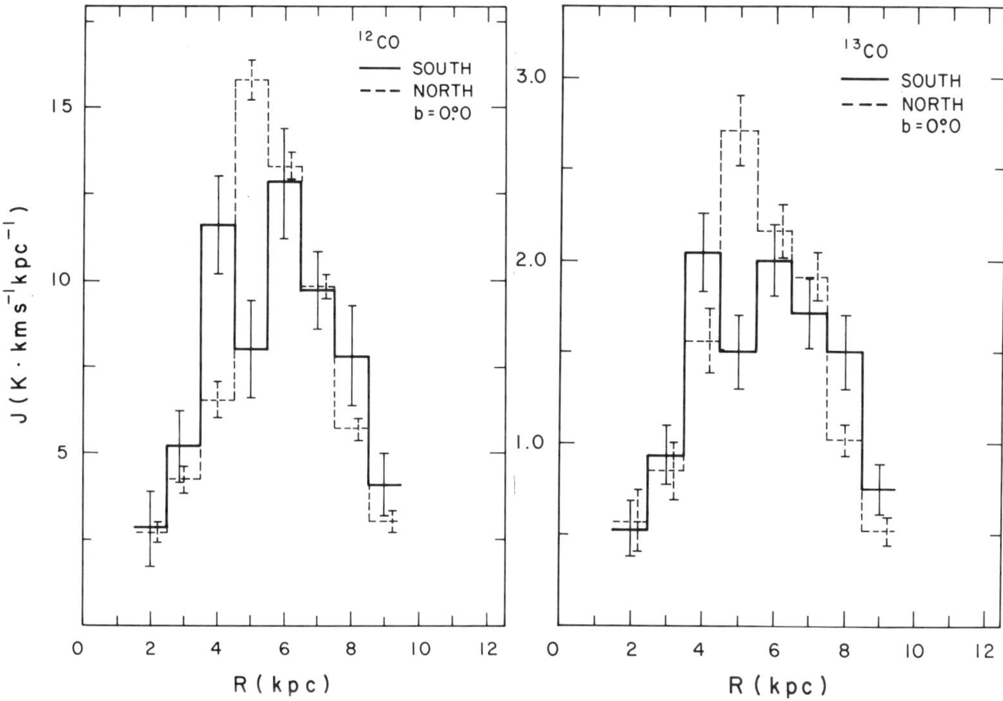

Figure 1. Radial distribution of CO emissivity, J, in the southern and northern hemisphere at b = 0°.

Scoville 1983) which reveal a gaussian z distribution of CO emission with a half width, $z_{1/2}$, of 40 pc at R=3 kpc, increasing to 75 pc at R=10 kpc approximately as $R^{0.5}$. Since the product, $z_{1/2}J_o$, is proportional to the face-on CO integrated intensity, I_{CO}, the H_2 surface density (from equation 1) is given by,

$$\sigma(H_2) = 5.82\ I_{CO} \qquad (M_\odot\ pc^{-2}) \qquad (3)$$

Figure 2 shows $\sigma(H_2)$ for the Milky Way disk from the galactic center out to R = 16 kpc, the maximum radius of significant CO emission. Between 6 and 16 kpc $\sigma(H_2)$ falls by two orders of magnitude while σ(HI) is virtually constant. The percentage of total disk mass in gas, as determined from the theoretical disk model of Caldwell and Ostriker (1981), is approximately 12 percent over the entire region, R = 5 - 13 kpc. Between 4 and 9 kpc the total mass of H_2 exceeds HI by at least a factor of four; thus, even if the lowest conversion factor ($2x10^{20}$) were used in equation 1, H_2 still dominates HI in the inner galaxy. Table 1 summarizes the values for the disk mass of H_2 and HI.

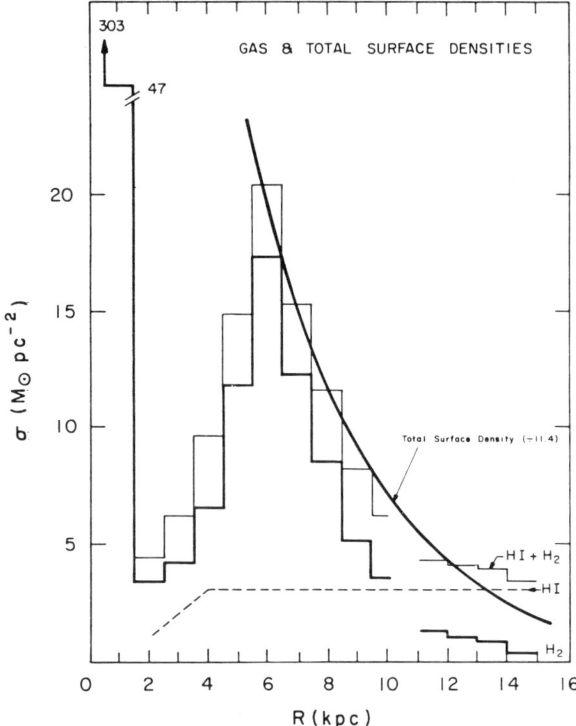

Figure 2. Comparison of gas surface density and total (gas+stars) surface density (Caldwell and Ostriker 1981) in the Milky Way disk

Table 1.
Total Disk Mass of HI and H_2 at R = 0 to 16 kpc
($R_\odot$ = 10 kpc, $V_\odot$ = 250 km s^{-1})

	R(kpc) 0-1.5	1.5-10	10-16	Total
$M(H_2)$	$.5x10^9 M_\odot$	$2.7x10^9 M_\odot$	$.3x10^9 M_\odot$	$3.5x10^9 M_\odot$
$M(H_2)/M_T(H_2)$	.14	.77	.09	1.0
M(HI)	$1x10^7 M_\odot$	$.9x10^9 M_\odot$	$2x10^9 M_\odot$	$3x10^9 M_\odot$
$M(HI)/M_T(HI)$	.003	.30	.67	1.0
$M(H_2)/M(HI)$	50	3.0	.15	1.17

3. CO EMISSION AND SPIRAL ARMS IN THE INNER GALAXY

We have taken close spaced CO spectra along the galactic midplane (see Figure 3) to obtain a detailed view of the ℓ-v distribution of molecular clouds on size scales of a few parsecs to many kiloparsecs. Previous ℓ-v surveys of HI and HII line emission in the first galactic quadrant have been interpreted as showing evidence for a global pattern of 2 or 3 narrow spiral arms (e.g. Yuan 1969; Lockman 1979). Some CO observers (Cohen et. al. 1980; Blitz and Shu 1980) have argued that the CO also shows evidence for confinement of the molecular clouds to these few arms.

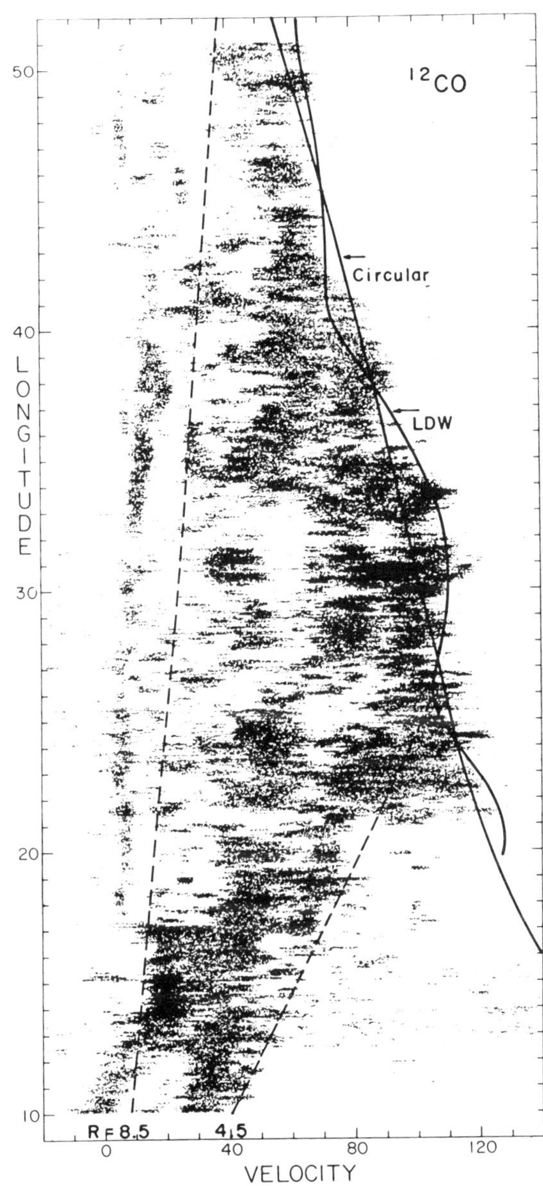

Figure 3. CO emission in longitude-velocity space near the galactic midplane between longitudes 10° and 51° ($\Delta\ell$ = 3'). The number of dots is linearly proportional to the CO antenna temperature with velocity resolution of 2 km s^{-1}. Saturation is at $T_A^* > 12$ K. No dots are plotted for $T_A^* < 1$ K. Most of the emission is confined to the R = 4-8 kpc region shown by the dashed lines. The CO terminal velocity is well represented in the mean by circular rotation as determined from HI, and in detail by the linear density wave model (e.g. Yuan 1969; Burton 1971)

Whether or not molecular clouds are confined to a few narrow spiral arms is an important astrophysical question which requires demonstrating the existence of self consistant spatial and kinematic patterns in the CO ℓ-v distribution. Here we will present evidence from an analysis of the spatial contrast in CO emission along the terminal velocity ridge which argues against confinement of molecular clouds to a few spiral arms. First, from simple inspection of Figure 3 one can see that CO emission extends to the terminal velocity defined by the ubiquitous HI at nearly all longitudes between 20 and 51 degrees. The absence of molecular clouds from interarm regions should show up as gaps in the CO emission along the terminal velocity ridge, similar to what is observed for the distribution of giant HII regions (e.g. Lockman 1979) which show concentrations along the terminal velocity ridge at spiral arm crossings near ℓ = 24° (4 kpc expanding ring), 30° (Scutum arm) and 50° (Sagittarius arm) and a large gap between ℓ = 36°- 47°.

We have quantified the variation in CO emission near the tangent point circle by integrating the CO emission within 10 km s^{-1} of v_T(HI). The CO emissivity, which is directly related to $n(H_2)$ (see equation 2), was determined using a linear density wave model (Burton 1971) which matches the terminal velocity in both CO and HI to compute the path-length corresponding to Δv. The resulting CO emissivity distribution (see Figure 4) shows 4 or 5 peaks superimposed on a 'background' of emission present at all observed longitudes between 20° and 51°. The peak-to-trough ratio is approximately 3:1 on scales of a few degrees, corresponding to regions of 1-2 kpc^2. Collectively the peaks occupy 30

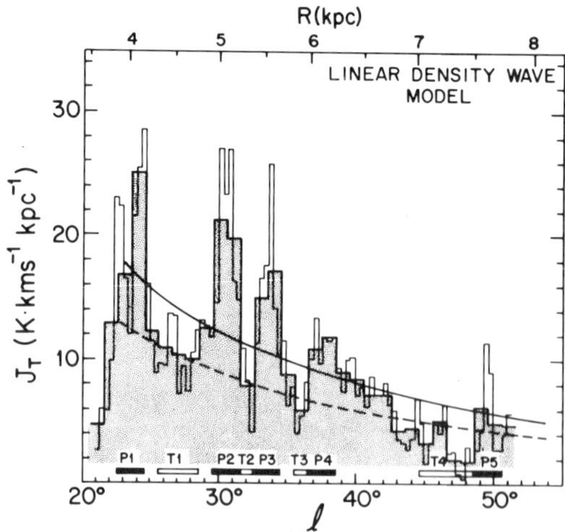

Figure 4. CO emissivity along the loci of tangent points. Peak (P) and trough (T) regions have been chosen to best reflect the variation of emissivity with longitude at a resolution of one degree (dark histogram). The smooth, solid curve represents an axisymmetric distribution of clouds with an exponential radial distribution normalized to give equal integrated emissivity compared to the data over the interval ℓ = 20°-50°. The dashed curve represents a similar model which best fits the data from trough regions (Sanders 1981).

percent of the effective area in the projected strip centered on the tangent point locus and contain approximately 50 percent of the total emission. This picture clearly is not consistent with spiral arm models tht locate all of the molecular clouds in only 2 or 3 narrow arms in the inner galaxy. It suggests a more widespread distribution, or possibly many 'arm segments'.

The exact nature of the cloud distribution will become more apparent when high resolution two dimensional maps of the galactic plane are available allowing one to identify large numbers of clouds and determine their distances. It should then be possible to tell whether patterns in the ℓ-v plane are due to physical associations of clouds or are only a superposition of clouds at different distances.

4. MOLECULAR CLOUD PROPERTIES

The use of large telescopes to study the galactic plane has allowed us to resolve and study the properties of individual clouds rather than just detect emission from a confusion of sources. Figure 5 shows a portion of our most recent high resolution survey of the galactic plane illustrating the clear breakup of emission into discrete molecular clouds. Figure 6 shows strips perpendicular to the plane where it is even easier to identify clouds owing to the smaller amount of near-far blending of emission at a given velocity as one moves away from b=0°.

Latitude-velocity strips like those shown in Figure 6 have been used by Sanders, Scoville and Solomon (1983) to identify a sample of clouds whose centroids are at large enough b such that they must be at the near-side kinematic distance. The distribution of cloud diameters is found to closely approximate a power law with spectral index -2.3 ±.25 for the range 10 - 75 pc (no discrete clouds larger than 75 pc were found while a 10 pc lower limit was imposed by the spacing between observations). By determining the ratio of the volume of the disk sampled to the total disk volume and weighting each annulus according to the radial distribution of Figure 2, we estimate that there are ~ 19,000 molecular clouds with diameters larger than 10 pc in the galaxy. Their average volume filling fraction at R = 2-10 kpc is 0.8%, and the mean free path between clouds is 3.5 kpc (at the R=6 kpc peak the numbers are 1.4% and 1.8 kpc).

A virial analysis of cloud masses using their measured diameters and CO line widths yields mean densities of 300 H_2 cm^{-3} for 15 pc clouds, systematically decreasing to 150 H_2 cm^{-3} for 50 pc clouds. Similar cloud densities are found from the the ^{12}CO integrated intensities using equation 1. From the size distribution and densities we find a mass distribution $N(m) \propto m^{-1.5}$. Most of the molecular mass in the galaxy is found in the high mass end of the cloud spectrum as shown in Figure 7. Over half of the mass is in approximately 1000 clouds with d > 50 pc and $M > 8x10^5$ $M_\odot$. Less than 20 percent of the H_2 mass is in clouds smaller than 20 pc.

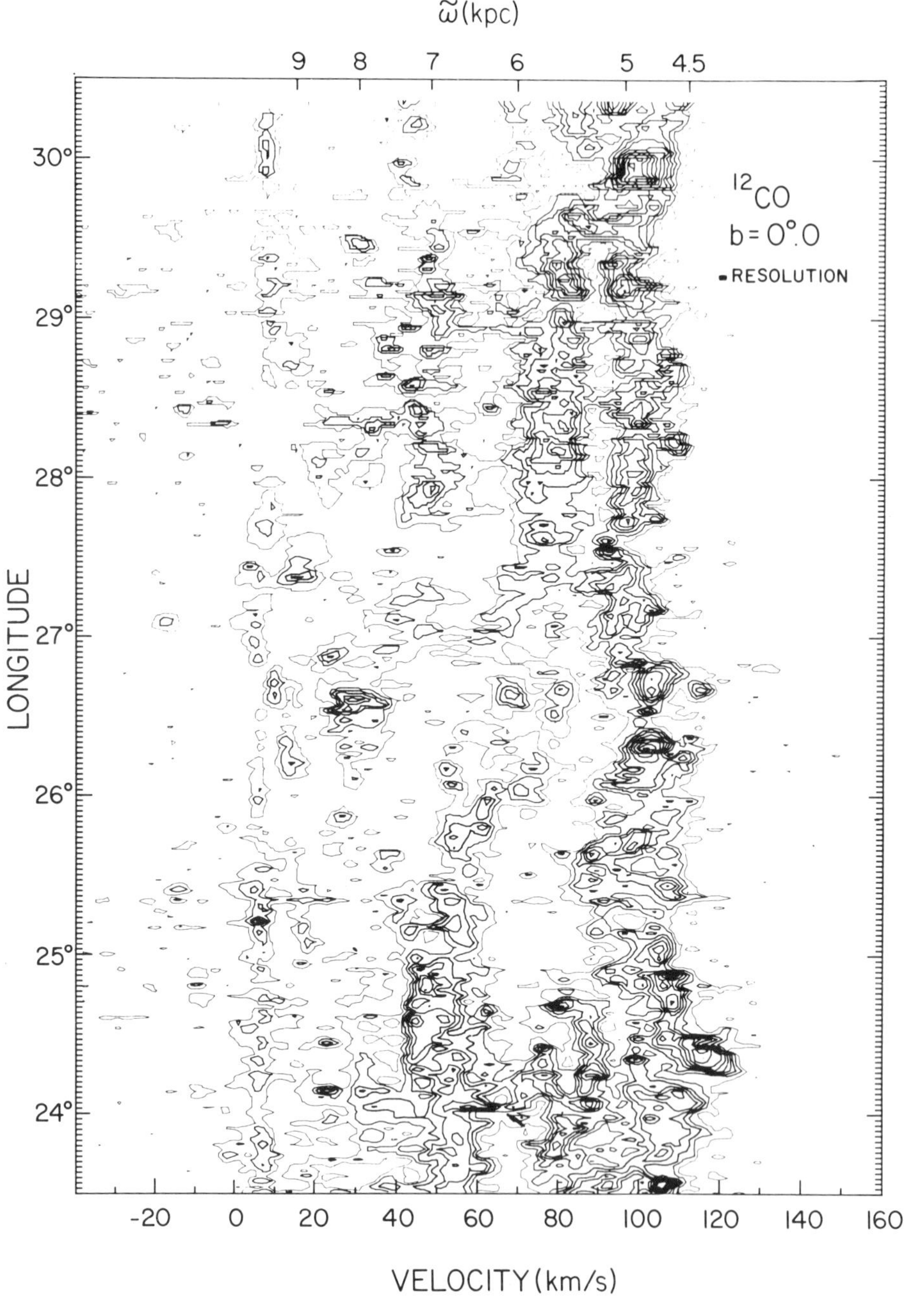

Figure 5. CO emission in the ℓ-v plane between ℓ = 10° - 51° at Δℓ = 3' resolution. Contour intervals are in steps of T_A^* = 1 K. The resolution is indicated by the black rectangle. The breakup into large molecular clouds is very striking above the third contour.

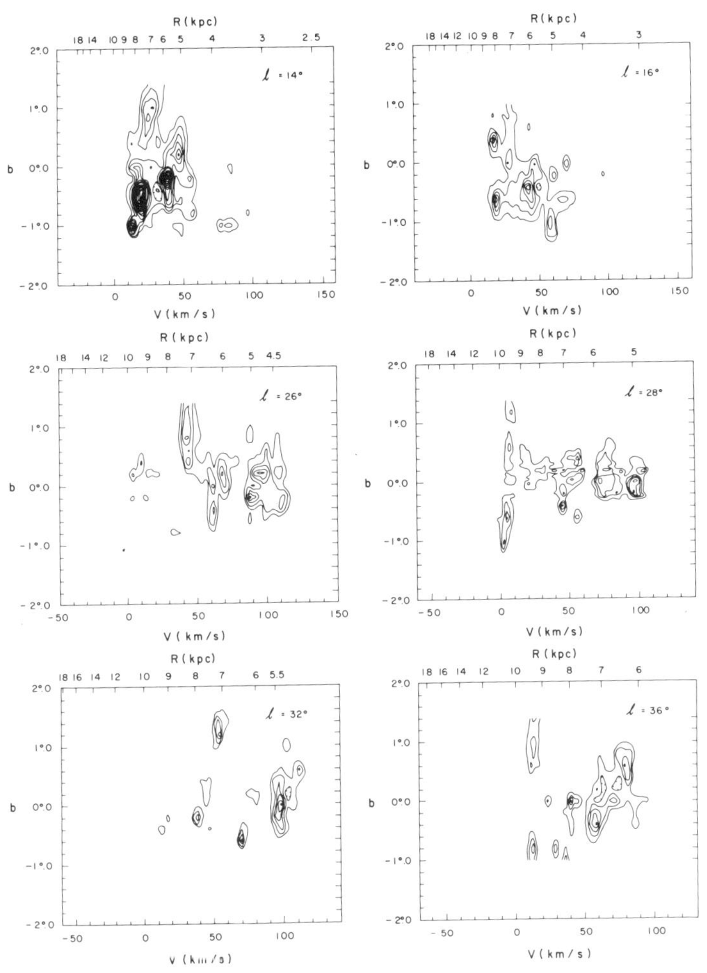

Figure 6. Sample latitude-velocity contour diagrams of CO emission. The top axis indicates galactocentric distance, R (kpc). Each contour is in units of $T_A{}^* = 1$ K.

The approximately 6,000 giant molecular clouds in the galaxy with individual masses greater than 10^5 $M_\odot$, diameters larger than 20 pc and total mass ~3×10^9 $M_\odot$ are the largest reservoir of interstellar matter in the galaxy. Our current observations with the 14-meter antenna are aimed at providing a more detailed sample of high resolution data to study the structure and evolution of these clouds and how their properties vary across the galactic disk.

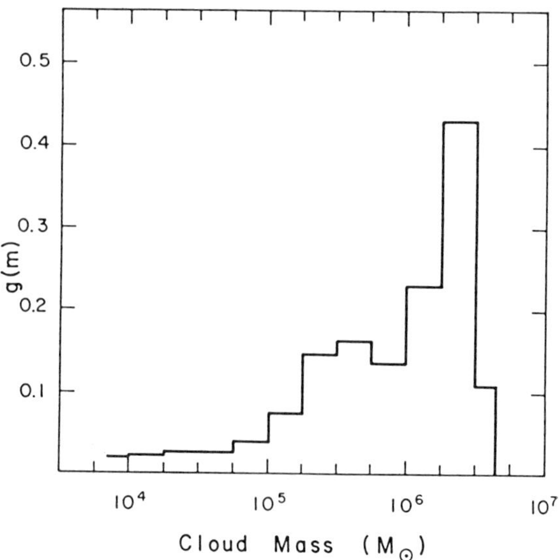

Figure 7. The fraction of H_2 mass in clouds per unit logarithmic mass interval. $g(m) = n(m)m^2 \int n(m)m\ dm$ where $\int g(m)\ d\ln m = 1$. n(m) is the cloud mass spectrum in units of $M_\odot^{-1}$.

REFERENCES

Blitz, L. and Shu, F.H. 1980, Ap.J., 238, 148.
Blitz, L. and Thaddeus, P. 1980, Ap.J., 241, 676.
Burton, W.B. 1971, Astr.Ap., 10, 76.
Burton, W.B., Gordon, M.A., Bania, T.M. and Lockman, F.J. 1975, Ap.J., 202, 30.
Caldwell, J.A.R. and Ostriker, J.P. 1981, Ap.J., 251, 61.
Cohen, R.S. and Thaddeus, P. 1977, Ap.J.(Letters), 217, L155.
Cohen, R.S., Cong, H.I., Dame, T.M. and Thaddeus, P. 1980, Ap.J., 239. L53.
Dickman, R.L. 1978, Ap.J. Suppl., 37, 407.
Gordon, M.A. and Burton, W.B. 1976, Ap.J., 208, 346.
Liszt, H.S. 1982, Ap.J., 262, 198.
Liszt, H.S., Xiang, D. and Burton, W.B. 1981, Ap.J., 249, 532.
Lockman, F.J. 1979, Ap.J., 232, 761.
Robinson, B.J., McCutcheon, W.H. and Whiteoak, J.B. 1982, Int. J. of Infrared and Millimeter Waves, 3, 63.
Sanders, D.B. 1981, Ph.D. Thesis, State Univ. of New York at Stony Brook.
Sanders, D.B., Scoville, N.Z. and Solomon, P.M. 1983, Ap.J. submitted.
Sanders, D.B., Solomon, P.M. and Scoville, N.Z. 1983, Ap.J. submitted.
Scoville, N.Z. and Solomon, P.M. 1975, Ap.J.(Letters), 199, L105.
Solomon, P.M., Sanders, D.B. and Scoville, N.Z. 1979, in The Large Scale Characteristics of the Galaxy, ed. W.B. Burton(Dordrecht:Reidel) p.35.
Solomon, P.M., Scoville, N.Z. and Sanders, D.B. 1979, Ap.J., 232, L89.
Solomon, P.M. and Sanders, D.B. 1980, in Giant Molecular Clouds in the Galaxy, ed. P.M. Solomon and M.G. Edmunds (Oxford:Pergamon), p 41.
Stark, A.A. 1979, Ph.D. Thesis, Princeton University.
Yuan, C. 1979, Ap.J., 158, 889.

LATITUDE DISTRIBUTION OF CO IN THE SOUTHERN HEMISPHERE

R.N. Manchester, J.B. Whiteoak, B.J. Robinson,
Robina E. Otrupcek
Division of Radiophysics, CSIRO, Sydney, Australia
and
C.J. Rennie
Mt. Stromlo and Siding Spring Observatories, Australian National University, Canberra, Australia

ABSTRACT

The latitude distribution of CO in the southern galactic hemisphere has been investigated using the 4-m telescope of the CSIRO Division of Radiophysics. Observations of the 115 GHz transition of CO were made over the latitude range ±0°.975 every 3° of longitude from 294° to 357°. Results show that the mean z-distance for the CO distribution deviates by ±60 pc from the galactic equator but that on the average the data are well fitted by a Gaussian curve of FWHM 130 pc.

1. INTRODUCTION

Any complete galactic plane survey of molecular clouds should cover a latitude range large enough to include both the nearby clouds, which are often large in angular size and centred at non-zero latitudes, and the more distant clouds, which, because of the warping of the plane, are displaced from zero latitude. Quite apart from the general value of complete sampling, the latitude extent of clouds within the solar circle is a useful parameter in resolving the ambiguity of distance estimates based on kinematic considerations. A limitation of the southern galactic plane CO survey (Robinson et al., 1983) is that the latitude extension is only ±4'.5 arc. Considerable observing time would be necessary to cover a larger range. To obtain some indication of the latitude distribution we have fully sampled latitude strips to ±0°.975 at intervals of 3° in longitude between 294° and 357°. This partial survey gives useful results on the mean z distribution of CO and on the continuity of spiral features discussed in the accompanying paper (Robinson et al. 1983).

W. B. Burton and F. P. Israel (eds.), Surveys of the Southern Galaxy, 137–141.

2. OBSERVATIONS

The observations were made in 1981 with the 4-m telescope at the CSIRO Division of Radiophysics. The equipment and observing procedure have been described by Robinson et al. (1982). As for the longitude survey, data taken on a nine-position grid were averaged to form a single spectrum having an effective angular resolution of 9' arc. Centres of adjacent observations were spaced by 9' arc and extended from $-0^{\circ}.9$ to $+0^{\circ}.9$ in latitude giving 13 spectra in each latitude strip. The final spectra covered a velocity range of -170 to +70 km s^{-1} (with respect to the local standard of rest) with an effective velocity resolution of 1.6 km s^{-1}. The temperature scale for the spectra was based on a "corrected" antenna temperature of 65 K for the CO associated with OMC1 (Davis and Vandenbout, 1973; Ulich and Haas, 1976). Typically the r.m.s. noise in the spectra was 0.5 K.

3. RESULTS

Figure 1 is a composite plot showing two contour levels of all 22 latitude-velocity maps and Figure 2 shows more complete contour data for selected longitudes. The first point to note is that many features, particularly at lower velocities, extend to higher latitudes than the cutoff of the survey, $\pm 0^{\circ}.975$. However, in general the latitude range surveyed does include the major CO features. At many longitudes the CO is concentrated off the galactic equator, for example, at $\ell = 345^{\circ}$, $v = -70$ and -120 km s^{-1}, and the feature extending from $\ell = 345^{\circ}$, $v = -20$ km s^{-1} to $\ell = 357^{\circ}$, $v = +10$ km s^{-1}. It is clear however that the major holes seen in the longitude survey (Robinson et al., 1983), for example, $\ell = 342^{\circ}$, $v = -110$ km s^{-1} to $\ell = 351^{\circ}$, $v = -40$ km s^{-1}, are not simply due to deviations of the CO distribution from the galactic equator and do in fact represent real minima in the galactic CO distribution. Figure 1, when viewed from a distance, gives a crude representation of the southern velocity-longitude diagram integrated over latitude from $-0^{\circ}.975$ to $+0^{\circ}.975$. It is somewhat more filled than the $b = 0^{\circ}$ survey (Fig. 1 of Robinson et al., 1983) but the overall appearance is very similar. In particular there is very little CO at low velocities.

To estimate the CO z-distribution, the latitude distribution at the tangential point was obtained by integrating in velocity across the highest velocity feature or clump for longitudes between 303° and 342°, e.g. from $v = -110$ km s^{-1} to $v = -75$ km s^{-1} at $\ell = 330^{\circ}$. These features were then assumed to lie at the tangential point ($R_{\odot} = 10$ kpc was assumed) and the corresponding z-distribution computed. Figure 3 shows the mean z and full width at half-maximum (FWHM) of the z-distribution for the resulting profiles. Systematic departures in mean z from the $b = 0^{\circ}$ plane are evident. The most prominent of these, a positive displacement of ~ 60 pc around $\ell = 327^{\circ}$, is probably related to the spiral feature tangential at that point (Robinson et al., 1983). It is interesting to note that for the corresponding longitude range in the northern hemisphere the mean z-values are displaced in the opposite

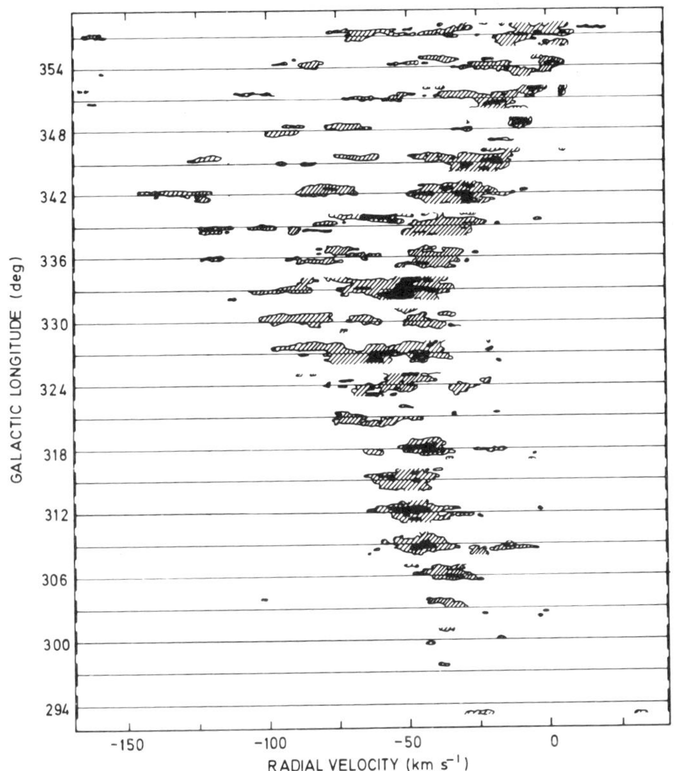

Fig. 1 - Composite plot showing the latitude-velocity distribution of ^{12}CO (J = 1-0) at 3° intervals of longitude from ℓ = 294° to ℓ = 357°. The thickened vertical bars on either side of each latitude-velocity plot show the latitude extent of each plot, ±0°.9. Two contour levels of corrected antenna temperature, 2 K and 8 K, are represented by hatched and filled shading respectively. The velocity scale is with respect to the local standard of rest.

sense by ~40 pc (Cohen and Thaddeus, 1977; Solomon et al. 1979), possibly indicating correlated or modal motions. FWHM values vary between 45 and 190 pc with a mean value of 103 pc, comparable to values found for the northern hemisphere. There is weak evidence for a positive correlation between galactocentric radius and FWHM, an effect seen at about the same level of significance in previous work.

The z-profile obtained by adding data for all tangential features across the longitude range 303° to 342° is shown in Figure 4. This distribution is accurately centred on the galactic plane ($\bar{z}$ = 0.7 pc) and has a FWHM of 133 pc. It is closely fitted by a Gaussian curve of FWHM 130 pc, as shown in Figure 4. There is some indication of bimodal behaviour near the peak of the curve. This is not simply due to distributions at different longitudes being centred above or below the plane,

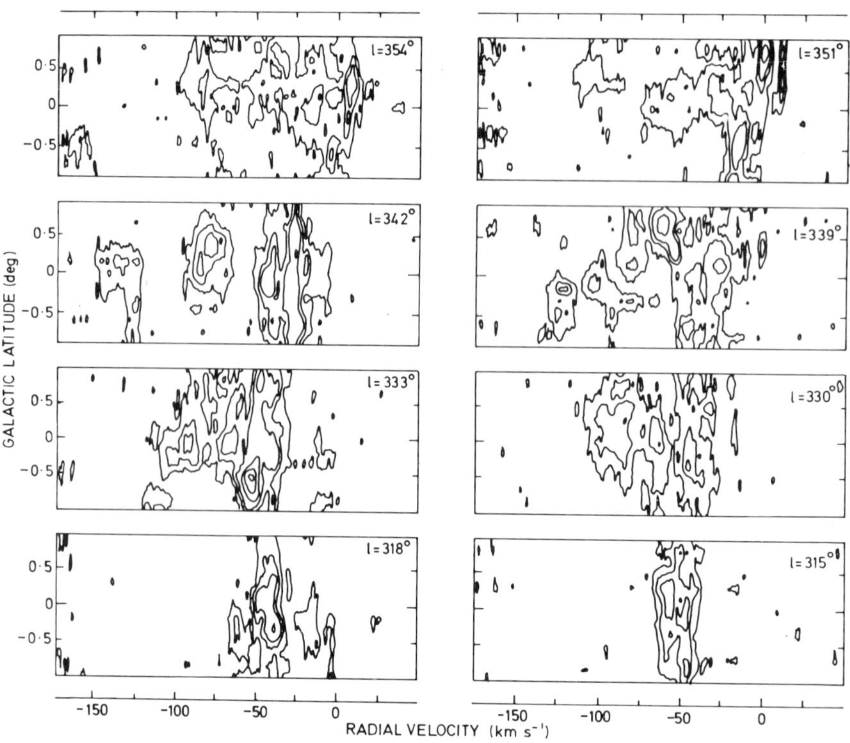

Fig. 2 - Latitude-velocity plots for eight longitudes selected from those observed. Contours are of corrected antenna temperature (T_A^*) with values of 1, 3, 5, 10, 15, 20 and 25 K. The velocity scale is with respect to the local standard of rest.

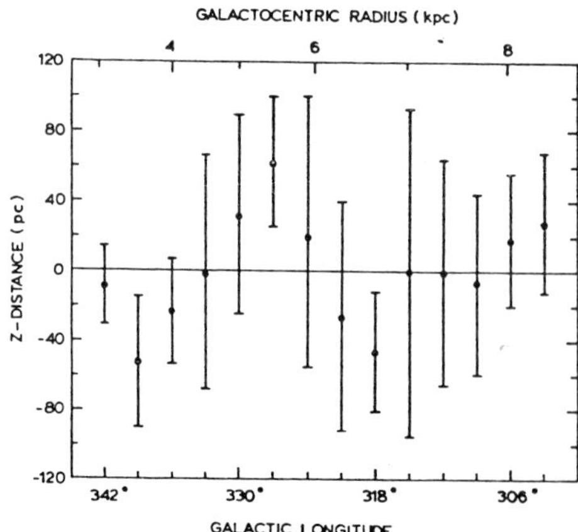

Fig. 3 - Galactic z-distributions of CO obtained from observations of gas at the tangential points at longitudes between 303° and 342°. At each longitude the point represents the mean z-value and the bars indicate the full width at half-maximum of the distribution. Corresponding galactocentric radii are indicated along the top of the figure.

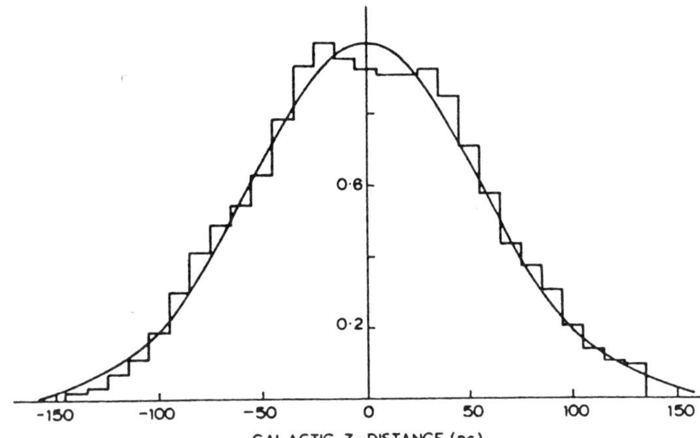

Fig. 4 - Galactic z-distribution of CO integrated over tangential points from longitude 303° to 342°. The sharp cutoff at high z is not real but results from the limited latitude extent of the survey. A Gaussian curve of FWHM 130 pc is a good fit to the observed distribution.

since evidence for it remains when smaller ranges of longitude, e.g. 324° to 330°, are averaged. It is possible that it results from self-absorption of background CO by cooler CO clouds concentrated near the plane.

We thank Anne Manefield, Isobel Goddard and Patrick Larkin for assistance with the observations and data analysis.

REFERENCES

Cohen, R.S., and Thaddeus, P.: 1977, *Astrophys. J.* 217, p. L155.
Davis, J.H., and Vandenbout, P.: 1973, *Astrophys. Lett.* 15, p. 43.
Robinson, B.J., McCutcheon, W.H., Manchester, R.N., and Whiteoak, J.B.: 1983, this volume.
Robinson, B.J., McCutcheon, W.H., and Whiteoak, J.B.: 1982, *Int. J. Infrared Millimeter Waves* 3, p. 63.
Solomon, P.M., Sanders, D.B., and Scoville, N.Z.: 1979, in "The Large Scale Characteristics of the Galaxy" (Proc. IAU Symp. 84) (W.B. Burton ed.) p. 35 (Reidel, Dordrecht).
Ulich, B.L., and Haas, R.W.: 1976, *Astrophys. J. Suppl. Ser.* 30, p. 247.

MOLECULAR CLOUDS IN THE OUTER MILKY WAY GALAXY

Marc L. Kutner
Physics Dept., Rensselaer Polytechnic Institute

1. INTRODUCTION

This is to follow up the report by Kutner and Mead (1981) of extensive CO emission from molecular clouds between R = 10 and 16 kpc. In the initial report, three strips were mapped, at b = 1.3°, 1.5°, and 1.7°, with l ranging from 55° to 95°, in steps of 0.1°. Maps of five giant molecular clouds (GMCs) were also presented. It was argued that CO emission from outer galaxy clouds is weaker than that from inner galaxy clouds partly because of a lower kinetic temperature in the outer galaxy clouds, and that this had to be taken into account in mass estimates of molecular material in the galaxy. It was estimated that the mass of H_2 in the outer galaxy is ~ 1/3 that in the inner galaxy.

Since that initial report, we have significantly extended our observations in a number of ways: (1) We have extended our longitude coverage, and now have at least some data taken over the range $45° \leq l \leq 230°$ (with no data taken within 20° of the anticenter). (2) We have extended our latitude coverage to enable us to estimate the location of the midplane and the thickness of the CO emitting layer. (3) We have greatly extended our mapping, now having at least partial CO maps of 55 clouds, along with ^{13}CO, $C^{18}O$, CO(2 → 1), and 2-mm H_2CO observations of some clouds.

2. LARGE-SCALE RESULTS

Our large-scale maps were all made on the NRAO 11-m telescope located on Kitt Peak. As part of this work, we have reobserved parts of the strips in our earlier study, under more favorable calibration conditions, and we find that the T_R*'s reported in Kutner and Mead should be reduced by ~ 15%. For the purpose of determining the latitude distribution of the outer galaxy molecular clouds, strips in b were done every 5° of l. Within the strips, points were 0.1° apart, except for several strips in the second and third quadrants where at 0.2° spacing was used since the material being observed was significantly closer to

W. B. Burton and F. P. Israel (eds.), Surveys of the Southern Galaxy, 143–148.

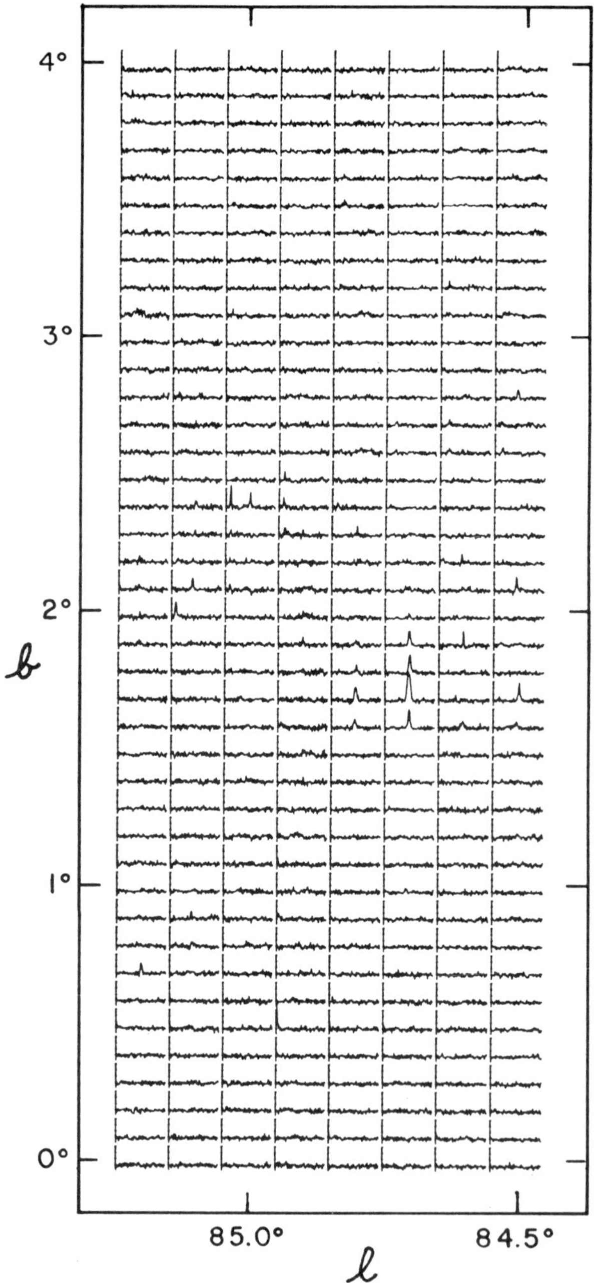

Fig. 1. In each spectrum, the full scale intensity corresponds to T_R^* = 6 K, and the rms noise level is 0.3 K. In each spectrum, distance from the galactic center increases from left to right, ranging from 12 to 20 kpc (corresponding to v_{LSR} from -42 to -125 km/s). Position-switching reference positions were at b = -1.5° and b = 5.5°.

the sun than that in the first quadrant. (The 0.1° sampling interval was chosen because GMCs 10 kpc from the sun subtend angles of a few tenths of a degree.) Our strips were 4° to 6° in extent. These observations indicate that the CO emission follows the warp evidenced in the HI.

To give an idea of the appearance of the outer galaxy clouds, Fig. 1 shows a number of spectra. The spectra are from a 4° (in b) by 0.8° (in l) rectangle, sampled every 0.1°. It can be seen that one cloud appears in several spectra, while others appear in only one or two spectra. Most of the emission is confined around b = 2°. To get a better idea of the latitude distribution, we have summed the spectra across in l, producing one spectrum at each b, and then used those to plot a b-v diagram, shown in Fig. 2. It should be noted that, when compared with the corresponding b-v diagram from the Weaver and Williams (1973) HI survey, the basic features appear to be the same. In Fig. 3, we sum over the velocity range -67 to -89 km/s, to estimate the

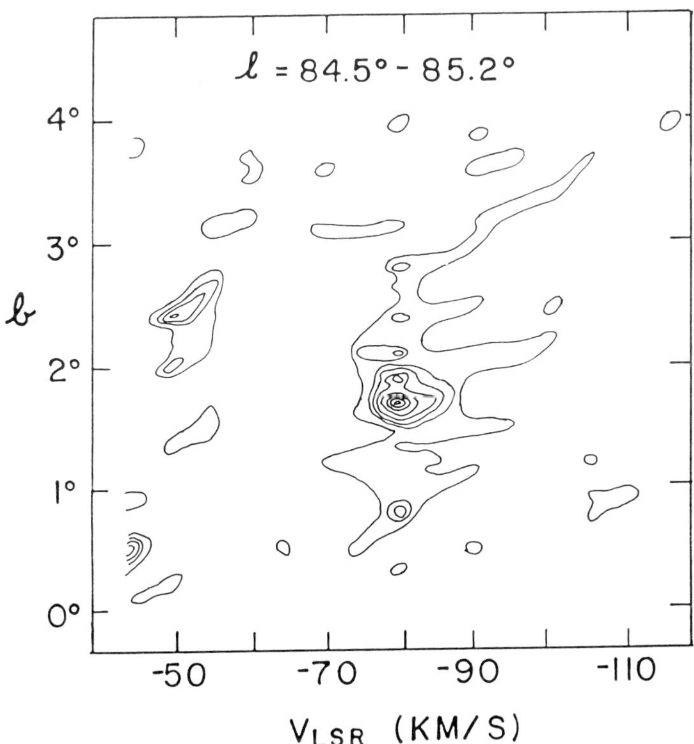

Fig. 2. A contour map made from the spectra in Fig. 1 after averaging across in longitude, producing a single spectrum at each b (with rms noise 0.1 K), and then taking $\int T_R^* dv$ in 5 km/s steps. Contours are of average $\int T_R^* dv$, in steps of 0.25 K-km/s.

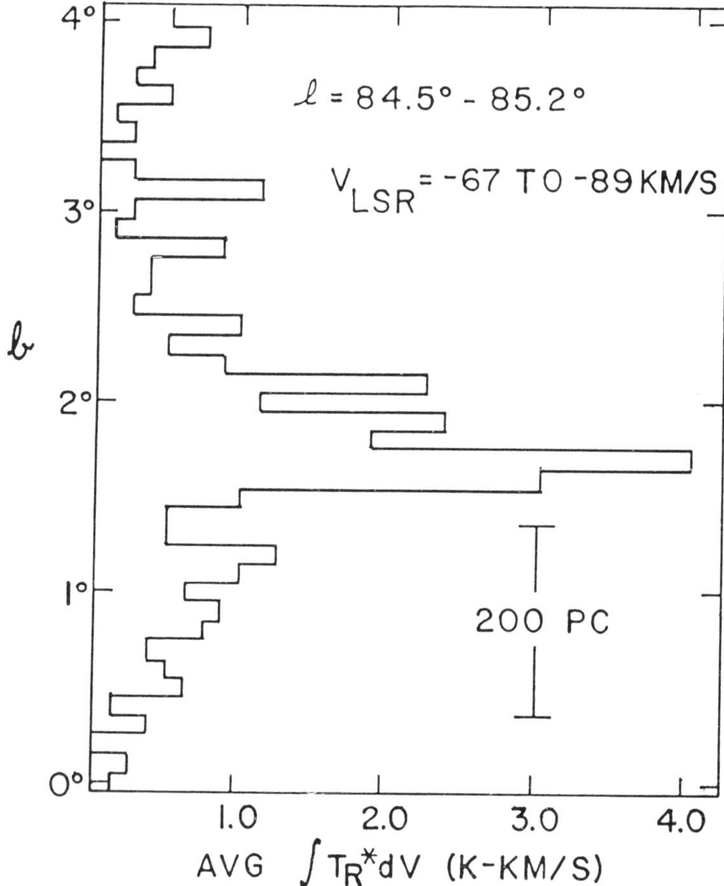

Fig. 3. Latitude distribution of CO emission from -67 to -89 km/s, from the longitude-averaged spectra described in Fig. 2.

thickness of the material emitting at these velocities. We estimate from this, and similar diagrams at other longitudes, that the molecular cloud layer is about three times thicker at R = 15 kpc than at 7 kpc. Thus, the molecular material "flares", just as the HI does. This flaring has also been noted by Fich and Blitz (1981) for HII regions in the outer galaxy.

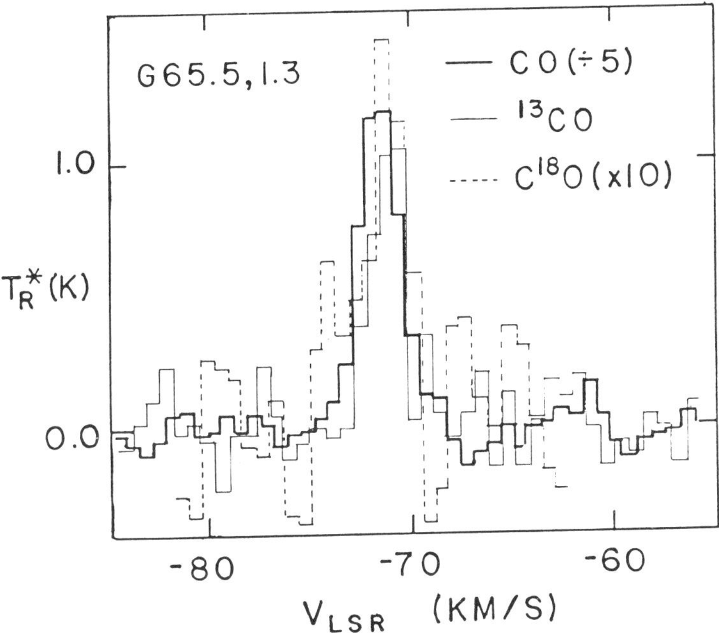

Fig. 4. Typical CO and isotopic spectra taken at FCRAO. The intensities have been scaled, as indicated, by the approximate intensity ratios observed in the inner galaxy.

3. INDIVIDUAL CLOUDS

Because of the sparse sampling, our basic survey has very little information on the nature of the clouds that are found in the outer galaxy. We have now used the 11-m telescope to make 1' sampled CO maps of 55 clouds. A number of them turn out to be tens of pc in extent, but we also find some smaller clouds. In general the emission from the smaller clouds is weaker than from the larger clouds, possibly explaining why Fich et al. (1982) found a relative paucity of small clouds in their study of two regions in the second and third quadrants.

As we have stated, a general feature of the outer galaxy clouds is that the CO lines are weaker than for the inner galaxy clouds. Kutner and Mead proposed that the lower cosmic ray flux in the outer galaxy can produce lower cloud kinetic temperatures. Kutner and Leung (1982, 1983) have investigated the question of the effect of cloud kinetic temperature on CO luminosity, and find that clouds with T_K = 13 K have three times the CO luminosity of clouds with T_K = 6 K (the clouds otherwise being identical).

These lower intensities, especially in the smaller clouds, has led to some misunderstanding on the part of other observers. For example, we attribute the failure of Solomon et al. (1982) to reproduce some of our earlier results to problems of calibration and sensitivity, arising primarily from the conditions under which their observations were done. We feel that the ultimate test of the reality of clouds is that they are reproduceable in our fully sampled maps.

To learn more about the nature of the clouds we observe, we have made ^{13}CO maps of 18 clouds (using the NRAO 11-m and FCRAO 14-m telescopes), and limited $C^{18}O$ maps of five clouds (using the FCRAO telescope). In addition, limited CO(2 → 1) maps have been made of five clouds (using the U. Texas MWO 5-m telescope) and 2-mm H_2CO observations have been made of seven clouds (using the 11-m telescope). (Details of the MWO and FCRAO observations will be given by Dickman et al. 1983.) A sample of CO and isotopic spectra is shown in Fig. 4. The important feature is that the relative intensities of the CO, ^{13}CO and $C^{18}O$ lines are the same as in inner galaxy clouds. This rules out the possibility that the outer galaxy cloud CO lines are weak due to low optical depths. Our CO(2 → 1) data also supports the idea of normal column densities with low kinetic temperatures.

4. CONCLUSIONS

Our large scale mapping of CO in the outer galaxy indicates that the CO emitting layer follows the HI warp in the outer galaxy and that the thickness of the layer increases by a factor of about 3 from R = 6 to 16 kpc. Detailed studies of individual clouds confirm the suggestion by Kutner and Mead that the kinetic temperatures in the outer galaxy clouds are significantly lower than those in the inner galaxy clouds. This difference means that, for a given amount of mass, inner galaxy clouds are ~ 3 times as luminous in CO as outer galaxy clouds.

This work was supported in part by NSF grant AST81-20900. The NRAO is operated by Assoc. Univ., Inc., under contract with the NSF. The MWO is operated by the EERL of the Univ. of Texas at Austin with support from the NSF and McDonald Obs. The FCRAO is operated with support from the NSF and with permission of the Met. Dist. Comm., Comm. of Mass.

REFERENCES

Blitz, L., Kulkarni, S., and Heiles, C.: 1981, BAAS, 13, p. 539.
Dickman, R. L., Mead, K. N., and Kutner, M. L.: 1983, (in preparation).
Fich, M., and Blitz, L.: 1981, BAAS, 13, p. 539
Fich, M., Terebey, S., and Blitz, L.: 1982, BAAS, 14, p. 617.
Kutner, M. L., and Leung, C. M.: 1982, BAAS, 14, p. 616.
Kutner, M. L., and Leung, C. M.: 1983, (in preparation).
Kutner, M. L., and Mead, K. N.: 1981, Astrophys. J. (Lett.), 249, p. L15.
Solomon, P., Stark, A. A., and Sanders, W.: 1982 (preprint).
Weaver, H., and Williams, D. R. W.: 1973, Astron. & Astrophys. (Suppl.), 17, p. 1.

A CO STRUCTURE NEAR THE GALACTIC CENTER WITH STRONG POSITIONAL AND KINEMATIC GRADIENTS

W.B. Burton
Sterrewacht Leiden

H.S. Liszt
National Radio Astronomy Observatory

ABSTRACT

We report here ^{12}CO observations made on a two-dimensional grid in the direction of the galactic center. The observations reveal an intense emission feature lying predominantly in the quadrant $\Delta\ell$, $\Delta b >$ 0', 0' and strongly inclined with respect to the galactic equator. The inclined feature shows gradients in velocity of order 100 km s^{-1} over angular scales of about a degree, continuity of intensity over length scales $\gtrsim$ 100 pc, and large apparent velocity dispersion. Lack of symmetry with respect to the direction and velocity of the galactic center seems to rule out interpretation in terms of a closed-orbit response to a general galactic potential.

The attention in our earlier work (e.g. Liszt and Burton, 1978) on the CO distribution in the galactic core focussed on the material characterized by obviously anomalous velocities and discussed by others as the "expanding molecular ring". Our interpretation placed this gas in the context of the large-scale tilted distribution also displayed by the HI in the inner one or two kpc. Study of this gas is facilitated by the relative ease with which it can be separated from line-of-sight contamination, either from emission from the core or from the galaxy at large. This ease of separation stems from the anomalous velocities of order 150 km s^{-1} and from the structural and kinematic regularities of the gas. These regularities allow gas tilted on a large scale to be studied using coarsely sampled data. Coarse sampling is necessary in any case because of the large angular extent of the tilted distribution studied earlier. In the HI data (e.g. Burton and Liszt, 1978) the tilted signature may be followed to $|b| \sim 4^{\circ}$, $|\ell| \sim 8^{\circ}$ from the center; there is no indication that CO does not pervade this same region, although fully sampling a region of such extent with a 1' beam is currently not feasible.

An important incompleteness in earlier work on the CO distribution in the galactic core is the relative lack of attention paid to gas at

W. B. Burton and F. P. Israel (eds.), Surveys of the Southern Galaxy, 149–157.

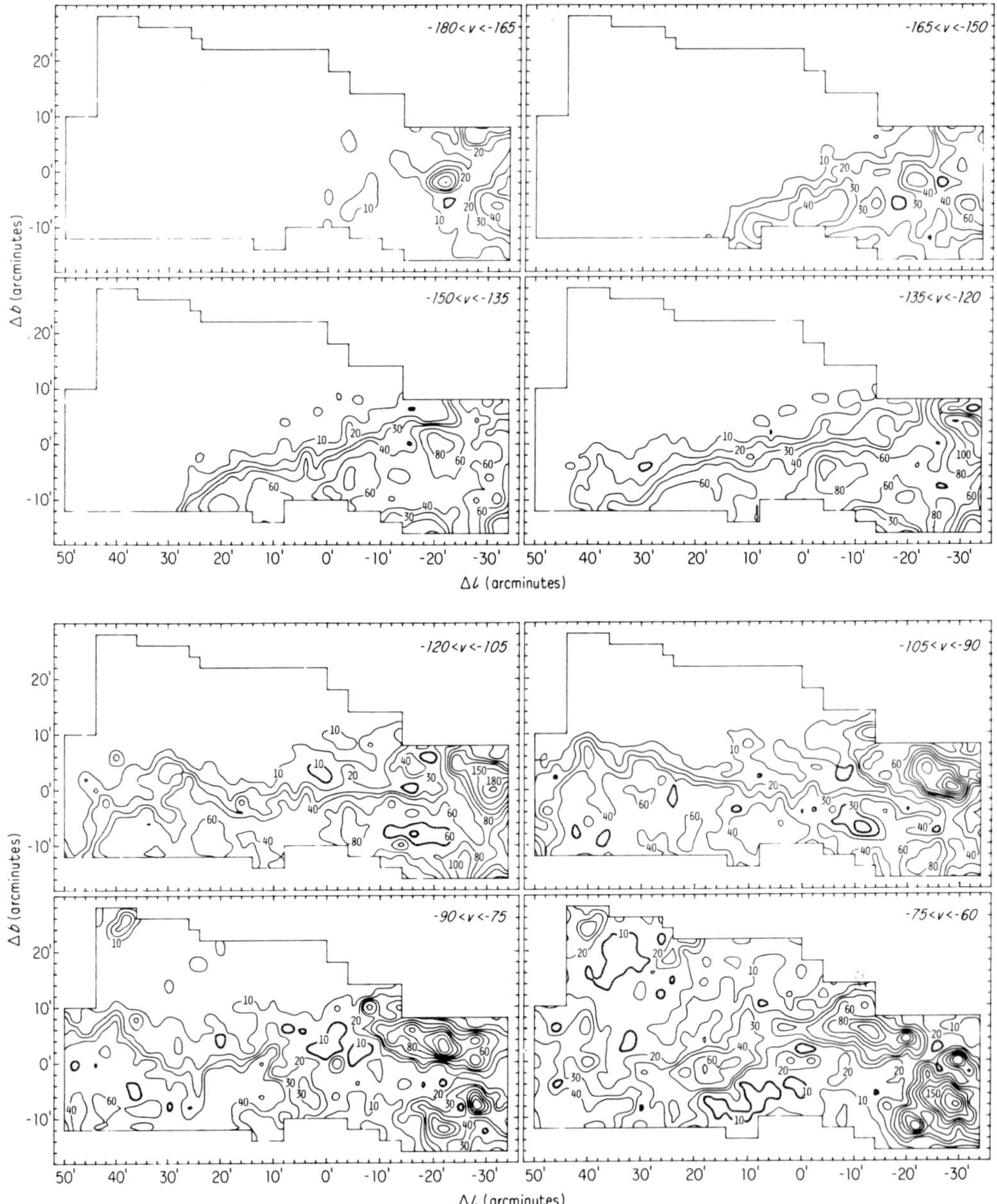

Figure 1. Contours of observed ^{12}CO temperatures T_A^* integrated over the indicated negative-velocity ranges and projected onto the plane of the sky. The contour units are K km s^{-1}. Emission from gas at v $\lesssim$ -100 km s^{-1} adheres to the kinematic and spatial signature of the large-scale tilted molecular structure described earlier (e.g. Liszt and Burton, 1978). This emission includes that associated with the negative-velocity branch of the feature often called (at b $\sim$ 0') the "expanding molecular ring".

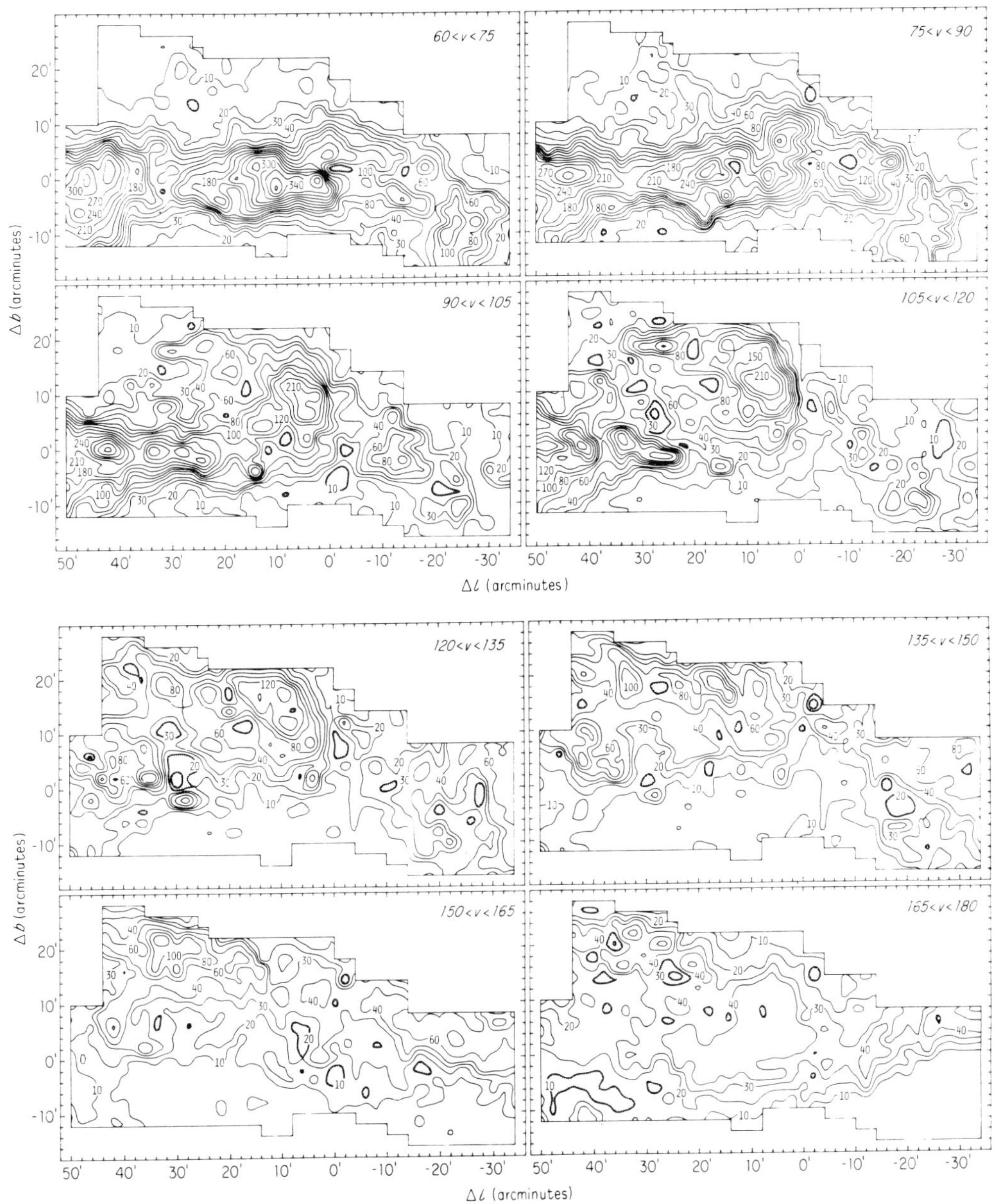

Figure 2. Contours of observed ^{12}CO temperatures T_A^* integrated over the indicated positive-velocity ranges and projected onto the plane of the sky. Emission patterns at v $\lesssim$ 75 km s^{-1} are dominated by the Sgr A and Sgr B molecular features located near the galactic equator. Our study of the present observations focusses on the emission pattern which separates from the Sgr A emission in the upper-left panel at Δb ∿ 6', Δℓ ∿ 0', and continues with rather regular gradients in Δb, Δℓ, and v to higher Δb and Δℓ as v increases. At the highest velocities, represented in the lower-right panel, the pattern is near the edge of the observed grid.

velocities less extreme than $|v| \sim 135$ km s^{-1}. Inspection of one of the longitude, velocity diagrams available for the galactic equator (e.g. figure 2 of Liszt and Burton, 1978) reveals that the most intense emission patterns within a few degrees of the center occur over the range $0 \lesssim v \lesssim 100$ km s^{-1}. This range includes the Sgr A and B molecular complexes, but it is clear even from the situation as only observed in the equator that the most intense emission is more extensive than these complexes. That the most-intense-emission patterns originate in the galactic core itself seems proven by the characteristics which they share with all other large-scale molecular emission features from the core: large ($\gtrsim 25$ km s^{-1}) velocity widths, continuity over lengths of at least a degree, and strong gradients in velocity and position. These are all characteristics which distinguish molecular emission from the core from emission from molecular clouds in the galaxy at large. The ^{12}CO observations reported here (and in more detail in a paper by Liszt and Burton, in preparation) were made to support investigation of the moderate velocity, but very intense, emission patterns.

The spectra were observed in four sessions during the past several years at the NRAO 36-foot telescope on Kitt Peak. The half-power beamwidth of the telescope is 1'at λ2.6 mm. The effective angular resolution of the present data is set, however, by the 2' spacing of the observed grid points. The grid positions were measured in terms of displacements from the peak of the 2 μm radiation in Sgr A West at $\ell = -3\overset{!}{.}3$, $b = -2\overset{!}{.}8$. We observed using a filter bank consisting of 256 channels, each of 1 MHz (2.6 km s^{-1}) width and separated by 1 MHz. The total velocity coverage of 650 km s^{-1} encompasses all known CO emission from the nucleus. Velocities are measured with respect to the local standard of rest. The intensity scale is $T_A/0.65$, giving the radiative temperature for a 65% beam efficiency.

Figures 1 and 2 show the arrangement on the plane of the sky of CO emission integrated over the indicated velocity ranges. Emission from the negative-velocity branch of the "expanding molecular ring" feature is represented in the upper panels of Figure 1. As expected, this emission displays the signature of the tilted gas distribution described in our earlier work. At the moderate negative velocities represented in the lower panels of the figure, emission patterns are not so clearly delineated. At $v \gtrsim -75$ km s^{-1}, several aspects of the observations compete in the same velocity region. The influence of the 3-kpc arm, CO self-absorption from very cold gas (Liszt <u>et al</u>., 1977), emission from clouds in the $4 < R < 9$ kpc annulus, as well as emission from the cone, together result in an observational situation which is rather confused in the present data.

An interesting, new emission pattern emerges at the positive velocities represented in Figure 2. In the upper panels of this figure there is substantial emission from the Sgr A and B complexes centered near $\Delta b \sim 0'$. There is a pattern which begins to separate from the Sgr A emission at $\Delta b \sim 6'$, $\Delta\ell \sim 0'$ in the upper-left panel. In successive panels this feature can be followed to higher Δb and higher $\Delta\ell$ as v in-

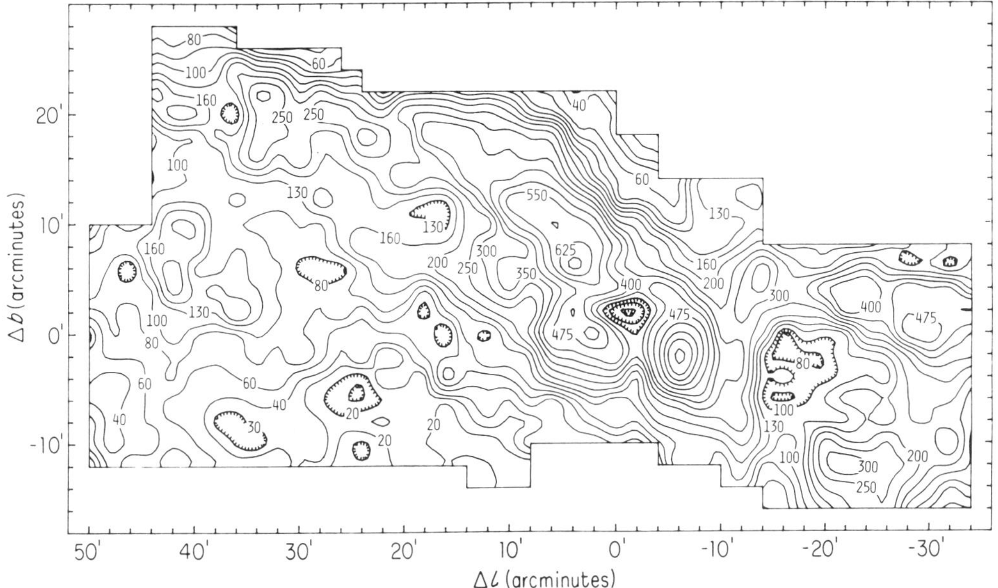

Figure 3. Gradient on the plane of the sky of the inclined ^{12}CO feature. The contours represent emission integrated over a range extending ±70 km s^{-1} from the velocity which we identify from inspection of the individual position, velocity maps as the central velocity of the inclined feature. At positive longitudes the feature extends from the region of Δb, Δℓ, v ∿ 0', 0', 85 km s^{-1} to Δb, Δℓ, v ∿ 25', 40', 170 km s^{-1}; beyond there it cannot be followed on our grid. Although the feature may extend to negative longitudes and latitudes, we have not been able to follow it with any certainty in the observations because of confusion with other emission at |v| $\lesssim$ 75 km s^{-} .

creases. The feature leaves the observed grid near Δb ∿ 22', Δℓ ∿ 44', at v ∿ 170 km s^{-1}.

The strong positional gradient of the inclined feature is shown in Figure 3. The velocity gradient is revealed by Figure 4. The various panels of this figure represent b, v cuts through the data cube at the indicated longitudes. The inclined feature can be seen separating from Sgr A emission in the Δℓ = 0' panel at Δb ∿ 8', v ∿ 60 km s^{-1}. At Δℓ = 6', the feature is centered near Δb ∿ 12', v ∿ 90 km s^{-1}. By Δℓ = 18', the feature has begun to blend with emission from the positive-velocity branch of the "expanding molecular ring". The velocity gradient of the feature over the range of position represented in Figure 4 is about 100 km s^{-1}. The very broad velocity width, evident in each of the panels of the figure, is obviously an important aspect of the signature of the inclined feature.

The continuity of the inclined feature in position and velocity, as well as the strong gradients in position and velocity, suggest that the emission is contributed from long lines-of-sight through a filled

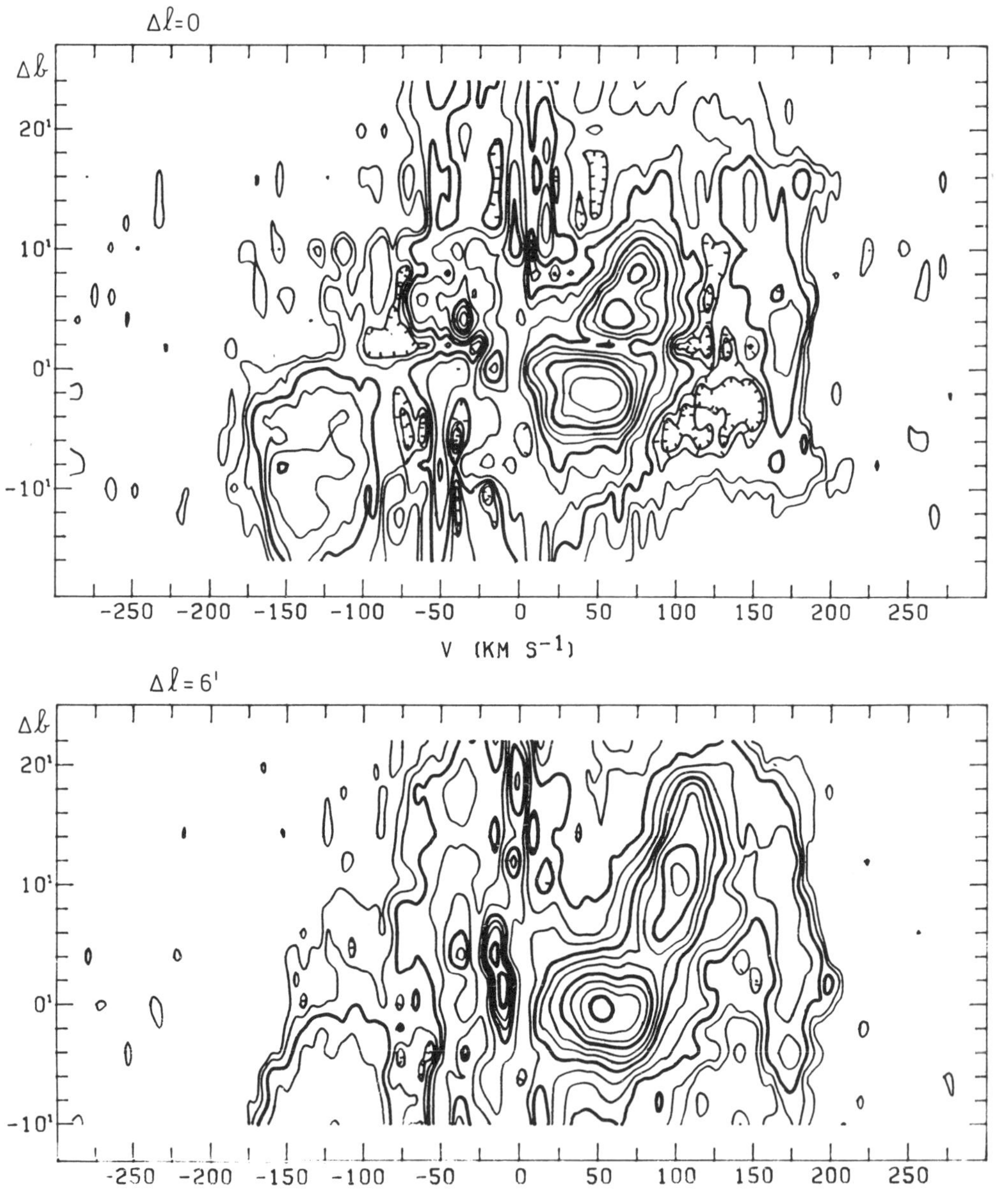

Figure 4. Latitude-velocity arrangement of ^{12}CO at the indicated longitudes. The $\Delta b/v$ gradients of the inclined feature are evident on the individual maps. The $\Delta \ell/v$ and $\Delta \ell/\Delta b$ gradients are evident from the changes from map to map of the centroid of the emission pattern. The emission from the inclined feature shares several characteristics with other emission patterns from neutral gas associated with the galactic core: 1) large apparent velocity dispersion ($\sigma_v \gtrsim 30$ km s^{-1}), 2) continuity of emission intensities over length scales $\gtrsim$ 100 pc, and 3) gradients in velocity of order 100 km s^{-1} over angular scales of less than a degree.

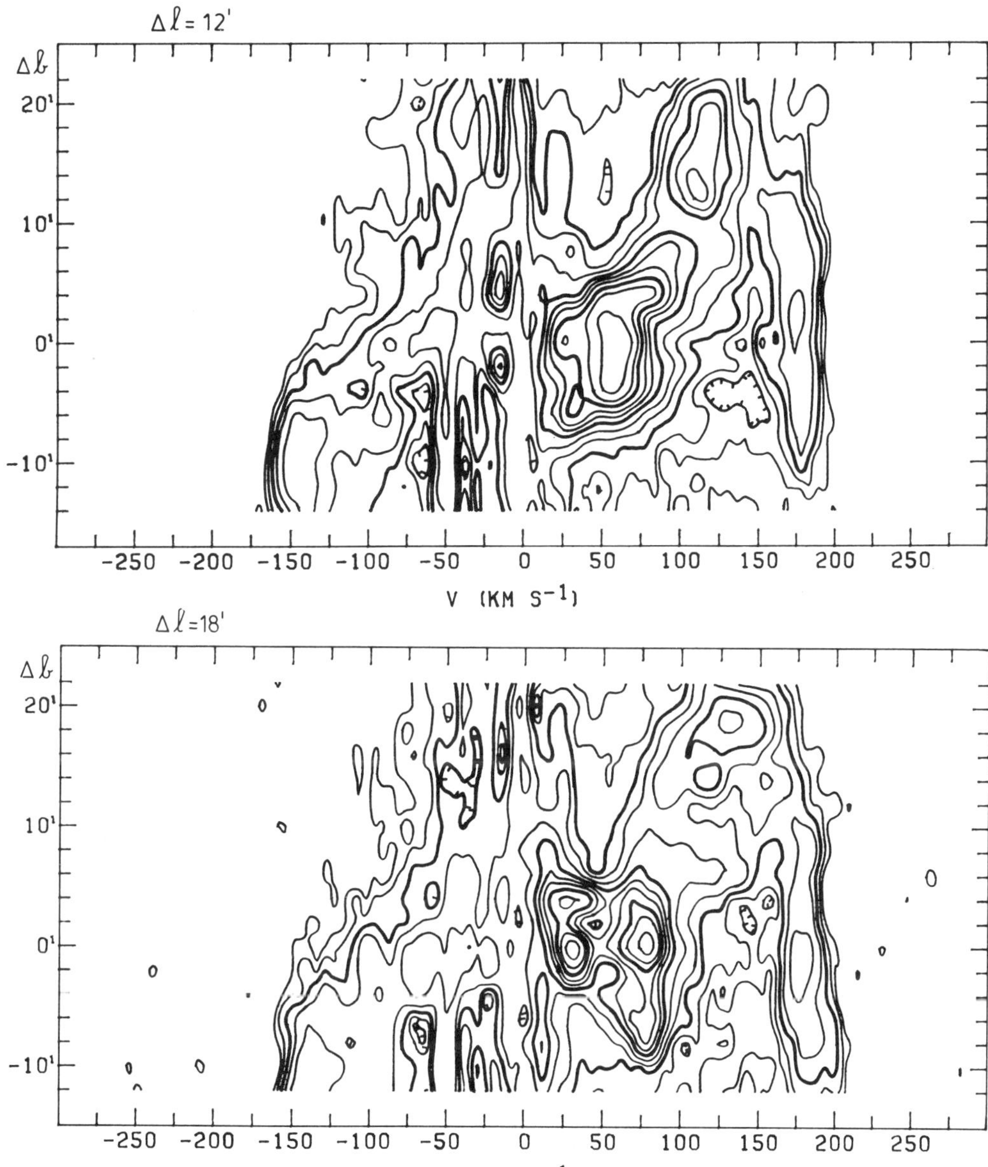

Figure 4, continued. Δb, v distribution of the inclined feature.

distribution. The velocity widths in particular would be difficult to account for in terms of a highly confined jet of molecular gas. Modelling of the "expanding molecular ring" by a filled, extended distribution can account for similar properties. But in our earlier modelling we were able to draw on the positional and kinematic symmetries of the "expanding molecular ring" feature and of the apparently anomalous HI features in the core to suggest, albeit in an *ad hoc* fashion, motions and a dis-

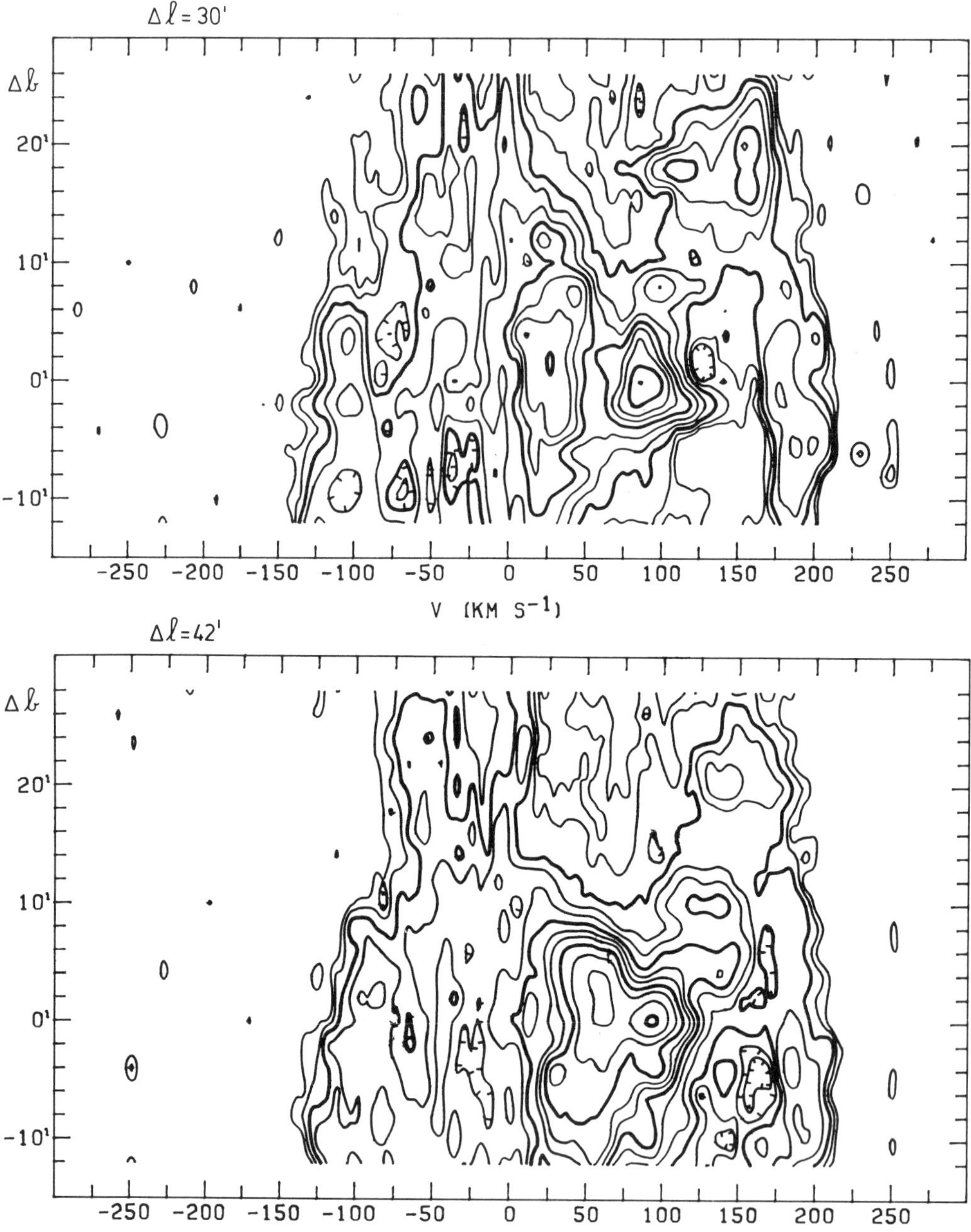

Figure 4, continued. Δb, v distribution of the inclined feature.

tribution governed by a general galactic potential. The newly studied inclined feature does not show sufficient symmetry with respect to Δℓ, Δb = 0', 0' and v = 0 km s^{-1} to justify a similar model.

The observed grid of the present data is not extensive enough to allow us to determine the total size of the inclined feature. Of particular importance is the extent to which the inclined feature continues to negative velocities and negative values of $\Delta\ell$ and Δb. Line-of-sight contamination at $|v| \lesssim 50$ km s^{-1} makes it difficult to follow the inclined feature there. Although our observations are very limited at $\Delta\ell < 0'$, $\Delta b < 0'$, they do seem to rule out a symmetric inclined feature in that quadrant. This apparent lack of symmetry places important constraints on the interpretation of the inclined feature. Definite conclusions in this regard should be supported by more extensive observations, preferably of tracers such as ^{13}CO or CS which - revealing higher densities - would suffer less than ^{12}CO from line-of-sight contamination.

Acknowledgement: The National Radio Astronomy Observatory is operated by Associated Universities, Inc., under contract with the National Science Foundation. We gratefully acknowledge support from the North Atlantic Treaty Organization through Research Grant No. 008.82.

REFERENCES

Burton, W.B., and Liszt, H.S.: 1978, Astrophys. J. 225, 815.
Liszt, H.S., Burton, W.B., Sanders, R.H., and Scoville, N.Z.: 1977, Astrophys. J. 213, 38.
Liszt, H.S., and Burton, W.B.: 1978, Astrophys. J. 226, 790.

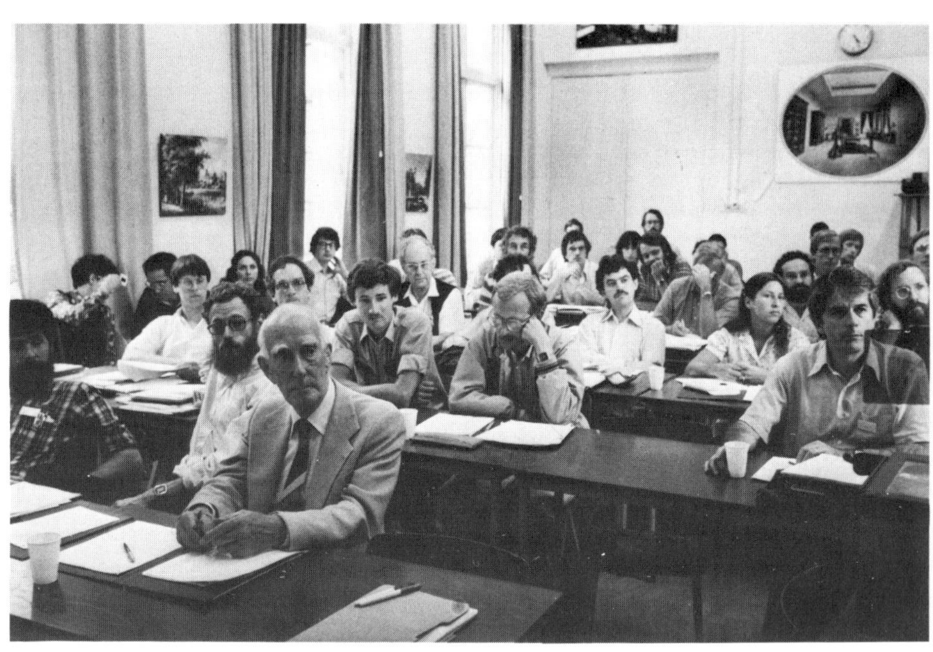

OH IN THE CENTRE OF THE GALAXY

R.J. Cohen and W.R.F. Dent
University of Manchester,
Nuffield Radio Astronomy Laboratories, Jodrell Bank,
Macclesfield, Cheshire SK11 9DL, England

ABSTRACT

Results are reported from a large-scale survey of OH in the galactic centre region. The molecular clouds are distributed in a bar-like structure at least 2 kpc in diameter. The outer parts of the bar are warped out of the galactic plane, and extend to z-distances of ± 200 pc.

1. INTRODUCTION

The galactic centre contains the largest concentration of dense molecular clouds in the Galaxy. These were first detected through their OH absorption lines at 18 cm wavelength, and the subsequent OH survey by Robinson and McGee (1970) has remained the best guide to the large-scale distribution and kinematics of molecular clouds in the galactic centre. A new survey of the region was carried out at Jodrell Bank last summer (Cohen 1982). This was not strictly a southern survey, but it would certainly have been easier in the south. In fact it was necessary to observe as low as elevation 2° in order to map the OH distribution completely. The OH main lines at 1667 and 1665 MHz were observed simultaneously in both hands of circular polarization. The lines occur primarily in absorption against the galactic continuum background, but a number of maser emission sources were also detected. The parameters of the survey are summarized below:

Area surveyed	$-6\overset{\circ}{.}0 \leqslant \ell \leqslant 8\overset{\circ}{.}6$
	$-1\overset{\circ}{.}0 \leqslant b \leqslant 1\overset{\circ}{.}0$ plus extensions to $b = 1\overset{\circ}{.}6$ and $b = -2\overset{\circ}{.}0$
Telescope beamwidth	$0\overset{\circ}{.}17$ (10 arcmin)
Sampling interval	every $0\overset{\circ}{.}2$ in ℓ and b
Velocity coverage	900 km s^{-1}
Velocity resolution	3 km s^{-1}
Detection level	0.2K

W. B. Burton and F. P. Israel (eds.), Surveys of the Southern Galaxy, 159–164.

2. RESULTS

OH absorption was detected over the region shown in Fig.1. The molecular layer as a whole extends to $\pm 6^{\circ}$ in galactic longitude, corresponding to projected distances of ±1 kpc from the centre. The densest part, found in the Parkes OH surveys, lies between $\ell = -1^{\circ}$ and $\ell = 3^{\circ}$ (Fig.1). As a whole this central part of the concentration is not inclined to the galactic plane, although some structures within it are inclined (Cohen and Few 1976). The outer distribution of molecular clouds found in the present survey lies almost entirely out of the galactic plane in a warped layer, reaching z-distances of ±200 pc. The warp or tilt in the molecular layer closely follows the warp seen in the HI in the centre.

Some preliminary longitude-velocity (ℓ-V) maps of the new data are given in Fig. 2. The two OH lines at 1667 MHz and 1665 MHz are shown, but velocities are given for the 1667 MHz line only. In the map at $b = 0\overset{\circ}{.}0$ there is overlap of the two lines near $\ell = 0$ because of the wide velocity range ($\sim$400 km s^{-1}) over which absorption occurs. Elsewhere the two lines are clearly separated and show a very good correspondence in general. The strongest OH features occur at low and

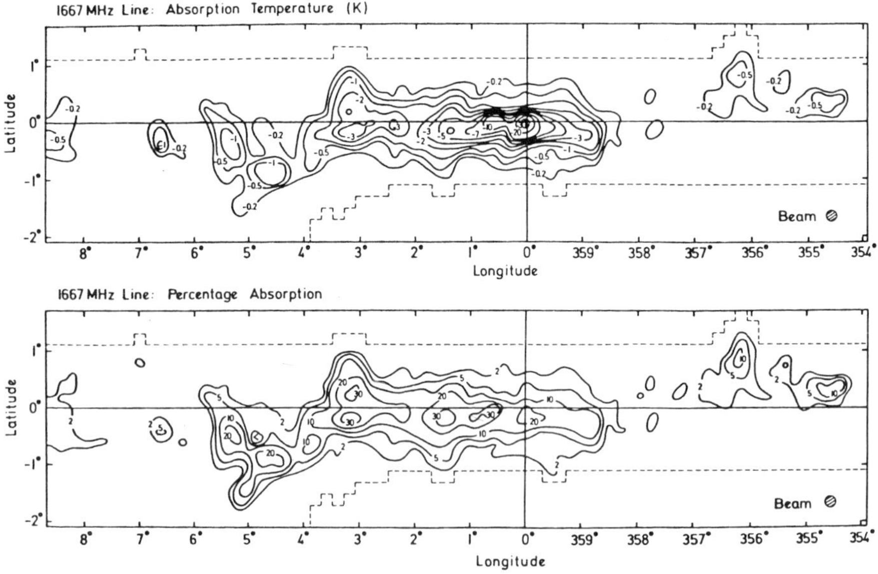

Fig. 1 The upper diagram shows the strongest absorption temperature measured in the 1667 MHz line, as a function of position on the sky. The lower diagram shows the absorption as a percentage of the 18 cm continuum. The dashed lines indicate the boundary of the region surveyed. The telescope beamsize is indicated at the lower right.

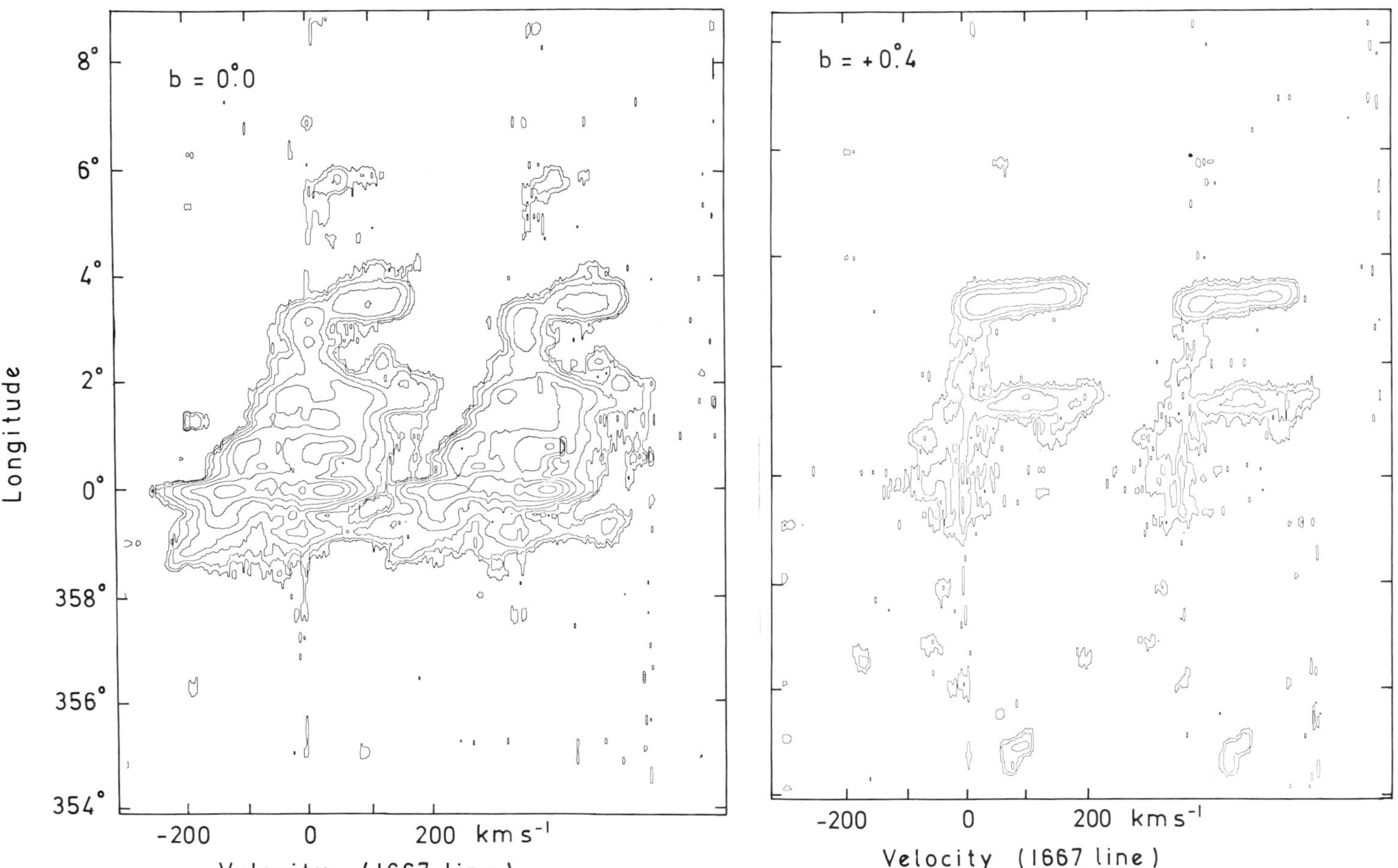

Fig. 2. Longitude-velocity maps of OH absorption at latitudes ℓ = 0°.0 (left) and b = +0°.4 (right). Contours give antenna temperature, and are plotted at -0.2,-0.4,-0.8,-1.6,-3.2,-6.4,-12.8 and -25.6K.

intermediate velocities, for example near $\ell = 3^\circ$, V = 0 km s^{-1} in the b = 0°0 map. The high-velocity nuclear disk which is so prominent in HI spectra is only a second rank feature in the OH spectra and is sometimes not even detected. The map at b = +0°4 shows two examples of very broad line profiles, at ℓ = 1°4 and 3°2. It also shows a molecular cloud near $\ell = 355^\circ$, V = 100 km s^{-1} which has a very large noncircular velocity.

3. INTERPRETATION

The molecular clouds have complex noncircular motions. In the past, attempts have been made to model these in terms of either

(i) rotation about the centre plus radial outflow, perhaps due to explosions at the centre, or

(ii) elliptical streamlines about a central bar.

Detailed models of both types have been constructed by Burton and Liszt (1978) and Liszt and Burton (1980). They have emphasized that kinematic effects can give rise to spectral features at certain positions and velocities, even when the molecular cloud distribution is random, because large path lengths can contribute to certain narrow velocity ranges. However the dominant OH features in our data occur at positions and velocities where the kinematic effects are working in the opposite way, namely where a small path length is contributing to a very wide velocity range. The fact that we nevertheless find the strongest features here is strong evidence that the OH distribution is not random but is highly structured.

We have used the velocity field of type (i) by Burton and Liszt (1978) to locate the molecular clouds in the central region according to their radial velocities. The results are shown in Fig. 3, which is a plan view of the central region as it might appear to an observer located outside the Galaxy. The molecular clouds are strongly concentrated in a narrow bar-like structure. The details of this picture may be model dependent but its general character is not. The bar appears to be rather lop-sided, being only fragmentary on the negative longitude side (the side furthest from the Sun). However this could be due to the relative distributions of the molecular clouds and the continuum sources against which the OH absorption lines are seen (Cohen and Few 1976). The bar might be more symmetric than it appears in OH. Observations of CO emission would be valuable in clearing up this point.

The bar lies in the orientation required by models of type (ii) which have been proposed to explain the noncircular motions of HI features out to and including the 3 kpc arm (Peters 1975; Liszt and Burton 1980). We are currently extending our analysis to velocity models of this type. Gas dynamical calculations by Sørensen et al. (1976) and others have shown that only a weak bar potential is necessary to excite a strong response in the gas flow. These calculations predict the formation of shock-waves along the major axis of

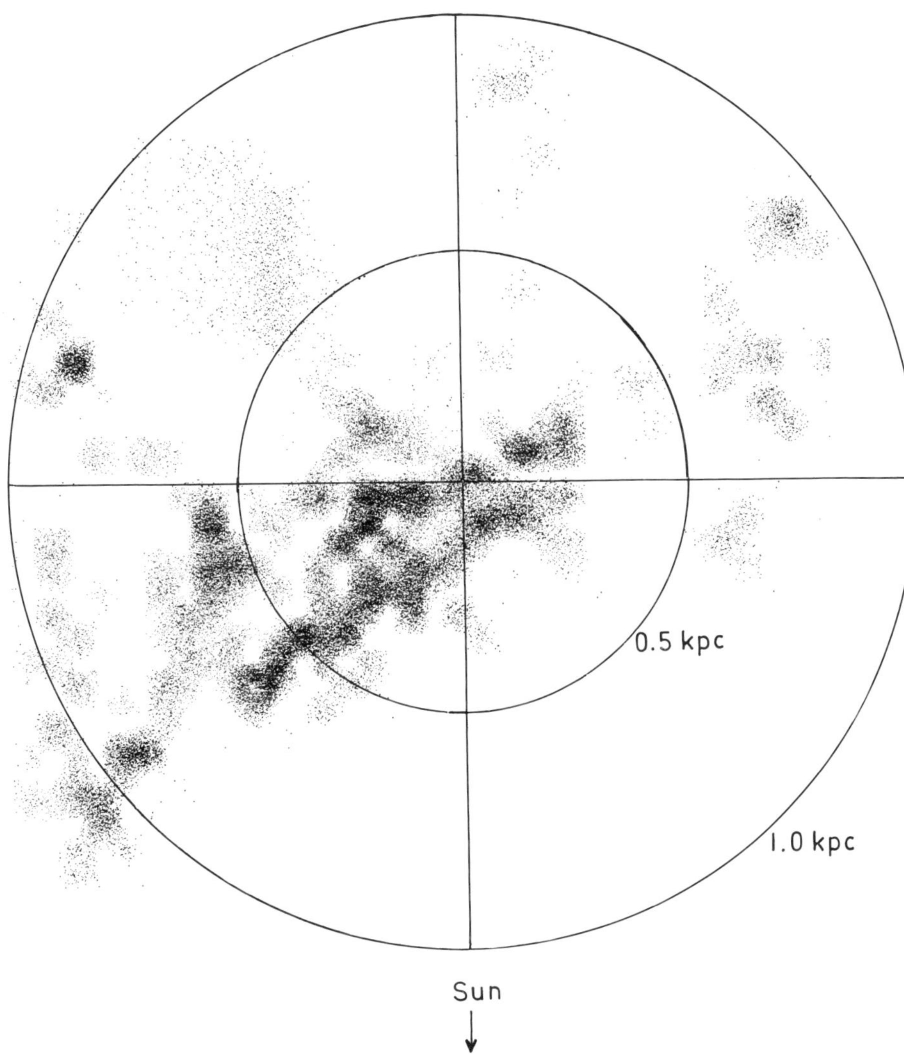

Fig. 3. Plan view of the galactic centre showing the distribution of molecular clouds. Data have been summed over all observed latitudes and smoothed to a resolution of 50 pc.

the bar, and it is natural to identify the bar-like concentration of molecular clouds with such a shock wave driven by a weak bar in our own galactic centre.

REFERENCES

Burton, W.B., and Liszt, H.S., 1978. Astrophys.J., 225, pp.815-842.
Cohen, R.J., 1982. "Submillimetre wave astronomy" pp.185-190. Eds. J.E. Beckman and J.P. Phillips, Cambridge University Press.
Cohen, R.J. and Few, R.W., 1976. Mon.Not.R.astr.Soc., 176, pp.495-523.
Liszt, H.S. and Burton, W.B., 1980. Astrophys.J., 236, pp.779-797.
Peters, W.L., 1985. Astrophys.J., 196, pp.617-629.
Robinson, B.J. and McGee, R.X., 1970. Aust.J.Phys., 23, pp.405-423.
Sørensen, S.-A, Matsuda, T. and Fujimoto, M., 1976. Astrophys.Space Sci., 43, pp.491-503.

OH/IR STARS IN THE CENTRAL REGION OF THE GALAXY

B. Baud
Laboratorium voor Ruimteonderzoek
Groningen

ABSTRACT

OH/IR stars offer a unique opportunity to study the distribution and kinematics of the stellar population in the nuclear region of the Galaxy. Systematic 1612 MHz surveys with the 100m Effelsberg telescope and the VLA have detected these stars as close as a few arcminutes from the Galactic Centre. Their density distribution in the inner 5 kpc follows that of the molecular species: a steep decrease from a maximum at R=5 kpc down to a minimum at 3 kpc and a very sharp peak in the inner 350 pc from the Centre. The similarity in radio characteristics with OH/IR stars in the galactic disk indicates that the stars in the nuclear region are of the same population type: a mix of population I and intermediate disk population stars. Nevertheless there is evidence for a paucity of both very old (10^9yrs) and very young (10^7yrs) stars in the inner 350 pc.

The distribution of stars in the inner 1° is flattened with an axial ratio of 2:1. There is clear evidence for rotation of 150 km s^{-1} deg^{-1} and the velocity dispersion is also 150 km s^{-1}.

1. INTRODUCTION

Studies in the Galactic Centre region have been carried out predominantly at radio and infrared wavelengths. Most of these have been limited to the distribution and kinematics of the gas in the nucleus. Optical and near infrared observations have traced the stellar component locally in the Baade windows of low interstellar absorption or very close to the Centre, but they are limited by the strong and variable extinction in the direction of the Galactic Centre. OH/IR stars do not pose this limitation. These stars, with their characteristic radio signature of a strong double peaked 1612 MHz OH emission profile, can be detected throughout the Galaxy unhindered by interstellar extinction. They have therefore proven to be very useful for statistical studies of the stellar distribution and kinematics on a galactic scale (Johansson et al. 1977, Bowers 1978, Baud et al. 1981a).

W. B. Burton and F. P. Israel (eds.), Surveys of the Southern Galaxy, 165–171.

In recent years several systematic OH radio surveys have been carried out, searching for OH/IR stars around the Galactic Centre. The most important are listed in Table 1.

TABLE 1

Strip Survey	$358<\ell<$	$\|b\|<1.25$	$\|V\|<200$ km s^{-1}	Effelsberg
Deep Survey	circle of 1° radius around $(\ell,b)=(0,0)$		$\|V\|<450$ km s^{-1}	Effelsberg
	0.5° field near Galactic Centre		$-224<V<-14$ km s^{-1}	VLA
Parkes Survey	$270<\ell<326$	$b=0$	$-230<V<130$ km s^{-1}	Parkes
	$340<\ell<355$	$\|b\|<0.2$	$-230<V<130$ km s^{-1}	Parkes
	$355<\ell<2$	$\|b\|<0.6$	$\|V\|<450$ km s^{-1}	Parkes

All surveys, except the one with the VLA, have been carried out at 1612 MHZ only, by integrating on a rectangular grid of points, separated by the HPBW of the telescope or somewhat less. Close to the Centre, within 0.5° from the galactic plane, strong an multiple absorption features affect the frequency baselines of the spectra and hamper the detection of weak emission features. In this case the VLA was used to resolve out the more extended background continuum sources. This gives a clean frequency baseline on which the narrow line emission is more clearly visible. Some preliminary results have been published by Baud et al. (1981b), Olnon et al. (1981) and Caswell et al. (1981).

2. OH/IR STARS WITHIN 2 KPC FROM THE GALACTIC CENTRE

I will only consider the results from the Strip Survey in the first quadrant (15°<ℓ<0°). This contains the most complete sample of CH/IR sources in the inner region of the Galaxy and has some overlap with the extensive surveys of the remainder of the Galaxy in the northern hemisphere.

Most OH/IR stars found in the radio surveys are close to the galactic plane and their radial velocities indicate large distances. Some can be detected at the far side of the Galaxy. The interstellar and circumstellar extinction obscures them optically. Hence, most of their properties are derived from the radio data. The large scale surveys in the galactic disk at ℓ>10° (e.g. Baud et al. 1981a) have shown that:

- The radial distribution of OH/IR stars follows that of the molecular species, with a pronounced maximum at R = 5 kpc and a decrease in density on either side. In particular, inside 5 kpc the density drop-off is very steep. Most OH/IR stars in the Galaxy are found in this so-called molecular ring.

- OH/IR stars represent a continuum of population types which can be distinguished statistically by the parameter ΔV, the velocity separation of the two peaks in the OH emission profile. Stars with $\Delta V < 30$ km s^{-1} are disk population stars with ages of 0.5 - 1×10^9 yrs and have a main sequence mass of 2 - 3 $M_\odot$. They are probably similar to the OH emitting Miras with long periods (>300^d), found in the solar neighbourhood, but their OH emission is two orders of magnitude more luminous. At $\Delta V > 30$ km s^{-1} a larger proportion of stars is probably young (10^7 - 10^8 yrs) and may resemble the OH emitting supergiants near the Sun.

The distribution of OH/IR stars in the first quadrant is shown in Fig. (1). This is derived from a combination of the Strip Survey and the surveys in the galactic disk. The maximum at ℓ = 25° reflects the peak in the density distribution at R = 5 kpc ($R_\odot$ = 10 kpc). The results of the strip survey confirm the low density inside 5 kpc, but the numbers steeply increase again within two degrees from the Galactic Centre, implying a density increase within 350 pc from the Centre. Note that the molecular gas distribution also has a local maximum within a few hundred pc from the Centre. This strengthens the conclusion from previous surveys that the OH/IR distribution follows that of the molecular gas. The link between the two is probably the star formation rate: a large density of molecular clouds implies a large rate of star formation and therefore a relatively high density of young disk stars that will evolve through the OH/IR phase. The maximum in the old stars near ℓ = 0° further implies that the star formation rate and the density of molecular gas in the galactic centre region must have been high for at least the last 5×10^8 yrs.

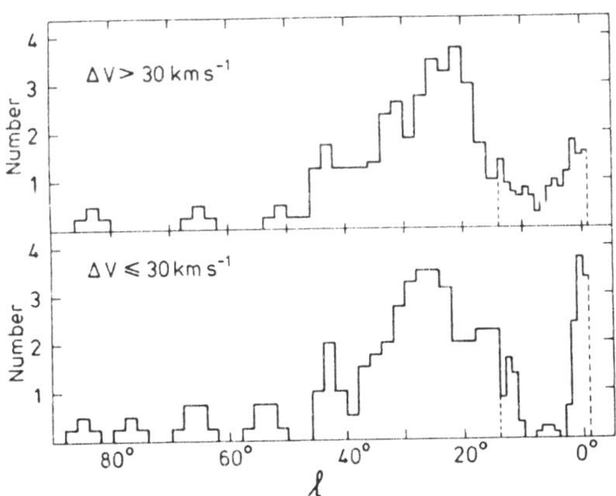

Figure 1. The number of OH/IR stars per 2° longitude interval in a 2.5° wide strip along the galactic equator, divided into two groups by peak separation ΔV. The data within the dashed lines are taken from the Strip Survey, those at ℓ>14° from the survey with the Dwingeloo telescope (Baud et al. 1981a). The numbers from the strip survey have been corrected in order to account for the higher sensitivity.(From Olnon et al. 1981)

Another result from a recent analysis of the Strip Survey is the different population mix in the nucleus. Fig. (2) shows the ΔV distribution of stars found in the Strip Survey in the Centre and away. For $\ell > 2°$ most stars are foreground objects, that are probably part of the molecular ring at R = 5 kpc. Here, the population distribution, as derived from the ΔV distribution, is similar to that found in the surveys of the galactic disk at larger longitudes. Within 2° from the Centre, the distribution is much narrower, with an apparent lack of stars with $\Delta V < 25$ and > 35 km s^{-1}, corresponding to a main sequence mass $< 2\ M_\odot$ and $4\ M_\odot$ respectively (Baud and Habing 1983). This is confirmed in the velocity distribution of OH/IR stars close to the Centre: only the intermediate mass stars with ΔV between 25 and 35 km s^{-1} show the large increase in velocity distribution, that is typical for matter close to the Galactic Centre.

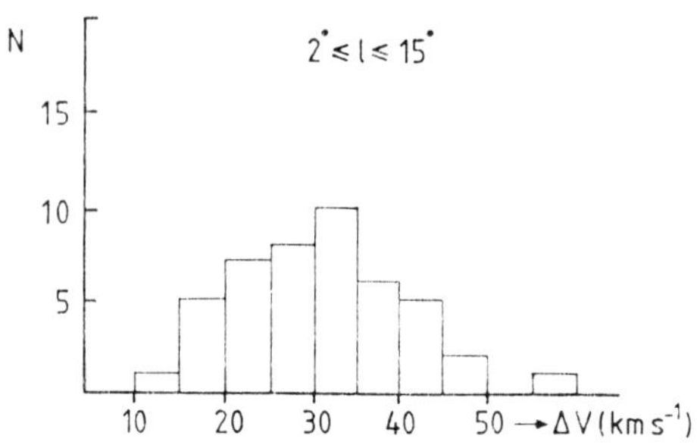

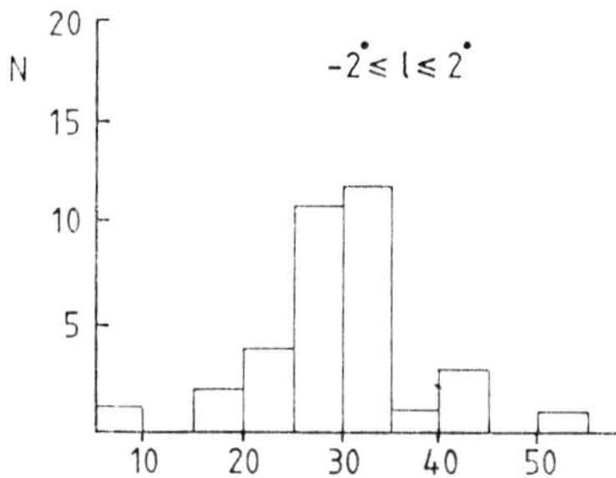

Figure 2. ΔV distribution of OH/IR stars in the galactic disk and in the galactic centre region.

Although the longitude distribution of OH/IR stars is similar to the distribution in CO, we find no detailed correspondance in the kinematics of the so called tilted disk model (Liszt and Burton 1980), which appears to explain some of the peculiar kinematic features of the gas around the Galactic Centre in a consistent way. Such a correspondance is not expected, however, because of the large velocity dispersion of the stars of about 150 km s^{-1} (see next section).

3. THE INNER 170 PC

The Deep Survey, a circular area of 2° diameter centered on $(\ell, b) = (0,0)$, was carried out to explore the gravitational potential near the Centre. Long integration times were required to get a sufficiently large sample of stars for a statistical study. Hence the name "Deep".

The distribution on the sky is shown in Figure (3). The flattening towards the plane is clear. This is further strengthened if one takes into account the strong absorption close to the plane ($< 0.5°$), that hampers detection of OH/IR stars in this region. This selection effect in the data must be taken into account when the mass distribution around the Galactic Centre is derived. This is also true for the velocity distribution in Figure (4).

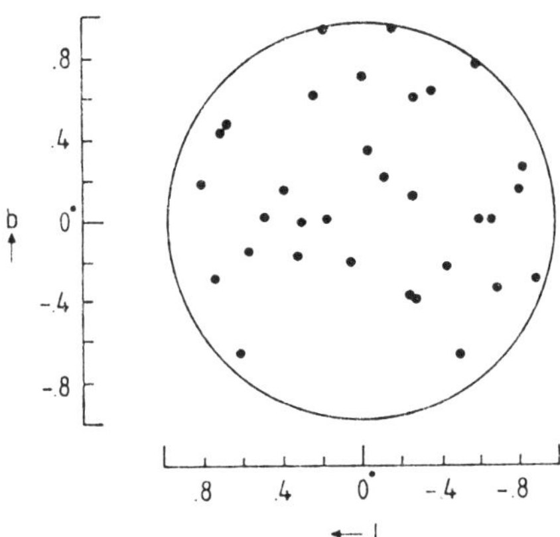

Figure 3. Distribution of OH/IR stars around the Galactic Centre in galactic coordinates. Results from the Deep Survey (Habing et al. 1983).

Nevertheless it is clear from the large observed velocity range that the majority of stars is actually associated with the Galactic Centre. Habing et al. (1983) estimate that only one or two stars may be foreground objects. Despite the large velocity dispersion (150 km s^{-1}), there is clear evidence for rotation of the whole sample, with a velocity of about 150 km s^{-1} deg^{-1}. This is to our knowledge the first actual detection of galactic rotation in a stellar population near the Galactic Centre.

By taking into account the above mentioned selection effect Habing et al. (1983) construct a model density distribution, based on the observed sky and velocity distribution. They find that the distribution inside ℓ = 1° is flattened, with an axial ratio of about 2:1 and that it has the same shape as the 2.4 μm distribution outside this region. The latter reflects the distribution of the K and M giants in the nuclear bulge. Since OH/IR stars are disk population objects this suggests that the stars in the bulge near the plane consist of a disk population and not of very old population II objects. Recently, Feast et al. (1981) have determined radial velocities of 25 Mira variable near the Galactic Centre near two Baade windows, about 3° below the plane. They found no net rotation, which is in apparent contradiction with the results from OH/IR stars. This suggests that rotation may only occur close to the plane. Another explanation would be the short periods ($<300^d$) of the Miras, which

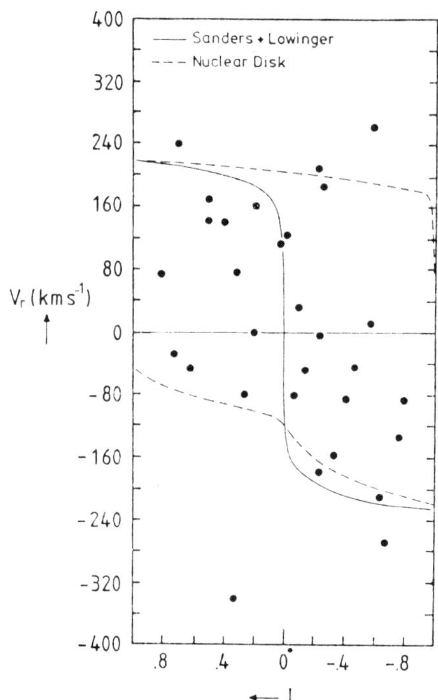

Figure 4. Longitude - radial velocity diagram of OH/IR stars around the Galactic Centre, from the Deep Survey. Indicated are the rotation curve in the nuclear region (continuous) and the extent of the CO emission (between the dashed lines)

indicates an older population than OH/IR stars.

An interesting speculation is given by Habing et al. (1983) on the nature of the Ne II nebulae discovered in the nucleus by Lacy et al. (1980). These may be ionized remnants of the circumstellar shells of OH/IR stars. Both the rotation and the large velocity dispersion of the Ne II clouds are also seen in the OH/IR stars, while the expected density of these objects within 5 pc from the Centre agrees with the observed number of Ne II clouds.

ACKNOWLEDGMENTS

Most of the work reported here is the result of a long-term effort together with Drs. H.J. Habing, H.E. Matthews, S. Tereby, R. Walterbos and A. Winnberg.

REFERENCES

Baud, B., Habing, H.J.: 1983, submitted to Astron. Astrophys.

Baud, B., Habing, H.J., Matthews, H.E., Winnberg, A.: 1981, Astron. Astrophys. 95, 156

Baud, B., Habing, H.J., Matthews, H.E., Winnberg, A.: 1981, Astron. Astrophys. 95, 171

Bowers, P.F.: 1978, Astron. Astrophys. 64, 307

Caswell, J.L., Haynes, R.F., Goss, W.M., Mebold, U: 1981, Aust. J. Phys. 34, 333

Feast, M.W., Robertson, B.S.C., Black, C.: 1980, Monthly Notices Roy. Astron. Soc. 190, 227

Habing, H.J., Olnon, F.M., Winnberg, A., Matthews, H.E., Baud, B.: 1983, in preparation

Johansson, L.E.B., Andersson, C., Goss, W.M., Winnberg, A.: 1977, Astron. Astrophys. 54, 323

Lacy, J.H., Townes, C.H., Geballe, T.R., Hollenbach, D.J.: 1980, Astrophys. J. 241, 132

Liszt, H.S., Burton, W.B.: 1980, Astrophys. J. 236, 779

Olnon, F.M., Walterbos, R.M., Habing, H.J., Matthews, H.E., Winnberg, A., Brzezinska, H., Baud, B.: 1981, Astrophys. J. Letters, 245, L103.

^{13}CO EMISSION FROM THE GALACTIC DISK IN THE RANGE $\ell = 40^{\circ} - 60^{\circ}$

Didier Despois and Alain Baudry
Observatoire de l'Université de Bordeaux, 33270 Floirac, France

ABSTRACT

We have surveyed the longitude range $40^{\circ} \leq \ell \leq 60^{\circ}$ in the J = 1-0 line of ^{13}CO with the 2.5-m telescope of the Bordeaux Observatory. At $b = 0^{\circ}$ the galactic disk was sampled at 15' spacings. Out of the plane, spectra were obtained at $b = \pm 0^{\circ}\!.5$ and at 5' spacings in b at $|b| \leq 0^{\circ}\!.25$. Some major features of the ℓ - v and b - v diagrams and variation with ℓ of integrated intensities and latitude profiles are briefly discussed.

1. INTRODUCTION

The carbon monoxide molecule is a major tracer of the morphology and kinematics of the Galaxy. CO large scale surveys (e.g. Burton and Gordon, 1978) reveal that the molecular material is distributed in many individual clouds but do not show the intercloud gas phase well detected in HI surveys. The large scale structure of the gas is clearly seen in the closely sampled data of Cohen et al. (1980); the CO molecule appears to be as good a tracer of galactic arms as the atomic hydrogen. The presence or not of a spiral structure is related to two fundamental questions: how were the molecular clouds formed? and, what is the lifetime of these clouds?

In this work we have used the J = 1-0 transition of ^{13}CO to investigate the galactic structure in the first quadrant. Despite the weaker emission, ^{13}CO was preferred to CO for several reasons. (i) An arm-interarm contrast and individual clouds are more easily detected in ^{13}CO than in CO because of narrower linewidths and less frequent blending of velocity features. (ii) In general, ^{13}CO intensities are related to column densities whereas CO intensities are not because this line is optically thick. (iii) The Earth's atmosphere is more transparent at the ^{13}CO line frequency than around 115 GHz and reliable calibration procedures can be expected. A relatively small number of large scale ^{13}CO maps have been published until now. Stark (1979), Solomon et al. (1979), Sanders (1981), and Liszt et al. (1981) have investigated the inner regions of the Galaxy

W. B. Burton and F. P. Israel (eds.), Surveys of the Southern Galaxy, 173–180.

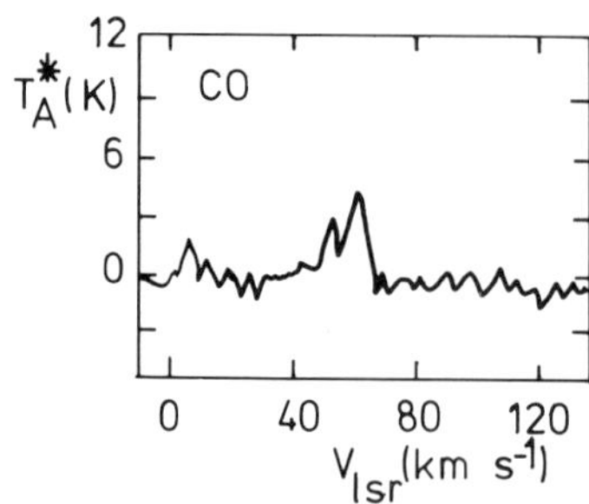

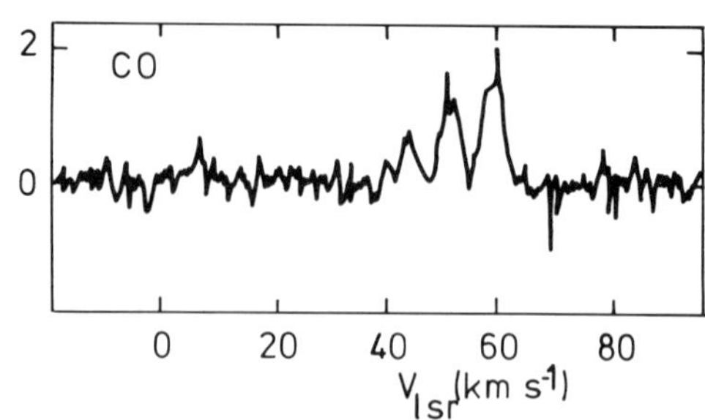

Figure 1. CO (Cohen, 1981) and ^{13}CO (this work) spectra obtained towards $\ell = 49\overset{\circ}{.}25$, b = 0°. Spectral resolution: CO = 500 kHz, ^{13}CO = 100 kHz. T_A^* is the antenna temperature corrected for atmospheric absorption and telescope efficiency.

at $\ell = 51^\circ$ and between $\ell \simeq 10^\circ$ and 44°.

In our project we have decided, as a first step, to survey the $\ell = 40^\circ - 60^\circ$ region at $|b| \leq 0\overset{\circ}{.}5$. This part of the Galaxy has not yet been extensively observed in ^{13}CO and contains the Sagittarius arm.

2. OBSERVATIONS

The observations were carried out during the 1981-1982 winter and the 1982 spring with the 2.5-m telescope of the Bordeaux Observatory. The properties of this telescope are described by Baudry _et al._ (1981). The half-power beamwidth of the antenna is 4.4 arc min at the J = 1-0 line frequency of ^{13}CO. The receiver had a double side band noise temperature of about 350 K at the beginning and about 250 K at the end of our observations. In order to have good spectral resolution we have used the 100 kHz filter bank.

Figure 1 displays three blended CO features (Cohen, 1981) which are completely resolved in ^{13}CO. In this program the local oscillator was frequency switched and the velocity coverage was split into two overlapping intervals extending from -20 to 50 km s^{-1} and from 28 to 97 km s^{-1}. Baseline curvature was removed by means of a polynomial fit to those regions of a spectrum free of ^{13}CO emission. At b = 0°, the $\ell = 40^\circ - 60^\circ$ region was regularly sampled at 15' spacings. Twice as many spectra have been obtained out of the galactic plane, at $|b| \leq 0\overset{\circ}{.}5$. The survey sensitivity was uniform and fixed to 0.2 K in one 100 kHz channel.

Four sources were regularly used as calibration sources: Barnard 335, L134N, G45.5+0.0, and G59.75+0.0. The integrated intensities, $\int T_A^*$ (^{13}CO) dv, towards G45.5+0.0 and G59.75+0.0 (Figure 2) are 15.4 and 11.5 K km s^{-1} respectively. For both sources, the standard deviation of the mean derived from about 53 distinct observations showing various sky qualities is 1.5 km s^{-1}. Nineteen individual features of the survey were measured on two separate occasions. The observed differences (for individual features up to $\sim$ 6 K km s^{-1}) are typically 0.7 K km s^{-1}, a number which is only slightly higher than expected from noise and calibration errors.

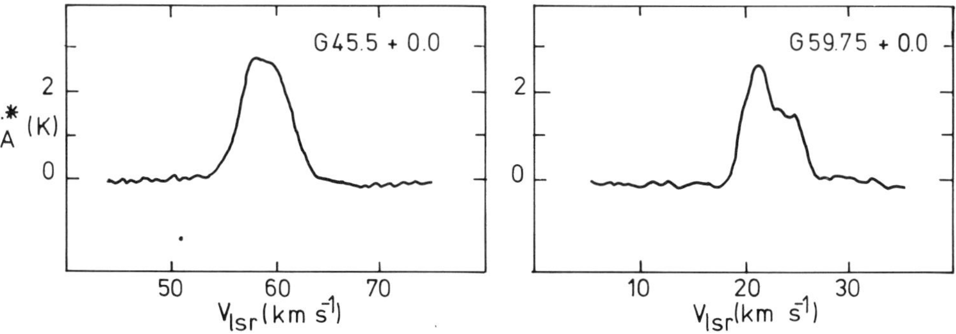

Figure 2. ^{13}CO spectra obtained towards $\ell = 45\overset{\circ}{.}5$, $b = 0^{\circ}$ and $\ell = 59\overset{\circ}{.}75$, $b = 0^{\circ}$. Spectral resolution = 100 kHz.

3. DISTRIBUTION OF ^{13}CO CLOUDS

3.1. Cloud distribution in the ℓ - v diagram at $b = 0^{\circ}$

Our ^{13}CO survey at $b = 0^{\circ}$ is presented in a longitude-velocity (ℓ - v) diagram (Figure 3). The data have been smoothed to a spectral resolution of 500 kHz for better comparison with the Columbia CO survey (Cohen et al., 1980). The large scale distributions of ^{13}CO and CO are very similar when one compares our Figure 3 with Figure 3 of Cohen et al. (1979). However, along the ridge of emission corresponding to maximum velocity allowed by circular motion the gas is more clumped in ^{13}CO than in CO. Enhancement of molecular emission is observed around $\ell = 45^{\circ}$ where the Sagittarius arm is seen tangentially, but in the region $\ell \simeq 46^{\circ} - 52^{\circ}$, $V_{LSR} \simeq 20 - 40$ km s^{-1}, we have not detected any significant molecular emission. Molecular clouds are very often observed in the vicinity of HII regions. All the H110α sources of Downes et al. (1980) lying near the galactic plane between $\ell = 40^{\circ}$ and 60° can be identified, as expected, with ^{13}CO complexes.

3.2. Variation with ℓ of integrated intensities

The scattered distribution of integrated ^{13}CO intensities (Figure 4) is not due to noise but to real clumpiness of the gas. The major peaks in Figure 4 are closely associated with or lie at the edge of well known sources such as W49, W51, or G45.5. The latter source, which has been mapped in CO by Israel (1982), deserves further comments.

Our ^{13}CO strip map made towards $\ell = 45\overset{\circ}{.}5$ at 5' spacings in ℓ and b shows that the ^{13}CO cloud has a total size of about 53 pc if it is placed at a distance of 7 kpc. Then, from a phenomenological relationship between integrated ^{13}CO intensities and visual extinction (Liszt, 1982) and from the observed gas to extinction ratio (Bohlin et al., 1978) we derive a total mass of about $1.5\ 10^5\ M_{\odot}$ and a mean H_2 density of ~ 40 cm^{-3}. Towards $\ell = 45\overset{\circ}{.}5$, $b = 0^{\circ}$, we obtain $\tau(^{13}CO) \simeq 0.48$. An LTE analysis gives $N(^{13}CO) \simeq 1.5\ 10^{16}$ cm^{-2} and $N(H_2) \simeq 7.5\ 10^{21}$ cm^{-2} if one assumes $[H_2]/[^{13}CO] = 5\ 10^5$. This value of the projected H_2 density is similar

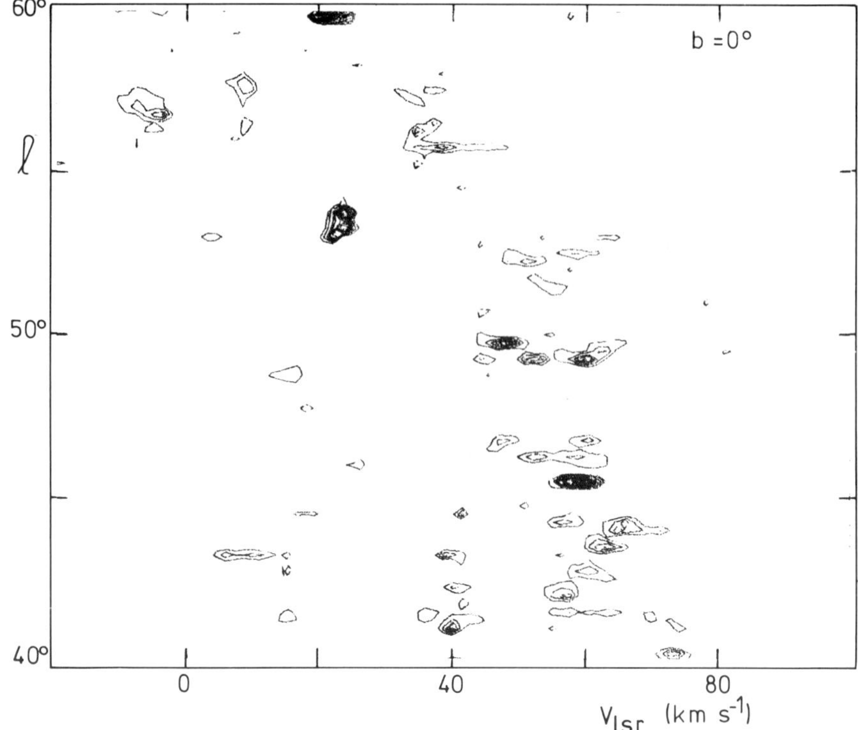

Figure 3. The longitude -velocity diagram of ^{13}CO intensities along b = 0° at 15' spacings. The ℓ-coverage is complete except for 5 directions where, because of negligible CO emission, one half of the total velocity coverage was not observed. Contour intervals are 0.3 K in units of T_A^*. The spectra have been smoothed to 500 kHz. The ℓ and v scale intervals are approximately two times larger than those used in Figure 3 of Cohen *et al.* (1979) for their CO, ℓ - v diagram.

to that derived in the same direction, $N(H_2) \simeq 10^{22}$ cm^{-2}, from the phenomenological relationship mentioned above.

Integrated intensities measured in the region ℓ = 40° - 60° are smaller than those measured by Liszt et al. (1981) below 40° (compare our Figure 4 to their Figure 2). Although the details of such a comparison have still to be refined, it is clear that ^{13}CO emission decreases rapidly at ℓ ≥ 40° - 45°. This general trend is well observed in CO (Cohen et al., 1980) or H_2CO (Few, 1978), in the radio continuum (Altenhoff et al., 1978), and in the infrared (Boissé et al., 1981) brightness profiles, and corresponds to a decline of neutral and ionized material outside the 4-8 kpc molecular ring.

3.3. The latitude distribution

^{13}CO spectra have been obtained out of the galactic plane between ℓ = 40° and 60° at b = ±0°.5 and at 5' spacings in b at |b| ≤ 0°.25. A ±0°.5 coverage in latitude may often be adequate since the CO scale height,

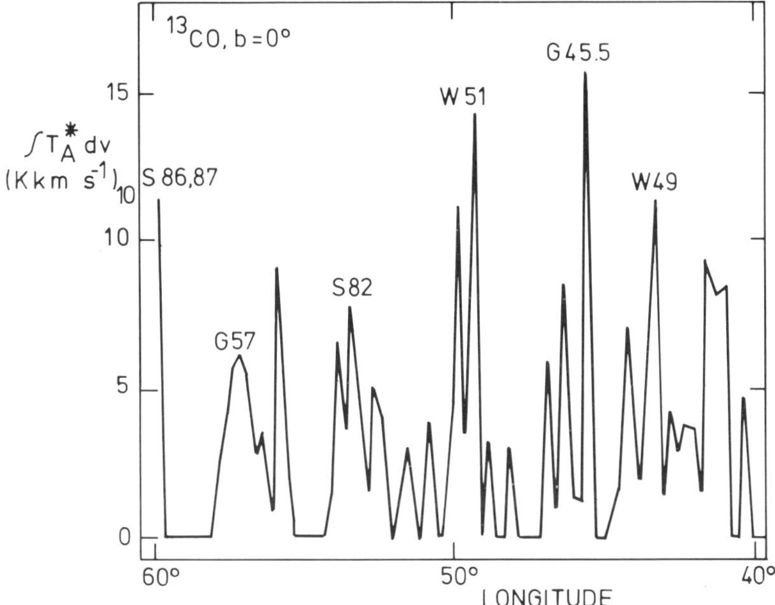

Figure 4. Longitudinal distribution of integrated intensities. Longitude intervals = 0°.25. Spectral resolution = 100 kHz.

$z_{\frac{1}{2}}$, between ℓ = 40° and 60° is ∿ 70 pc (e.g. Sanders, 1981): at a distance of 7 kpc, this scale height corresponds to $b_{\frac{1}{2}} \simeq 0°.5$.

The latitude distribution of the ^{13}CO gas layer is given in Figure 5. No clear systematic displacement of peak integrated intensities above or below the galactic plane is observed in the longitude range ℓ = 40° - 60°. However the total integrated intensity at $-0°.5 \leq b < 0°$ is about 3/2 larger than that measured at $0° < b \leq 0°.5$. Towards ℓ = 43°.5, 45°.5 and 52°.5 the latitude profiles are nearly symmetrical.

Individual clouds whose properties will be discussed elsewhere are shown in a latitude-velocity diagram (Figure 6). The diagram is restricted to $|b| \leq 0°.25$ where the gas was sampled at 5' spacings in b. This type of sampling is well adapted to the distribution of cloud sizes and numbers as determined by Liszt et al. (1981).

REFERENCES

Altenhoff, W.J., Downes, D., Pauls, T.A., and Schraml, J.: 1978, Astron. Astrophys. Suppl. 35, pp. 23-54.

Baudry, A., Cernicharo, J., Pérault, M., de La Noë, J., and Despois, D.: 1981, Astron. Astrophys. 104, pp. 101-115.

Bohlin, R.C., Savage, B.D., and Drake, J.F.: 1978, Astrophys. J. 224, pp. 132-142.

Boissé, P., Gispert, R., Coron, N., Wijnbergen, J.J., Serra, G., Ryter,

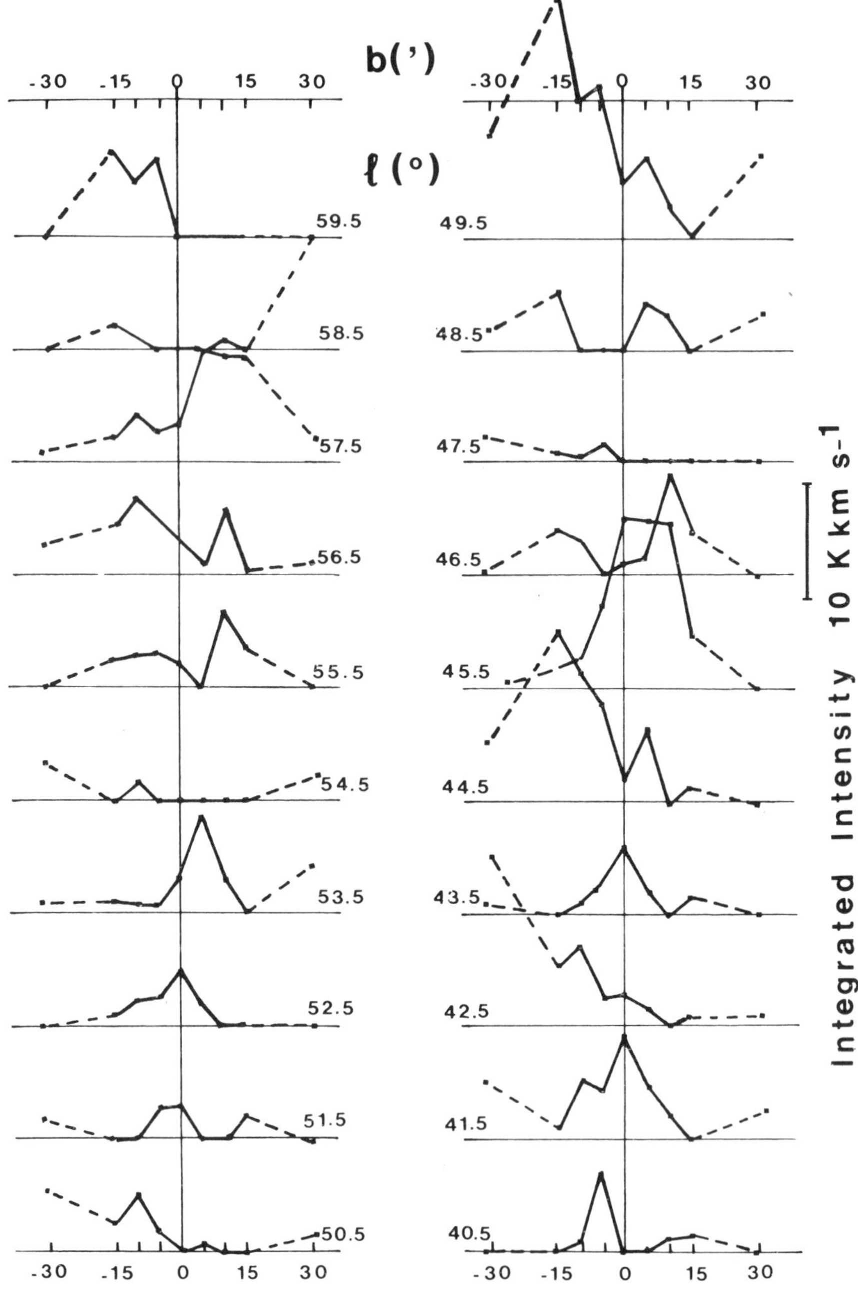

Figure 5. The latitude profiles of integrated intensities at 5' spacings in b (continuous lines at $|b| \leq 0^\circ.25$) and at $b = \pm 0^\circ.5$ (dotted lines).

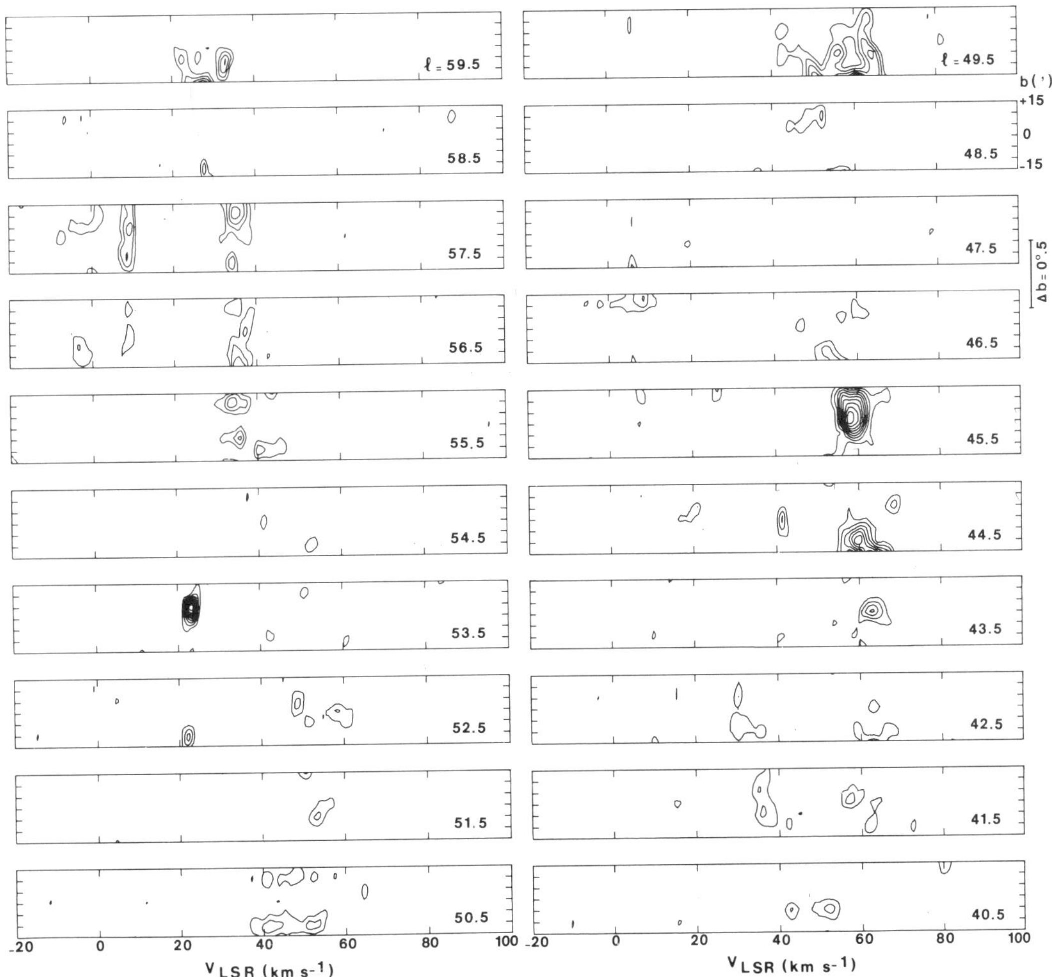

Figure 6. The latitude-velocity maps of ^{13}CO emission obtained at $|b| \leq 0^{\circ}.25$, between $\ell = 40^{\circ}.5$ and $59^{\circ}.5$. Latitude spacing = 5'. Spectral resolution = 500 kHz. Contour intervals are 0.3 K in units of T_A^*.

C., and Puget, J.L.: 1981, Astron. Astrophys. 94, pp. 265-271.

Burton, W.B., and Gordon, M.A.: 1978, Astron. Astrophys. 63, pp. 7-27.

Cohen, R.S.: 1981, private communication.

Cohen, R.S., Cong, H., Dame, T.M., and Thaddeus, P.: 1980, Astrophys. J. Letters 239, pp. 53-56.

Cohen, R.S., Tomasevich, G.R., and Thadeus, P.: 1979, IAU Symposium No. 84, edited by W.B. Burton, D. Reidel Publishing Co., pp. 53-56.

Downes, D., Wilson, T.L., Bieging, J., and Wink, J.: 1980, Astron. Astrophys. Suppl. 40, pp. 379-394.

Few, R.W.: 1979, Monthly Notices Roy. Astron. Soc. 187, pp. 161-178.

Israel, F.P.: 1982, Astrophys. J. 255, pp. 475-488.

Liszt, H.S.: 1982, private communication.

Liszt, H.S., Xiang, Delin, and Burton, W.B.: 1981, Astrophys. J. 249, pp. 532-549.

Sanders, D.B.: 1981, Ph.D. Thesis, SUNY Stony Brook.

Solomon, P.M., Scoville, N.Z., and Sanders, D.B.: 1979, Astrophys. J. Letters 232, pp. 89-93.

Stark, A.A.: 1979, Ph.D. Thesis, Princeton University.

^{13}CO OBSERVATIONS TOWARDS THE 2nd GALACTIC QUADRANT MADE WITH THE BORDEAUX TELESCOPE

F. Casoli, F. Combes, and M. Gérin
Observatoire de Meudon, 92190 Meudon, France

A ^{13}CO survey has been carried out in the 2nd galactic quadrant. The emission is expected to be fainter here than in the first quadrant, because it originates outside the molecular ring; but it is also expected to be easier to interpret because there is neither distance ambiguity nor line overlap. Indeed the region $\ell = 108^o - 128^o$, already surveyed in ^{12}CO by Cohen et al. (1980) consists of two main spiral features, the Perseus and Orion arms, and the Lindblad local ring, all of which are well separated either in radial velocity or in latitude. The Perseus arm is near v = -50 km s^{-1}; the Orion arm is near 0 km s^{-1}. Due to its smaller optical thickness, the ^{13}CO molecule yields a better estimate of column densities than the existing ^{12}CO observations, and allows discrimination between several small clouds that often appear blended in the same CO feature.

The giant molecular complex found in the Perseus arm between ℓ = 108° and 113° and covering more than 12 square degrees, has been entirely observed, and also two regions around ℓ = 111° and ℓ = 126° in the Orion arm, each covering about 6 square degrees. These observations were made in 1981-82 with the 2.5m millimeter antenna in Bordeaux. The cooled receiver had a single-sideband noise temperature of 600 K, at 110 GHz. The spatial resolution is 4.5 arc min. A velocity resolution of 0.26 km s^{-1} was obtained with 256 filter banks, each 100 kHz wide. The sampling was every 8 arc min (a little less than two beams) and the noise level was 0.1 K in the 100 kHz filters (about 40 minutes actual observing time for each spectrum). More than a thousand spectra were obtained.

Comparison of the ^{12}CO (Cohen et al. 1980) with the ^{13}CO observations, convolved to the same-spatial resolution, reveal that the ^{13}CO line is indeed optically thin: the optical thickness is always between τ = 0.1 and τ = 0.7.

If a distance of 3 kpc is adopted for the region $\ell \sim 111^o$ of the Perseus arm, the observed giant molecular complex has a characteristic dimension of 100 pc. We estimate a total mass of 10^6 $M_\odot$ for this complex, assuming a ^{13}CO/Hz ratio of 2 x 10^{-6} (Dickman 1978) and calculating ^{13}CO

W. B. Burton and F. P. Israel (eds.), Surveys of the Southern Galaxy, 181–187.

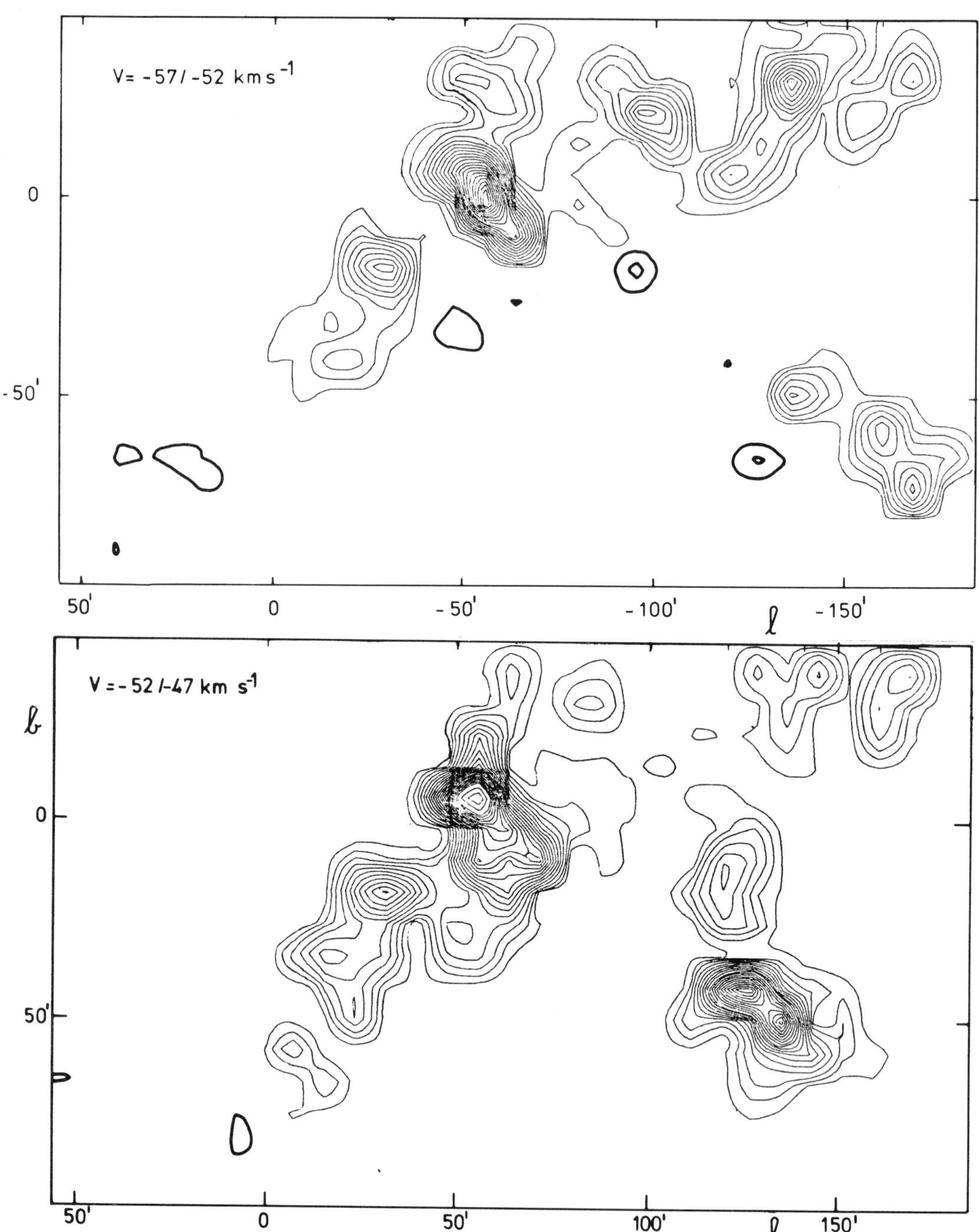

Figure 1. Perseus arm: 2 channel maps of the region L111 P centered at $\ell = 111^o$, $b = -0^o.1$.

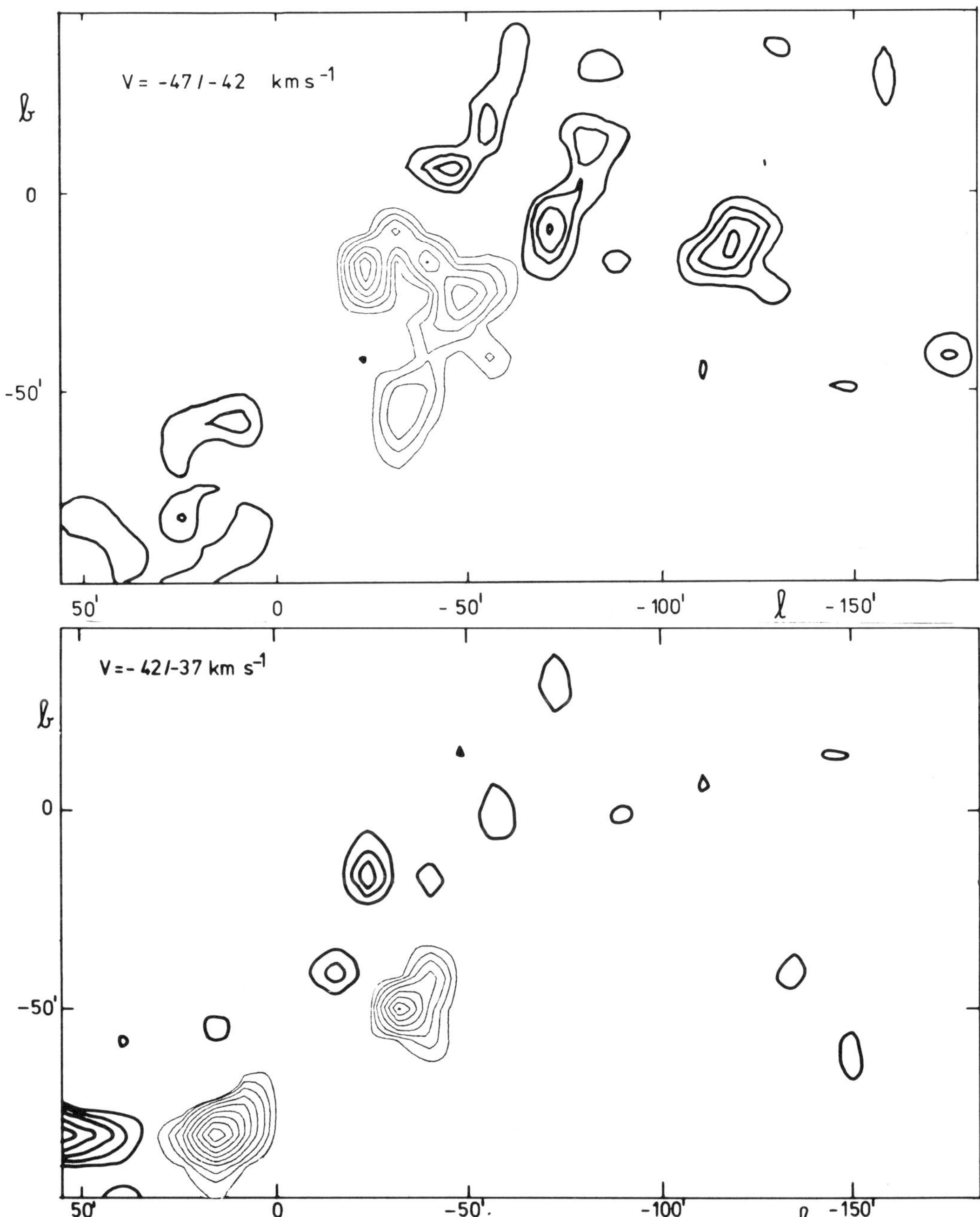

Figure 1. (cont.) Perseus arm: 2 channel maps of the regions L111 P centered at ℓ= 111°, b= -0°.1.

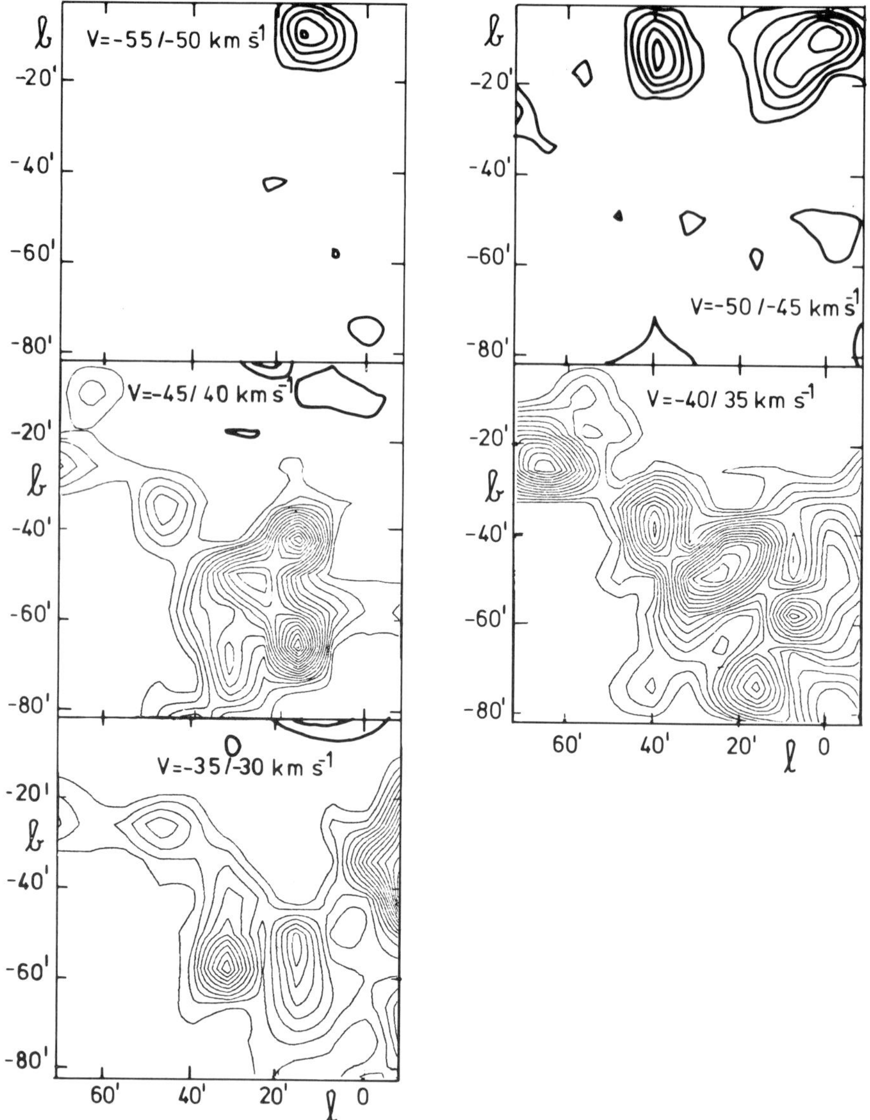

Figure 1. (cont.) Perseus arm: 5 channel maps of the region L 111 PB-2 centered at $\ell = 111^{\circ}$, $b = -2^{\circ}.1$.

column densities under the usual LTE assumptions. If the line of sight dimension is of the same order as the mean apparent diameter, the mean molecular density is $n \sim 20$ H_2 cm^{-3}.

A general characteristic of the observations is the clumpiness of the medium, at the spatial resolution used here (4 pc). Most of the spectra contain several velocity features. Some can be followed over a few beams. There is also a large velocity gradient along the region (about 30 km s^{-1} over 100 pc), as is easily seen on the channel maps in figure 1.

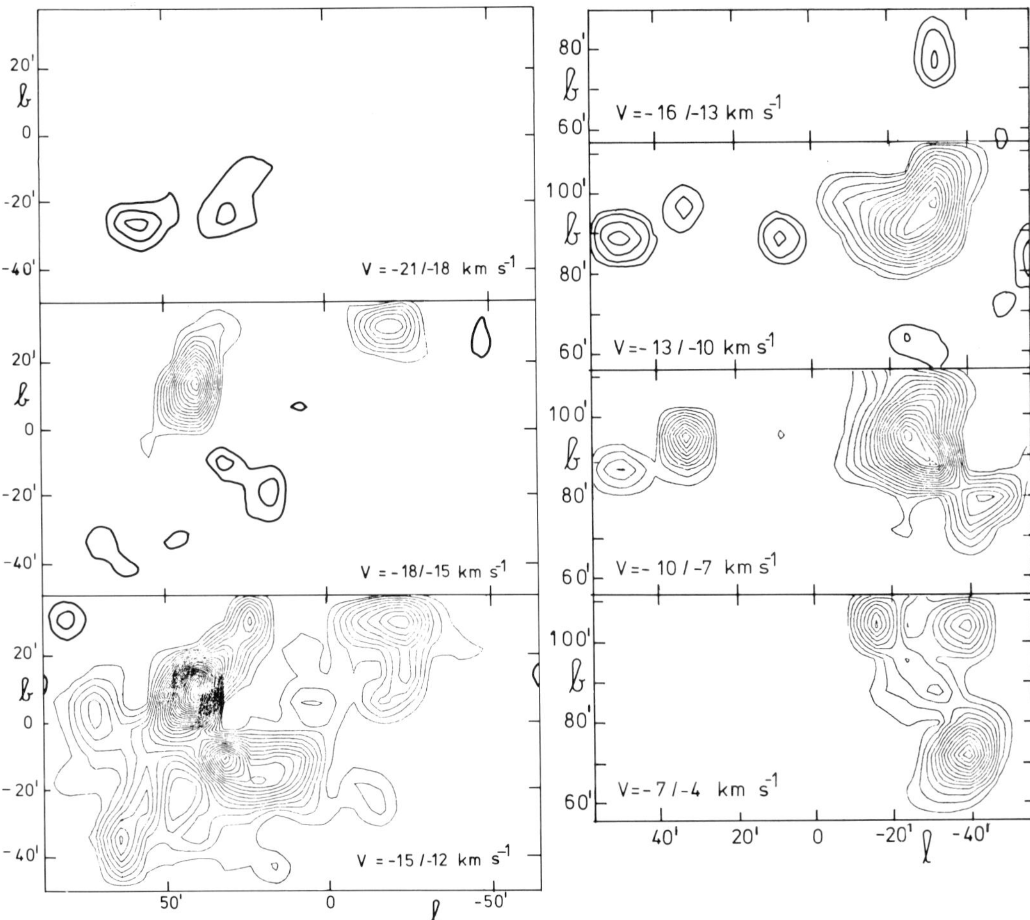

Figure 2. Orion arm: left) 3 channel maps of (ℓ = 126° b = -1°). right) 5 channel maps of the region ℓ = 111°, b = 0°.

We have tried to evaluate the cloud mass spectrum, with two "definitions" of the individual clouds. First, several features are considered to belong to the same cloud if they are geometrically associated (i.e. found at consecutive positions), and if their central velocities are equal within 0.5 km s^{-1}, (which is about twice the sound velocity in the GMC). More than two hundred clouds were found in this manner; they approximately lie on a power-law mass spectrum of index -1.35 ± 0.1. Second, we considered as one individual molecular cloud several spatially associated features (also located at consecutive positions), with velocities within 4 km s^{-1}. The two cloud distributions were combined, and were found to obey a power-law mass spectrum of index -1.25 ± 0.1.

The error bar on the index was obtained by varying the manner in which clouds were classified in mass bins to compute a power-law fit. This flat distribution of cloud masses indicates that the spectrum is

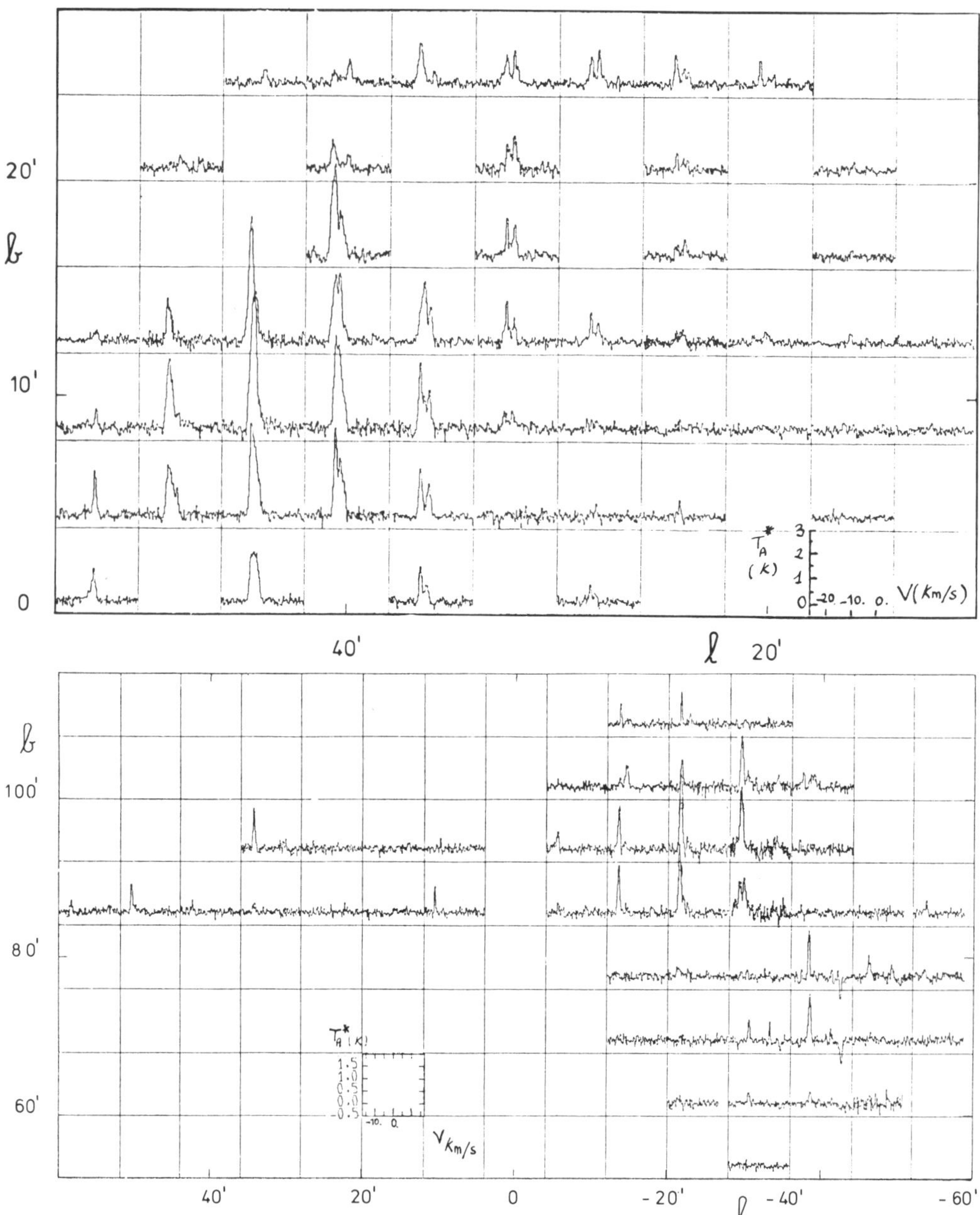

Figure 3. ^{13}CO spectra of the same regions as in Figure 2.

dominated by the large clouds, and is compatible with results for the first quadrant previously published by Stark (1979) and Solomon et al. (1979): power-laws of -1.4 ± 0.2. The molecular cloud spectrum is definitely flatter than the diffuse atomic clouds spectrum of index -1.8 ± 0.2 (e.g. Hobbs 1974). This is in agreement with a collisional model of the formation of these clouds (cf. Casoli and Combes 1982). The mean values of this cloud distribution are the following: mean radius 6 pc, mean mass 5 x 10^3 $M_\odot$, mean molecular density 100 H_2 cm^{-3}.

Most of the ^{13}CO maxima correspond to well known HII regions, for example S147-8-9, S152-3, S156-7-9, and N7538 in the Perseus arm complex, the ^{13}CO antenna temperatures reach here $\sim$ 5 K. The radial ^{13}CO velocities are in good agreement with the optical determinations.

In the Orion arm, at $\ell = 126^o$, the 13 CO observations suggest collision between two molecular clouds. The velocities of the two components shift towards each other until a certain position where they blend, and where the ^{13}CO emission shows a remarkable maximum (Figures 2 and 3). This particular position correspond to the HII region S187. To investigate this collision in more detail, better sampled observations (each beam) are now being carried out.

REFERENCES

Casoli, F., and Combes, F.: 1982, Astron. Astrophys. 110, 287.

Cohen, R.S., Cong, H., Dame, T.M., and Thaddeus, P.: 1980, Astrophys. J. 239, L53.

Dickman, R.L.: 1978, Astrophys. J. Suppl. 37, 407

Hobbs, L.: 1974, Astrophys. J. 191, 395.

Solomon, P.M., Sanders, D.B., and Scoville, N.Z.: 1979, IAU Symp. 84, 35.

Stark, A.A.: 1979, Ph.D. Thesis, Princeton University.

A COMPARISON OF ^{12}CO AND ^{13}CO GALACTIC SURVEYS

Antony A. Stark and Arno A. Penzias
Bell Laboratories

Brian Beckman
Princeton University Observatory

ABSTRACT

Five square degrees of the first galactic quadrant have been mapped on a square 3'x3' grid in both ^{12}CO and ^{13}CO. A comparison of the two sets of maps reveals the nature and extent of the influence on survey parameters caused by line saturation in the common species. Gross features of the ^{12}CO and ^{13}CO maps were found to be strikingly similar when the integrated line intensities of the two species were respectively plotted with linear and quasi-logarithmic contour intervals. In general, the two sets of maps yielded similar statistics, although the ∿25% of clouds that are blended together were easier to separate in ^{13}CO. The generally used rule-of-thumb that ^{12}CO lines are 5X brighter than ^{13}CO lines fit most of the data reasonably well. However, a quadratic fit: $T_A^*(^{13}CO) = 0.037[T_A^*(^{12}CO)]^2 + 0.056T_A^*(^{12}CO)$ was a significantly better representation of the data. The curvature in this relation, induced by saturation of the ^{12}CO line, explains the similarity of the quasi-logarithmic ^{13}CO maps and the linear ^{12}CO maps.

SURVEY PARAMETERS

We have surveyed five square degrees of the first Galactic quadrant, as a pilot project for a more extensive survey (cf. Stark 1982). Survey parameters are as follows:

1) Both ^{12}CO and ^{13}CO are observed.
2) Sensitivity - 0.1 K rms noise level in 0.68 km s^{-1} wide channels for ^{13}CO, 0.3 K rms noise level in 0.65 km s^{-1} wide channels for ^{12}CO.
3) Sampled on a square grid in l and b, 3' between grid points.
4) Calibrated within 5% of the absolute temperature scale in T_A^*.
5) Accurate beam pattern - the Gaussian central lobe contains $\geq$87% of the antenna response; less than 5% of the antenna response is from Milky Way emission not in the central lobe.

W. B. Burton and F. P. Israel (eds.), Surveys of the Southern Galaxy, 189–193.

6) Coverage - five squares, 1° on a side, centered at (l,b) = (34,0); (35,0); (36,0); (51,0); and (53,0). Velocity coverage was v_{LSR} = -30 to 130 km s^{-1}.
The survey yielded $T_A^*(^{12}CO)$ and $T_A^*(^{13}CO)$ at each of $5x10^5$ points in (l,b,v) space.

RELATION BETWEEN ^{12}CO AND ^{13}CO ANTENNA TEMPERATURES

A systematic comparison (Fig. 1) between ^{12}CO and ^{13}CO line intensities was made at a set of representative locations (plotted as circles) as well as at cloud centers (triangles). Although there is considerable scatter in the data, the dashed line plot of the general rule-of-thumb (e.g. Solomon, Sanders and Scoville 1979) relation $T_A^*(^{13}CO)=0.2T_A^*(^{12}CO)$ is seen to provide a useful approximation for all but the most intense features. The solid curve is an emperical relationship: $T_A^*(^{13}CO) = 0.037[T_A^*(^{12}CO)]^2 + 0.056T_A^*(^{12}CO)$ derived from a least-squares-fit to several thousand randomly-selected points. The error in the coefficients of this relationship is ∿5%, and is dominated by systematic calibration uncertainties.

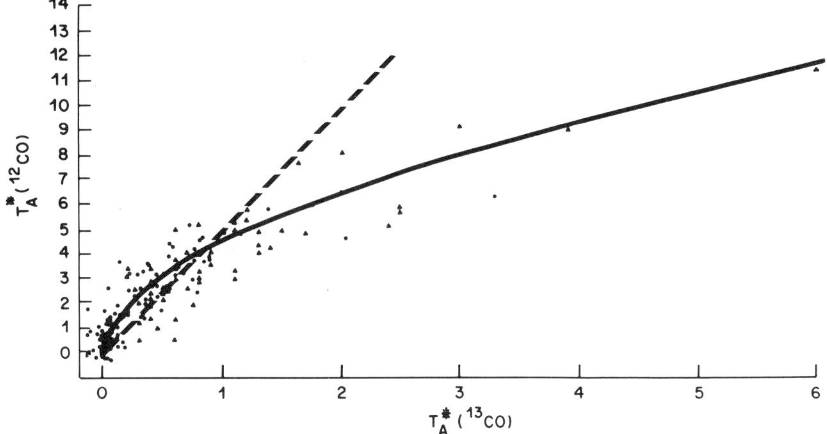

Fig. 1. $T_A^*(^{12}CO)$ vs. $T_A(^{13}CO)$ for selected points in (l,b,v) space. The rms noise is 0.1K on the horizontal axis and 0.3 K on the vertical axis. The solid curve is fit to many more points than those shown here.

CATALOGING CLOUDS

Because ^{13}CO and ^{12}CO brightnesses are well correlated, it is not surprising to find that most molecular clouds appear in both surveys. Even small clouds that have central ^{12}CO brightnesses under 2 K appear in the ^{13}CO survey at the ~0.3 K level. Essentially none of the observed ^{12}CO emission is optically thin: the ratio $T_A^*(^{12}CO)/T_A^*(^{13}CO)$ is always less than the $^{12}C/^{13}C$ isotope ratio. Figures 2 and 3 are position-position maps of T_A^* integrated over a 5 km s^{-1} wide velocity slice. When allowance is made for differences in contouring, it can be seen that almost all features appear similarly in both data sets. For most molecular clouds, the ^{13}CO and ^{12}CO surveys are equally useful. About a quarter of the clouds, however, are blended or self-absorbed, and for these we found that the ^{13}CO survey contained useful information not present in the ^{12}CO survey.

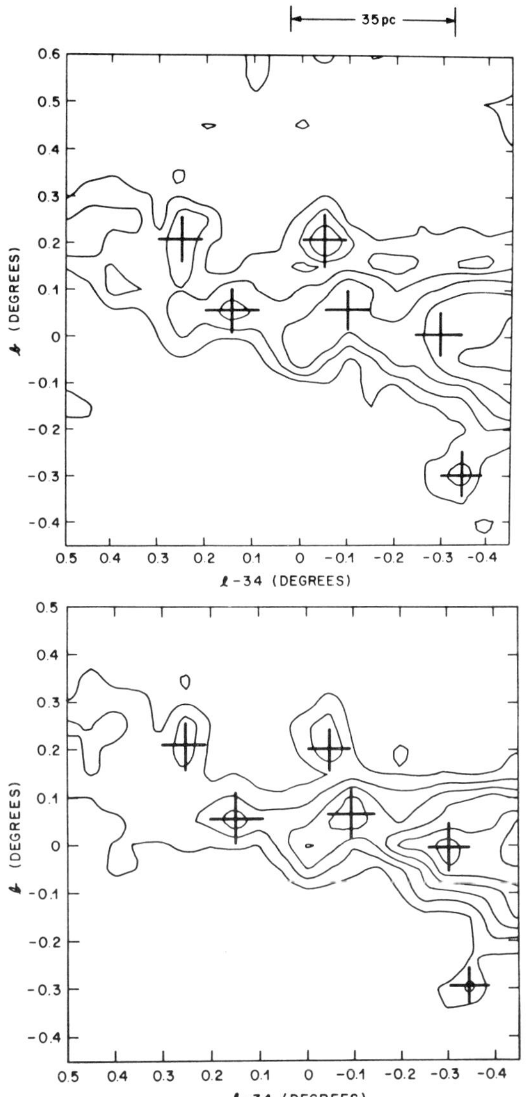

Fig. 2. $\int_{100\text{km s}^{-1}}^{105\text{km s}^{-1}} T_A^* dv$ over a square degree centered at (l,b)=(34,0). The ^{13}CO data (top) has contours at 0.4, 1, 2, 4, 10 and 20 K km s^{-1}; the ^{12}CO data (below) has contours at 2, 4, 6, 8, 10, 12 K km s^{-1}. Crosses denote cloud centers determined from the ^{12}CO data alone.

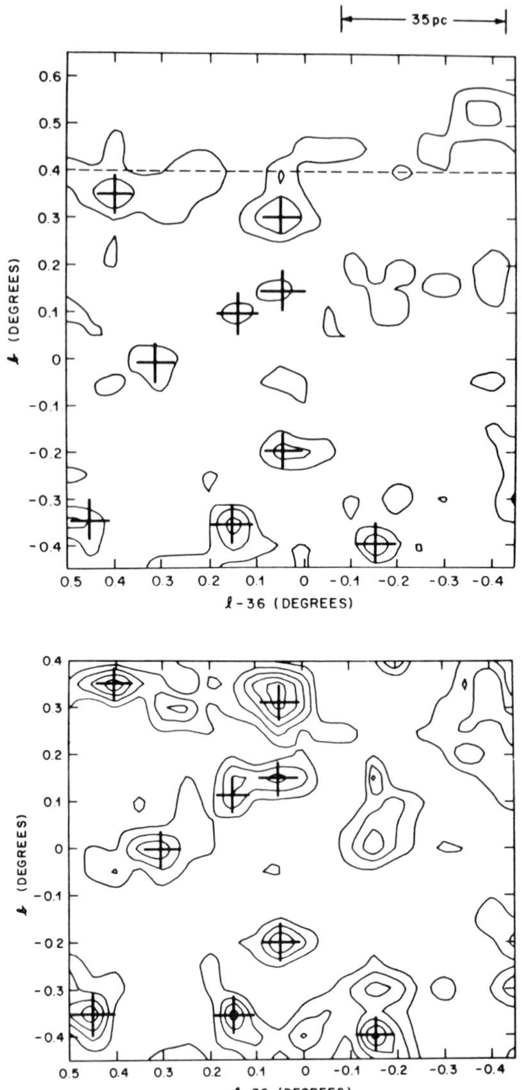

Fig. 3. $\int_{87.5\text{km s}^{-1}}^{92.5\text{km s}^{-1}} T_A^* dv$ over a square degree centered at (l,b)=(36,0). ^{13}CO data is at top; ^{12}CO data, below. Contours are as in Fig. 2.

We compared cloud number and size distribution data derived from the two surveys for various cloud-definining algorithms; for example the cloud boundary was taken to be the lowest intensity contour in one case and 0.5 of the brightest enclosed peak in another. In general, both surveys gave similar results. Only some 10% of the ^{13}CO and 25% of the ^{12}CO catalog entries exhibited appreciative changes when the algorithms were changed.

DISCUSSION

Molecular clouds are generally better delineated in ^{13}CO. In the spectrum in Figure 4 for example, the blended clouds between 50 and 100 km s-1 are an undifferentiated lump in ^{12}CO, but are fairly distinct in ^{13}CO.

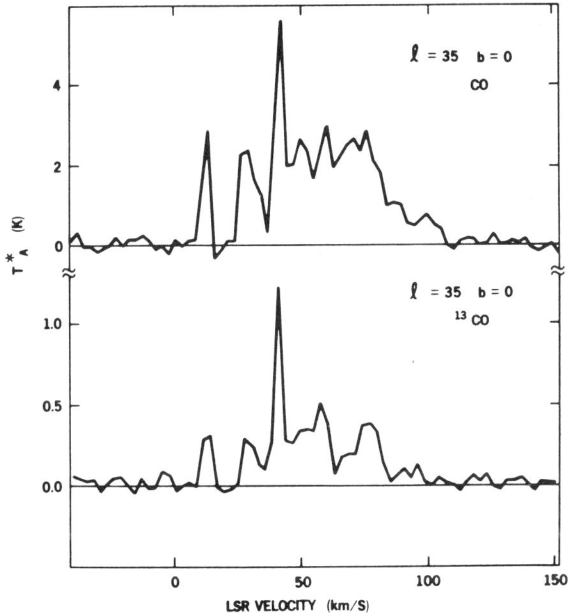

Fig. 4. ^{12}CO and ^{13}CO spectra towards (l,b)=(35,0).

Except for the observing time required, the ^{13}CO spectrum is preferable to the ^{12}CO spectrum. Our present receiver in average weather takes 36 seconds for each ^{13}CO spectrum and 16 seconds for each ^{12}CO spectrum in the survey. Atmospheric opacity is higher for ^{12}CO than for ^{13}CO, so that improved receivers will make ^{13}CO relatively easier to observe. Given a sufficiently sensitive receiver, galactic surveys should be made in ^{13}CO. Almost all molecular clouds, even those that are faint in ^{12}CO, are readily detected in ^{13}CO. Essentially no information is lost and some freedom from confusion is gained in ^{13}CO surveys compared to ^{12}CO surveys. At the same time the large body of existing ^{12}CO data as well as the continued use of this line for extragalactic studies compels us to improve our understanding of its properties.

REFERENCES

Solomon, P. M., Sanders, D. B., and Scoville, N. Z. 1979 in The Large-Scale Characteristics of the Galaxy, ed. W. B. Burton (Reidel: Dordrecht) pp. 35-52.

Stark, A. A. 1982 conference proceedings of Vancouver meeting on Galactic Structure, ed. W. L. Shuter (Reidel: Dordrecht).

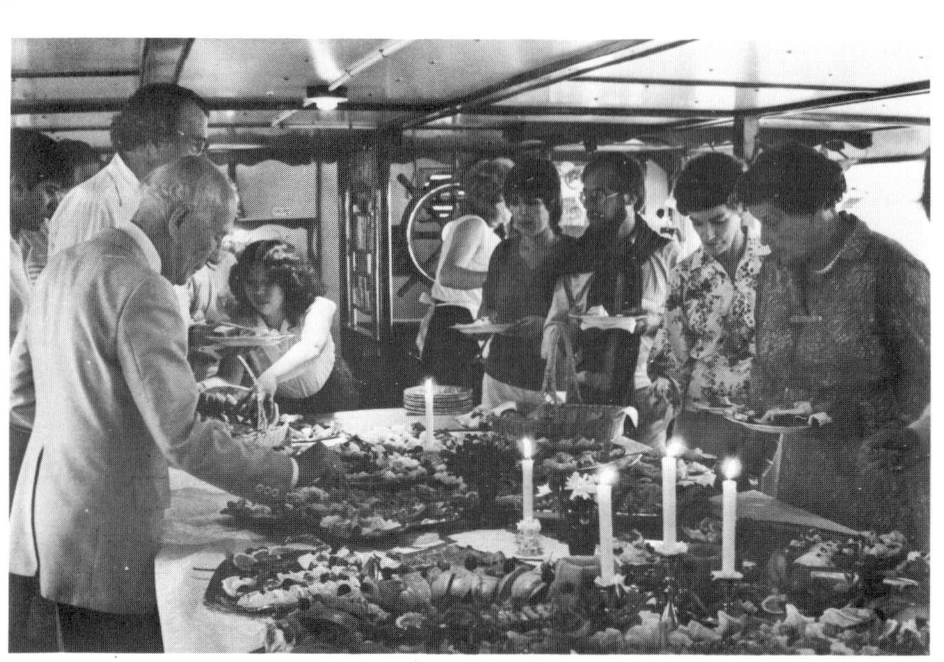

ARTIFICIAL BOUNDARIES *vs*. HI SHELLS AND SUPERSHELLS

Carl Heiles
University of California, Berkeley

I. KNOCKING DOWN ARTIFICIAL BOUNDARIES

At a meeting that concerns itself with the amalgamation of northern and southern hemisphere data, I think it is appropriate to emphasize the effects that artificial boundaries such as this can have on the interpretation of astronomical data. HI shells and supershells are large-scale features, so that they often cross boundaries of any kind; their study is particularly susceptible to the existence of artificial boundaries.

In HI survey data there are two types (really three - see section II) of artificial boundaries. One is the boundary between northern and southern hemisphere data near declination -25 degrees. The other is the boundary between "galactic plane" and "intermediate latitude" surveys at $|b| = 10^o$.

At intermediate latitudes, HI surveys exist in both hemispheres: the Hat Creek survey (Heiles and Habing, 1974; Heiles, 1975) in the north, and the Australian (Cleary, Heiles, and Haslam, 1979; Heiles and Cleary, 1979) and the Argentine (Colomb, Poppel, and Heiles, 1980, hereafter CPH) surveys in the south. In the south, the Argentine survey is much preferred for kinematical studies because of its better velocity resolution. The presentation of data by CPH is photographic, with HI column density in narrow velocity intervals pictures on a galactic coordinate grid; it includes both the Hat Creek and Argentine data.

The amalgamation of CPH is necessary to clearly show the most prominent HI shell in the sky. This shell is associated with radio loop I (the North Polar spur) and it is centered near $(1,b)=(330^o,+14^o)$. It is so large that only a fraction of it is visible in either the northern or southern hemisphere data alone.

On photographs at different velocities, its approaching hemisphere changes size in the way expected for a shell expanding at about 25 km s^{-1}. The HI filaments increase monotonically in diameter from velocity of about -30 km s^{-1} to about 14 km s^{-1}, to a maximum of about 120 degrees.

W. B. Burton and F. P. Israel (eds.), Surveys of the Southern Galaxy, 195–203.

At positive velocities there are other filaments, which are either smaller in diameter or are centered at different positions; these may be related to the expanding shell so easily seen at negative velocities, or may instead be parts of different structures. Nevertheless, the behavior of the filaments at negative velocities implies that the shell is expanding at about 25 km s^{-1}. This is in contrast to Weaver's (1979) statement that the shell is expanding at only 2 km s^{-1}, which was based on an examination of northern hemisphere data alone.

Shells at intermediate galactic latitudes are so large that they are easily resolvable at distances up to 10 kpc in the galactic plane. Many such shells are visible in the galactic plane data of Weaver and Williams (1973, 1974); they are discussed by Heiles (1979), who again presents photographs of HI column density in small velocity intervals on a galactic coordinate grid. But inspection of these photographs also reveals portions of many structures that extend outside the survey limits $|b| = 10^o$.

This artificial boundary at $|b| = 10^o$ has been eliminated by amalgamating the northern hemisphere intermediate-latitude and galactic plane data presented in a paper currently in preparation (Heiles, 1982b). The features visible only in part on the individual surveys are now visible in full. An important, and the most unusual, example is the large shell in the galactic anticenter, discussed below in section II. This forthcoming paper will also present again some of the photographs of CPH, which were reproduced poorly and on too small a scale to recognize some of the weaker HI structural features.

Expanding shells tend to show only one hemisphere. That is, only the approaching or the receding hemisphere is seen, and in those few cases in which both are seen one of the two hemispheres is very much weaker and more difficult to recognize. The implication is that most shells are not fully complete, a point which also emerges from the simple fact that the circular filaments on the sky are usually only partial circles and not equally discernable over their full circumferences.

There are many curved HI filaments visible on photographs of HI data that do <u>not</u> change size with velocity. The proper interpretation of such filaments is not certain. I assume that such filaments are portions of shells that are no longer expanding. In a homogeneous interstellar medium, such a shell would be complete and the filament would be that part of the shell seen tangentially; this would also be true for a reasonably homogeneous interstellar medium. Alternatively, such a filament might be a real filamentary condensation within a large shell - or it might be simply a filamentary condensation in the ambient medium, and not part of a formerly expanding shell at all. Detailed analysis of HI data, never yet performed, should enable these questions to be resolved.

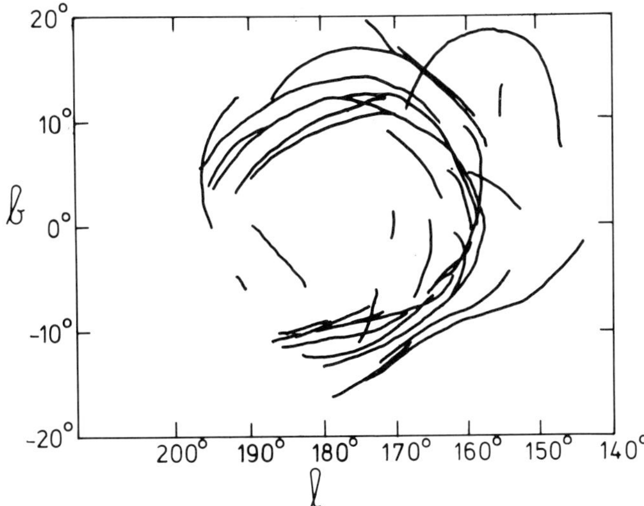

Figure 1. Ridge lines of large HI shell in galactic anticenter at velocities from -90 to -45 km s^{-1}.

II. ARTIFICIAL BOUNDARIES AND THEIR EFFECTS ON HI STRUCTURE OF PECULIAR MORPHOLOGY

Most circular filaments that appear on HI photographs fall into two classes: either they change size with velocity in roughly the way expected for an expanding shell, or they do not change at all with velocity. However, there are some cases in which the velocity structure does not mimic that expected for an expanding shell. Even though the filaments in these cases look circular on the sky, they might be parts of some different type of structure.

One spectacular example is a large shell, about 30 degrees in diameter, located in the galactic anticenter. It is discernable over a larger velocity range, from -90 to -20 km s^{-1} possibly over an even larger range. There is velocity structure, but it does not mimic that expected for an expanding shell. Figure 1 shows a sketch that depicts the locations of the ridges at velocities between -90 and -45 km s^{-1}. Portions of this shell contain very high negative velocity gas between -200 and -100 km s^{-1} (Mirabel, 1982).

This shell is so large that it crosses the artificial HI survey boundaries of $|b| = 10^{o}$. Historically this has led to its being studied over only very limited portions of its area, and the interpretations reflect this inadequacy. Kepner (1970) described it as part of feature Q, and portions of 0_i and possibly 0_o; some of these are large-scale galactic features. Dieter (1971) described it as part of an infalling rim of gas that supposedly defines the outer boundary of our galaxy. Verschuur (1973) described it as spiral arms, which of course had to have noncircular motions. Weaver (1974) described it as a "jet".

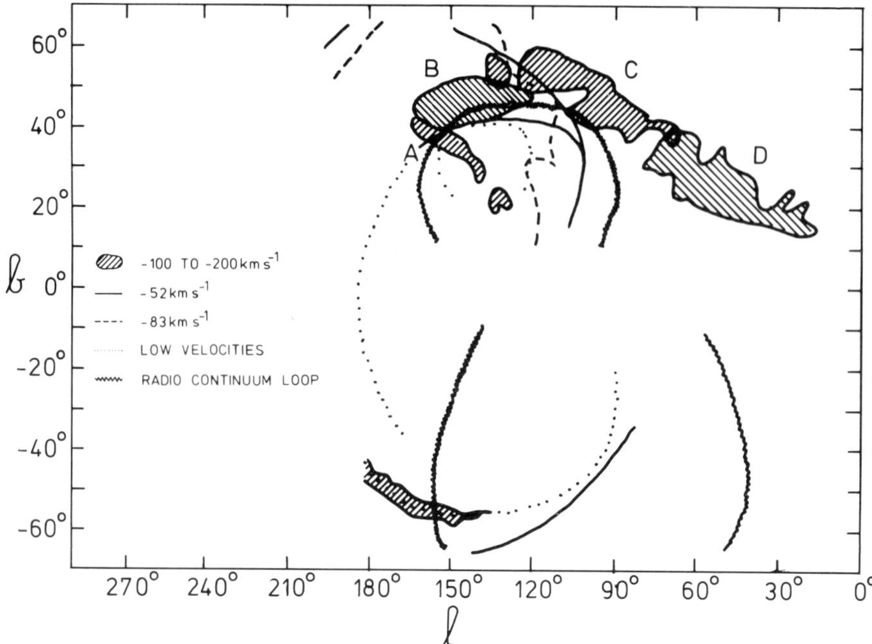

Figure 2. HI filaments at high, intermediate, and low velocities, together with radio continuum Loops II (at negative latitudes) and III (at positive latitudes).

Simonson (1975) described it as an external dwarf galaxy! Other authors, referenced by Heiles (1982b), describe portions of this shell as being local features, with the large velocity spread resulting from supernova remnants or interactions with high velocity gas.

In my view, the circular structures sketched in Figure 1 strongly argue that all of the gas discussed by the above authors is part of a single large shell. It has never been recognized before because of the artificial boundary of $|b| = 10^{\circ}$ imposed by the HI survey data. There is no reason to believe that the shell is a significant aspect of galactic structure. The shell is extraordinary, but probably not unique (see below), in having a large spread in velocity and in having a systemic velocity that is totally different from that expected from galactic rotation.

Another example of peculiar morphology are HI filaments of very large angular size that overlap radio loops II and III, sketched in Figure 2. These radio loops are centered roughly near longitude 130 degrees, with loop II centered at a negative latitude and loop III at a positive latitude. For radio loop II, at negative latitudes, there is a low-velocity ($\simeq$ -10 km s^{-1}) HI filament, an "intermediate-velocity" ($\simeq$ -50 km s^{-1}) filament, and a "high-velocity" ($\simeq$ -110 km s^{-1}) filament in the vicinity. The intermediate-velocity filament parallels the low-

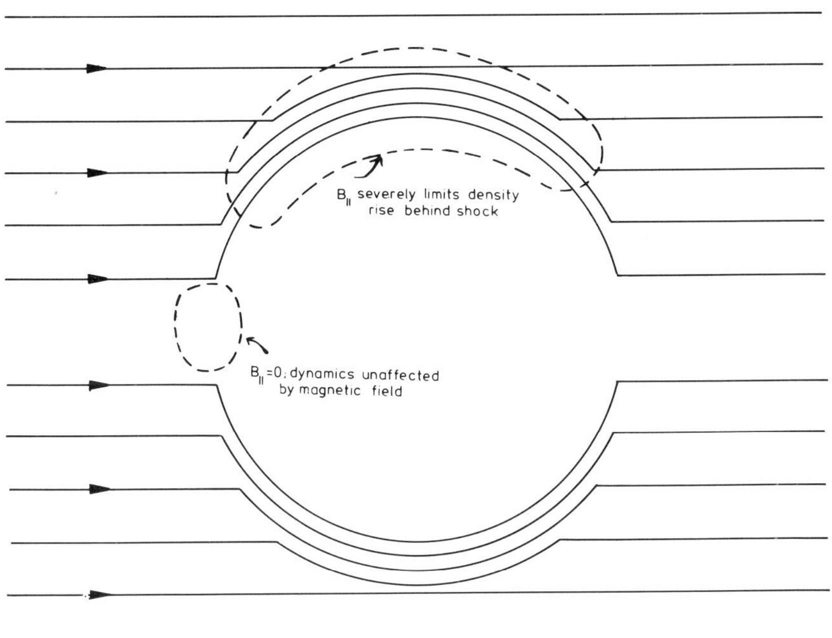

Figure 3. In a spherical shock with cooling, the magnetic field severely limits the density enhancement except at the "magnetic poles".

velocity filament. Surprisingly, the high-velocity filament overlies quite precisely a significant portion of the low-velocity filament (Cohen, 1981). For the radio loop III region, at positive latitudes, there is a similar paralleling of high- and intermediate-velocity HI filaments, plus a region where the end of an extensive low-velocity ($\simeq$ +17 km s^{-1}) filament abuts against the end of an intermediate-velocity ($\simeq$ -50 km s^{-1}) filament (at the letter "A" in Figure 2).

All of the features in Figure 2 are located in the northern sky, outside the limits $|b| = 10^{o}$. Thus, it might be supposed that artificial boundaries have played no part in the interpretation of these structures. However, this is not true: in this case the artificial boundary is velocity. Owing to instrumental considerations, HI surveys have concentrated on either the velocity range within about 100 km s^{-1} of the LSR, or the range outside this interval. Thus, the -110 km s^{-1} filament of Cohen (1981) was not discovered until recently. The juxtapositions of high-velocity (< -100 km s^{-1}) and intermediate-velocity (> -100 km s^{-1}) gas filaments at positive latitudes were recognized, but interpretations of gas in the two velocity ranges have tended to be independent and different.

Should the high-velocity gas and intermediate-velocity gas be regarded as being part of the same structure, or are they intrinsically different? It remains an open question, but it is one whose study should not be affected by the artificial boundary of -100 km s^{-1} imposed by instrumental considerations. The biggest problem with artificial boundaries is that they tend to attain significance in people's minds.

A final example is the "bisymmetric" HI remnant near $(1,b)=(209^{o},2^{o})$ discovered by Lockman and Ganzel (1982). This object consists of two arcs which join together to form a circle. The arcs have velocities which differ by 52 km s^{-1}, and the velocity within each arc is constant to within ±2 km s^{-1}. If interpreted as expanding pair of blobs, the kinetic energy is 1.6×10^{50} erg. However, it is hard to envisage a mechanism that would produce two blobs with such strange kinematic properties.

The juxtaposition of HI filaments at widely differing velocities, and the appearance of filaments over a very wide velocity range with velocity structure but no systemic change in size, are not expected if the filaments are parts of expanding shells, except coincidentally in the case of a nonuniform shell. The above cases are unlikely to be coincidental nonuniform shells, because they are the only ones in which HI filaments are visible over such wide velocity ranges; in the former two cases, the expansion velocities would have to be greater than 50 km s^{-1}, much higher than for other known shells. It seems to me that even though some of the filaments appear to be circular on the sky, their velocity structures imply that they are not parts of classical shells. My guess is that they represent another completely different physical phenomenon.

III. ENERGIES OF SHELLS AND SUPERSHELLS

Examination of the shells in the galactic plane (Heiles, 1979) proved to be very surprising. Large shells, such as that in Eridanus (Heiles, 1976) or that associated with the North Polar spur discussed above, should subtend angular diameters of a couple of degrees at 10 kpc distance. Many such shells were seen. But the largest shells at such distances were much bigger, and were wholly unexpected.

Such "supershells" are not very numerous, but their properties are spectacular. With diameters up to nearly 3 kpc and expansion velocities up to about 25 km s^{-1} they have very large kinetic energies - over 10^{53} erg. They seem to be concentrated at fairly large galactic radii, outside the solar circle. This concentration might be an observational bias, reflecting the fact that shells are easier to recognize outside the solar circle, where the distance is uniquely related to velocity through galactic rotation. However, this is unlikely because large HI shells have been observed in external galaxies - in M101 by Allen, Goss, and van Woerden (1973) and Allen and Goss (1979), and in M31 by Brinks (1983) - and the same tendancy is observed.

The majority of shells have relatively modest energies and sizes, and are adequately explained by the conventional picture of energy input from stellar winds and supernovae from a large star cluster (Bruhweiler et al., 1980). However, the largest supershells are probably not adequately explained by this picture. The energy requirements for supershells are large and require a star cluster much, much larger than the standard cluster adopted by Bruhweiler et al. (1980), which is similar to the Sco OB1 association and has 28 B0 and earlier stars. There is no independent evidence that such large star clusters exist outside the solar circle; indeed, current opinion is that star formation activity is weaker there.

For the largest supershells, then, another explanation seems to be called for. The currently most prominent alternative is collisions of high-velocity clouds with the galactic disk (Tenorio-Tagle, 1980, 1981). An infalling cloud transfers a large fraction of its kinetic energy to the ambient gas in the disk; energy requirements are met fairly easily for clouds similar to those observed. The fact that only one hemisphere of expanding shells is visible is a prediction of this theory!

IV. THE EFFECTS OF MAGNETIC FIELDS

Troland and Heiles (1982) have measured the magnetic field in two expanding shells to be 7 μGauss, using Zeeman splitting of the 21-cm line seen in emission. A measurement of Zeeman splitting yields only the longitudinal component of the field, i.e. that component directed along the line of sight to the observer. From geometrical considerations alone, with a probability of 1/2, the field is at least twice as large as the measured value. If the field strength is 10 μGauss, it exerts a pressure of 3×10^4 cm^{-3} K, much larger than the gas pressure in the shell.

Temperatures in shells are generally low; Heiles (1982a) has analyzed HI absorption data and found that the shells are in fact the low temperature "cloud" component. Volume densities in shells are lower than those we have come to accept as characteristic of the cloud component; the shell geometry forces a change in the derived densities because the line of sight through the tangential portion of a shell is much longer than the extent of the cloud across the line of sight, contrary to the usual assumption that these two distances are equal. For the Eridanus shell the HI volume density is less than 3 cm^{-3} at the distance of 500 pc (see Reynolds and Ogden, 1979, and Heiles, 1976).

Thus the gas pressure in the Eridanus shell is only about 300 cm^{-3} K, two orders of magnitude smaller than the magnetic pressure and much smaller than the ram pressure of shell expansion. The magnetic field must have a profound effect on the gas dynamics of the shell. For example, Troland and Heiles (1982) assume that the Eridanus HI shell is produced by cooling behind an adiabatic shock with Mach number M=20 or so. In the absence of a magnetic field, the density increase behind the shock is equal to M^2, a factor of several hundred. However, for the Eridanus shell the magnetic field limits the density increase to only a factor of about three. Even if their assumption of an isothermal shock is incorrect, the magnetic field must play a crucial role in the gas dynamics.

However, the magnetic field cannot affect the gas dynamics everywhere in the shell. It only has an effect if it is at least partially orthogonal to the shock velocity. For the case of a spherical shock in a uniform magnetic field, there are two regions - the "magnetic polar caps" - where the field is strictly parallel to the shock velocity. This is illustrated in Figure 3. In these regions, the magnetic field cannot inhibit the large increase in density that is expected behind an isothermal shock. Models of supernova-induced star formation (e.g. Herbst and Assousa, 1978) or of formation of dense molecular clouds behind supernova shocks should be meaningful only in these "polar-cap" regions.

Is there any evidence for such magnetic effects in real interstellar shells? There is certainly no direct evidence. There are two very meagre possibilities for indirect evidence. One is the Cygnus Loop. The bright filaments, which are the densest parts of the shell, lie on opposite ends of a diameter. This diameter is oriented roughly parallel to the galactic equator. The large-scale galactic magnetic field also lies parallel to the galactic equator. Thus, if the magnetic field in the vicinity of the Cygnus Loop is parallel to the average galactic field, the bright filaments lie at the magnetic poles - as they should if the magnetic field is important. A second meagre possibility for indirect evidence is the fact that sequential star formation in an OB association tends to occur along a line (Lada, Blitz, and Elmegreen, 1978). Is this line the direction of the local magnetic field? It would certainly be nice to find direct evidence for magnetic effects in expanding shells.

REFERENCES

Brinks, E.: 1983, paper presented at this meeting.

Bruhweiler, F.C., Gull, T.R., Kafatos, M., and Sofia, S.: 1980, Astrophys. J. 238, L27.

Cleary, M., Heiles, C., and Haslam, G.: 1979, Astron. Astrophys. Suppl. 36, 95.

Cohen, R.J.: 1981, Mon. Not. R. Astron. Soc. 196, 835.

Colomb, F.R., Poppel, W.G.L., and Heiles, C.: 1980, Astron. Astrophys. Suppl. 40, 47.

Dieter, N.H.: 1971, Astron. Astrophys. 12, 59.

Heiles, C.: 1975, Astron. Astrophys. Suppl. 20, 37.

Heiles, C.: 1976, Astrophys. J. 208, L137.

Heiles, C.: 1979, Astrophys. J. 229, 533.

Heiles, C.: 1982a, Astrophys. J., in press.

Heiles, C.: 1982b, in preparation.

Heiles, C. and Cleary, M.: 1979, Aust. J. Phys. Ap. Suppl. No. 47.

Heiles, C. and Habing, H.J.: 1974, Astron. Astrophys. Suppl. 14, 1.

Herbst, W. and Assousa, G.E.: 1978, in T. Gehrels (ed.), Protostars and Planets, 368.
Kepner, M.: 1970, Astron. Astrophys. 5, 444.
Lada, C.J., Blitz, L., and Elmegreen, B.G.: 1978, in T. Gehrels (ed.), Protostars and Planets, 341.
Lockman, F.J. and Ganzel, B.L.: 1982, Astrophys. J., submitted.
Mirabel, I.F.: 1982, Astrophys. J. 256, 112.
Reynolds, R.J. and Ogden, P.M.: 1979, Astrophys. J. 229, 942.
Simonson, S.C.: 1975, Astrophys. J. 201, L103.
Tenorio-Tagle, G.: 1980, Astron. Astrophys. 88, 61.
Tenorio-Tagle, G.: 1981, Astron. Astrophys. 94, 338.
Troland, T.H. and Heiles, C.: 1982, Astrophys. J. 260, L19.
Verschuur, G.L.: 1973, Astron. Astrophys. 22, L39.
Weaver, H.F.: 1974, in F.J. Kerr and S.C. Simonson (eds.), Galactic Radio Astronomy, IAU Symp. No. 60, 573.
Weaver, H.F.: 1979, in W.B. Burston (ed.), The Large-Scale Characteristics of the Galaxy, IAU Symp. No. 84, 295.
Weaver, H.F. and Williams, D.R.W.: 1973, Astron. Astrophys. Suppl. 8, 1.
Weaver, H.F. and Williams, D.R.W.: 1974, Astron. Astrophys. Suppl. 17, 1.

SOUTHERN OB ASSOCIATIONS: NEW CLUES TO STAR FORMATION MECHANISMS?

Anneila I. Sargent
Owens Valley Radio Observatory
California Institute of Technology

ABSTRACT

The advantages of examining southern OB associations, and the molecular clouds from which they are born, for clues to the way in which luminous stars form are considered. A number of problems which may be addressed more readily when southern observations are available are reviewed. Methods of identifying star-forming clouds for detailed study are discussed, taking into account the current instruments in the southern hemisphere.

1. INTRODUCTION

Star formation may be regarded as a quite virulent disease which afflicts some galaxies and not others. Observational evidence suggests that, once contracted, this disease is highly contagious and spreads rapidly. Among the advanced symptoms are HII regions, their evolution described fairly appropriately as blistering (Icke, Gatley and Israel 1980) or, more romantically, as a champagne-type flow (Tenorio-Tagle 1982 and references therein). The CO molecule is frequently employed as a tracer of the early stages, since stars presumably form from the molecular gas. Observations of strong emission from the ^{13}CO isotope, or of far infrared radiation from dust, may, however, be better signposts of the denser regions of molecular clouds where star formation begins (cf. Beichman 1979, Sargent et al. 1981).

It is to these very early stages that attention must be directed, since it has not yet been possible to distinguish unambiguously between theories of how the formation of molecular clouds and, by implication, stars is initiated. The competing hypotheses rely on either global processes, such as the passage of a galactic shock (Blitz and Shu 1980, and references therein), or on more local coagulation effects (Scoville and Hersh 1979; Cowie 1980 and references therein). Their relative merits are probably best evaluated by comparing our Galaxy to others, where the operating formation mechanisms appear more obvious. Available

W. B. Burton and F. P. Israel (eds.), Surveys of the Southern Galaxy, 205–215.

molecular cloud observations are, however, insufficient for this comparison to be made. More measurements are also necessary to resolve the question of whether molecular clouds trace out spiral structure in the Galaxy (Cohen et al. 1980 and references therein).

In this paper I should like to consider what we can learn about these and other star formation problems from examining southern OB associations and their related molecular clouds. I will also discuss which of the necessary observations can most readily be made in the immediate future.

2. WHY OB ASSOCIATION MOLECULAR CLOUDS?

OB associations are optimum sites to search for clues to how star formation is initiated and proceeds. Because of their relative youth, $\sim$ a few $\times$ 10^6 yrs., member stars have moved very little from their birthplaces and their kinematic properties reflect those of the primeval clouds (cf. Sargent 1979). Indeed, it has been demonstrated that molecular clouds exist in conjunction with a cluster or association only if there are stars of spectral type earlier than B0 present (Bash, Green and Peters 1977). Studies of these OB association molecular clouds should provide information about the earliest stages of stellar evolution, whereas observations in regions more obviously associated with well-developed radio or optical HII regions can result in some degree of bias toward later phases of star formation.

The characteristics of OB associations which can be determined from optical studies have been described by Blaauw (1964). The distributions of many of these associations in the northern and southern skies are shown schematically in Figures 1a and 1b respectively. The positions and assignments to a particular spiral arm are taken from Humphries (1978) and from Alter, Ruprecht and Vanysek (1970). Most of the molecular clouds connected with these associations have been detected and studied. Like the stellar associations, they are frequently extended parallel to the galactic plane, with lengths of 80 - 100 pc and masses of order $10^5 M_\odot$ (Blitz 1980 and references therein). In some, star formation appears to be contagious and sequential, in the sense that a number of epochs of stellar birth are evident, each apparently triggered in some way by the adjacent and immediately preceding outbreak (cf. Lada, Blitz and Elmegreen 1978). In others, the presence of a number of recently formed stars, or protostars, over a relatively large area of the primeval cloud eliminates any mechanism requiring proximity (Sargent 1979; Jaffe and Fazio 1982). It seems likely that the few associations which are accessible only from the southern hemisphere will be similar in their intrinsic properties to their northern counterparts.

For example, the NGC 6334 molecular cloud, at a distance of 1.7 kpc (Neckel 1978) and galactic longitude l = 351.4°, appears to be extended along the plane for about 20 pc and to have a total mass of $10^4 M_\odot$ (Dickel, Dickel and Wilson 1977). Since the published ^{12}CO antenna

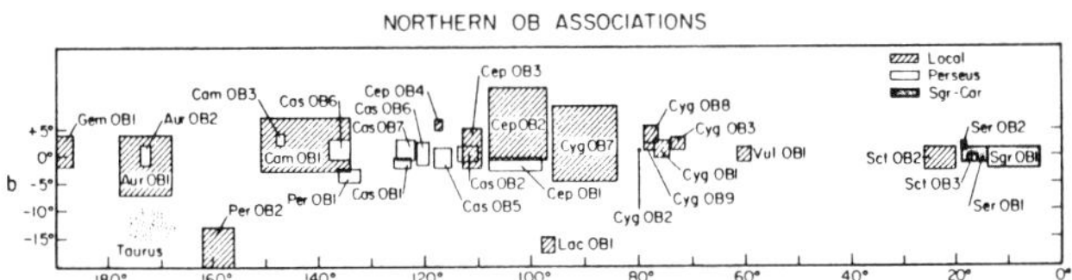

Fig. 1a The distribution of "northern" OB associations. Those in the local arm are indicated by hatched rectangles. Those in the Perseus and Sgr-Car arms are represented by blank and stippled rectangles respectively. The Taurus dark cloud is shown as an irregular dotted region.

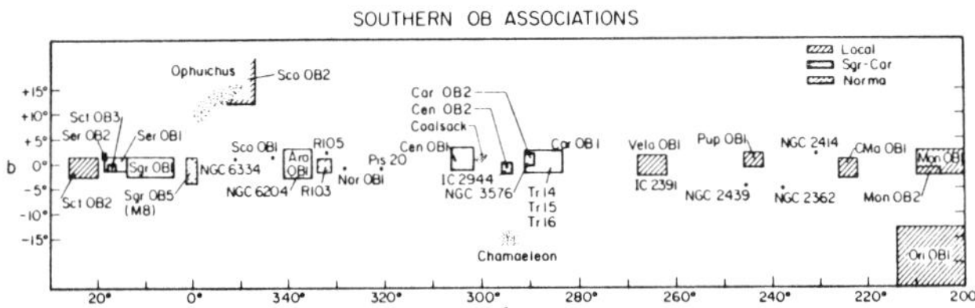

Fig. 1b As Fig. 1a, but for "southern" OB associations. Cross-hatched rectangles represent OB groups in the Norma arm. There is some overlap between the figures. The Ophiuchus, Coalsack and Chamaeleon dark clouds are shown as irregular dotted regions.

temperature contours do not extend below 10K, it is reasonable to assume that lower intensity emission is present, reaching to the 80 - 100 pc typical of individual star-forming clouds, and increasing the mass estimate correspondingly. At each of the molecular line-far infrared maxima shown in Figure 2 either compact HII regions or H_2O and OH masers have been detected (McBreen et al. 1979 and references therein; Rodriguez, Canto and Moran 1982). In addition, unusual far infrared emission at 400 μm has been detected from NGC 6334 I (Gezari 1982), while molecular hydrogen emission and characteristic high velocity CO wings are associated with NGC 6334 V (Fischer et al. 1982). All these characteristics are accepted indications of active star formation and, together with recent multi-wavelength infrared observations (Harvey and Gatley 1982) demonstrate that NGC 6334 is yet another cloud where stars are being born simultaneously over an extensive area. Thus, assuming NGC 6334 is a typical example, it seems that observations of southern OB associations will certainly contribute to the statistics of star forming regions. In the next sections, it will be shown that such measurements will also shed light on the formation mechanisms at work.

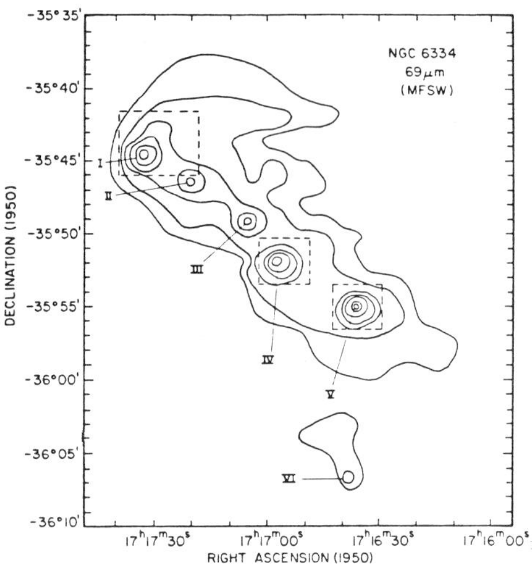

Fig. 2 The 69 μm map of NGC 6334 reproduced from McBreen et al. (1979). Far infrared maxima referred to in the text are denoted by roman numerals. Dashed rectangles indicate the regions studied by Harvey and Gatley (1982).

3. THE LARGE SCALE

If Figures 1a and 1b are re-examined, recalling the OB association clouds studied (Blitz 1980 and references therein), it is evident that observations have been confined, for the most part, to what are known as the local, Perseus and Sagittarius arms of our Galaxy. No extensive observations have been made of the Carina portion of the Sagittarius arm or of the Norma and inner arms. A major advantage of the southern hemisphere is the accessibility of these regions of the Galaxy. This is perhaps best illustrated by reference to Figure 3, where the spiral structure deduced using HII regions as tracers (Georgelin and Georgelin 1976) is shown. Not only can several putative spiral arms be more readily observed in the fourth quadrant, but the spacing between them is large. A variety of problems can therefore be studied with correspondingly less confusion than in the other three quadrants.

At the distances typical of these spiral arms, however, optical identification of OB associations will quickly become difficult. The number of molecular clouds available for study would be severely limited if the sample were confined to those related to visible stellar groups. The presence of massive star-forming clouds can, however, be inferred from the observation of H_2O masers, OH masers, far infrared peaks or enhanced molecular line emission, not necessarily accompanied by compact HII regions. Throughout this paper references to OB association clouds implicitly include such clouds, where visible, young stars may not have been detected. Optimum methods of searching for these will be discussed later.

Observations of molecular clouds in the southern hemisphere will obviously result in a more complete picture of the Galaxy and permit comparison with other galaxies in the throes of stellar birth. It

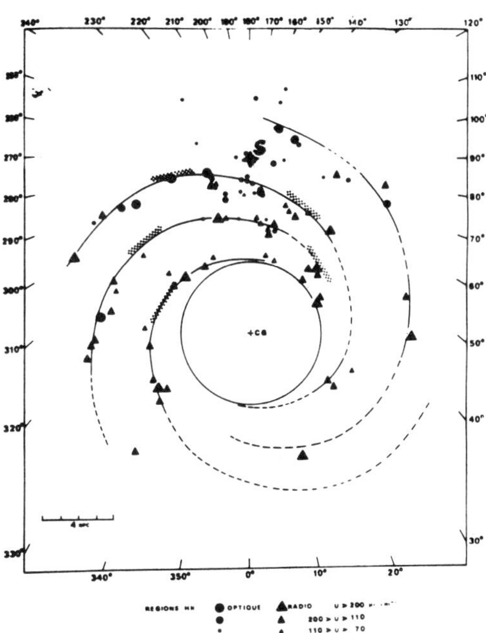

Fig. 3 The spiral structure of the Galaxy as suggested by the distribution of HII regions, reproduced from Georgelin and Georgelin (1976). The position of the sun is indicated by S.

should be possible to establish whether, like M31, spiral arms are discernible in all the typical population I tracers including CO (Linke 1982, and references therein). Such structure could be produced by density-wave induced star formation. Similarities to the galaxies studied by Young and Scoville (1982) would make such an interpretation less likely. In any case, if the molecular gas traces out spiral structure, the expected wide separation of any arms in the fourth quadrant should render their presence more readily discernible. The resultant well-defined distinctions in velocity between these arms will help in ascertaining whether molecular clouds exist in the interarm region. Such a situation would have profound implications for the lifetimes of the clouds and their star-forming behavior (cf. Blitz and Shu 1980).

Properties which may be galactocentric distance dependent can also be studied in more detail when southern hemisphere observations are available. Thus, observations of a small sample of northern molecular clouds suggest that the deuterium abundance increases with increasing galactic radius (Penzias 1979), as might be expected from the big bang model (cf. Ostriker and Tinsley 1975). On the other hand, it has been pointed out that variations in the DCO^+/HCO^+ ratio, used to trace deuterium abundance, can be caused by effects such as chemical fractionation and that measurements of a variety of deuterated molecules are required to determine unambiguously whether a deuterium abundance gradient exists (Wootten, Loren and Snell 1982). The search for and study of new southern molecular clouds should increase the statistical sample on which to base such a judgement.

Likewise, recent recombination line measurements of southern HII regions (McGee and Newton 1981) provide strong confirmation of a trend noted in northern studies for nebular electron temperatures to increase with distance from the Galactic center (Churchwell et al. 1978 and references therein). Variations of helium and heavier element abundances across the Galaxy bear strongly on interpretations of galactic evolution. It is therefore disturbing to find that, when the sample of HII regions is thus expanded, helium abundance seems independent of galactic radius, contrary to the conclusions of northern observers (Thum, Mezger and Pankonin 1980). Clearly, more observations of the HII regions representative of later stages of proto-stellar evolution are needed to resolve this question.

4. INDIVIDUAL STAR-FORMING REGIONS

Southern observations will naturally increase the known number of molecular clouds related to optical OB associations, permitting more extensive examination of the relative positions of the stars and their primeval clouds. If, along one spiral arm, there is a preferred orientation of stars and gas, or if the kinematics of the earliest formed stars are similar, star formation could well have been caused by a global event, such as the passage of the galactic density wave shock. For the nearby clouds, proper motion measurements of the association members using Hipparchos will allow more detailed studies of the relative kinematics of the stars and gas.

A larger sample of clouds will also provide a better statistical base from which to investigate the total linear extent of star-forming complexes. Although an individual association cloud is typically 80 - 100 pc in length, there is some evidence that groups of these clouds may be tenuously connected to form large aggregates of extent $\sim$ 300 pc (Elmegreen, Lada and Dickinson 1979; Sargent 1979), similar to the size of star formation regions in other galaxies.

5. LOW MASS STAR FORMATION

As mentioned previously, OB association molecular clouds, which are intrinsically less disturbed than clouds where HII regions already exist, should lead more easily to insights into early proto-stellar evolution. The Taurus region provides more extreme examples of quiescent clouds. There is not even evidence of an embedded OB star population (Elias 1978) capable of producing disruptive HII regions at a later date. The observation of a number of "pedestal" sources thought to be produced by stellar winds from very young, low mass objects (cf. Frerking and Langer 1982; Lichten 1982) does, however, indicate some star-forming activity. Similar dark clouds accessible from the South are the Southern Coalsack and Chamaeleon, both at about 170 pc distance (Hyland 1981; Jones et al. 1980). Searches at far infrared and millimeter wavelengths for emission from these regions should help establish whether star formation in such dark clouds is qualitatively different from that in OB association molecular clouds.

6. SEARCHES FOR SOUTHERN REGIONS OF STAR FORMATION

With the exception of the galactic plane CO work described at this workshop (Robinson et al. 1983; Manchester et al. 1983), most molecular line observations in the southern sky have been restricted to optical or radio HII regions (cf. Gillespie et al. 1977; Whiteoak 1983), which represent a relatively late stage of proto-stellar evoluation (Yorke and Shustov 1981). Limited searches in the J = 2 - 1 transition of CO in the directions of other types of young sources, such as Herbig-Haro objects, have not been successful (Brand 1983). Until a millimeter-wave telescope capable of large-scale survey work is available in the southern hemisphere, the quest for incipient OB star forming clouds may best be carried out using tracers other than the CO molecule. Southern H_2O maser catalogs (Batchelor et al. 1980; Braz and Scalise 1982 and references therein) can already assist in singling out possible proto-stars. In the immediate future, the IRAS survey in the far infrared should be particularly useful in revealing regions where stars may be forming and which should be investigated further at millimeter wavelengths.

This is well illustrated by recent far infrared observations of the NGC 2264 molecular cloud, part of the Mon OB1 complex (Blitz 1978), obtained using the University of Gronigen balloon infrared platform (Sargent

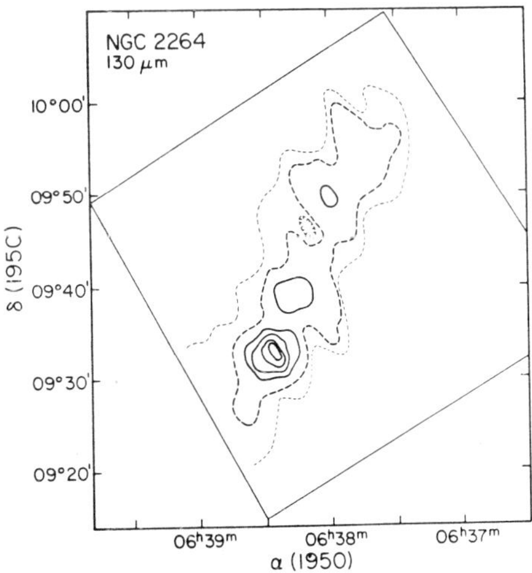

Fig. 4 The 130 μm surface brightness distribution in the NGC 2264 molecular cloud. Contours are at 0.9, 0.7, 0.5, 0.3 and 0.1 of the peak value of 2514 Jy into a 3' beam. The area searched in the far infrared survey (Sargent et al. 1982) is outlined by a rectangle.

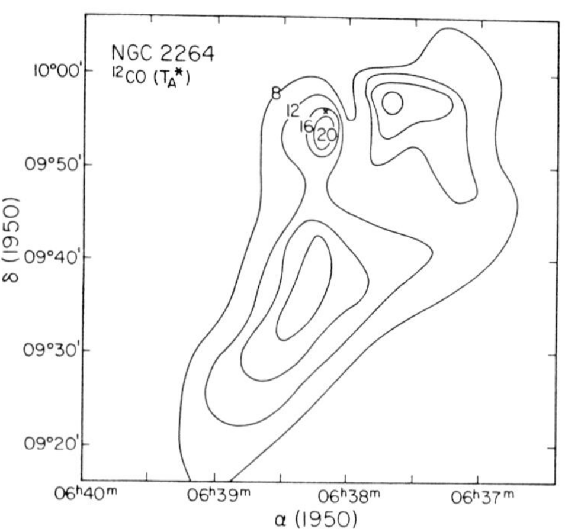

Fig. 5a A ^{12}CO antenna temperature map of the NGC 2264 cloud, reproduced from Crutcher, Hartkopf and Giguere (1978). An X marks the position of 15 Mon.

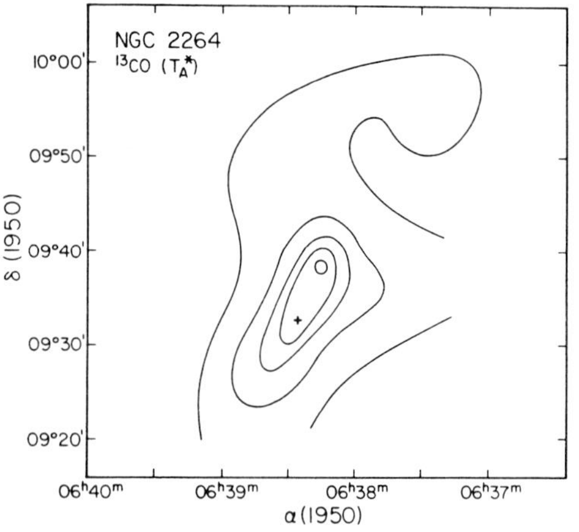

Fig. 5b As Figure 5a, but for ^{13}CO. A cross represents NGC 2264IR.

et al. 1982). the surface-brightness contour map of this cloud at 130 μm is shown in Figure 4. Contour maps of ^{12}CO and ^{13}CO antenna temperatures are reproduced from Crutcher, Hartkopf and Giguere (1978) in Figures 5a and 5b respectively. All three figures have similar general characteristics, and details of the ^{12}CO and ^{13}CO maps are reflected in the far infrared diagram. There is a maximum of far infrared emission at the location of NGC 2264IR (Allen 1972; Harvey, Campbell and Hoffmann 1977), shown as a cross in Figure 5b. A secondary far infrared peak which, our analysis suggests, is relatively dense and cold, occurs at the position of highest ^{13}CO antenna temperature. This may be a new site of star formation in the cloud. Adjacent to 15 Mon, represented by an X in Figure 5a, another enhancement in far infrared emission, probably due to external heating by the visible star, can be seen. The cloud orientation is the same as that of the larger complex, although the full 88 pc extent is evident only in the large-scale, 8' beam measurements (Blitz 1978). As suggested above, it is clear that the existence of the cloud and its salient properties are indicated by the far infrared observations alone. Higher resolution molecular line measurements are, of course, necessary to a comprehensive understanding of the structure and energetics of the cloud.

7. SUMMARY

It is clear that there are a number of advantages to be accrued from southern OB association molecular cloud observations using the new millimeter telescopes together with the Hipparchos and IRAS satellites. In particular, the reality of spiral structure in the molecular gas, and the variation of a number of properties with distance from the Center, can be investigated further. A complete picture of our Galaxy can be derived, permitting more detailed comparison with other galaxies than has hitherto been possible. Studies of individual clouds will be useful from a statistical viewpoint.

Initially, at least, it may be necessary to invert the standard, northern hemisphere, observing procedures and employ far infrared or H_2O surveys to locate these clouds, before examining their structure at millimeter wavelengths. Nevertheless, observations such as are discussed here should considerably improve our understanding of star formation processes and the next few years may see many advances in this field.

REFERENCES

Allen, D. A.: 1972, Ap. J. (Letters), 172, pp. L55-L58.
Alter, G., Ruprecht, J., and Vanysek, V.: 1970, "Catalog of Clusters and Associations," Akademiai Kiado, Budapest.
Bash, F. N., Green, E., and Peters, W. L.: 1977, Ap. J. 217, pp. 464-472.
Batchelor, R. A., Caswell, J. L., Goss, W. M., Haynes, R. F., Knowles, S. H., and Wellington, K. J.: 1980, Aust. J. Physics, 33, pp. 139-157.
Beichman, C. A.: 1979, Ph.D. thesis, University of Hawaii.
Blaauw, A.: 1964, Ann. Rev. Astr. Ap. 2, pp. 213-246.
Blitz, L.: 1978, Ph.D. thesis, Columbia University.
Blitz, L.: 1980, "Giant Molecular Clouds in the Galaxy," ed. P. M. Solomon and M. Edmunds (Oxford:Pergamon), pp. 1-18.
Blitz, L., and Shu, F. H.: 1980, Ap. J. 238, pp. 148-157.
Brand, J.: 1983, this workshop.
Braz, M. A., and Scalise, E.: 1982, Astron. Astrophys. 107, pp. 272-275.
Churchwell, E., Smith, L. F., Mathis, J., Mezger, P. G., and Huchtmeier, W.: 1978, Astron. Astrophys. 70, pp. 719-732.
Cohen, R. S., Cong, H., Dame, T. M., and Thaddeus, P.: 1980, Ap. J. (Letters) 239, pp. L53-L56.

Cowie, L.: 1980, Ap. J. 236, pp. 868-879.
Crutcher, R. M., Hartkopf, W. I., and Giguere, P. I.: 1978, Ap. J. 226, pp. 839-850.
Dickel, H. R., Dickel, J. R., and Wilson, W. J.: 1977, Ap. J. 217, pp. 56-67.
Elias, J. H.: 1978, Ap. J. 224, 857-872.
Elmegreen, B. G., Lada, C. J., and Dickinson, D. F.: 1979, Ap. J. 230, pp. 415-427.
Fischer, J., Joyce, R. R., Simon, M., and Simon, T.: 1982, Ap. J. 258, pp. 165-169.
Frerking, M. A., and Langer, W. D.: 1982, Ap. J. 256, pp. 523-529.
Georgelin, Y. M., and Georgelin, Y. P.: 1976, Astron. Astrophys. 49, pp. 57-79.
Gezari, D.: 1982, Ap. J. (Letters) 259, pp. L29-L33.
Gillespie, A. R., Huggins, P. J., Sollner, T. C. L. G., Phillips, T. G., Gardner, G. G., and Knowles, S. H.: 1977, Astron. Astrophys. 60, pp. 221-225.
Harvey, P. M., and Gatley, I.: 1982, in preparation.
Harvey, P. M., Campbell, M. F., and Hoffmann, W. F.: 1977, Ap. J. 215, pp. 151-154.
Humphries, R. M.: 1978, Ap. J. Suppl. 38, pp. 309-350.
Hyland, A. R.: 1981, "I.A.U. Symposium No. 96, Infared Astronomy," ed. D. P. Cruikshank and C. G. Wynn-Williams (Dordrecht:Reidel), pp. 125-151.
Icke, V. Gatley, I., and Israel, F. P.: 1980, Ap. J. 236, pp. 808-822.
Jaffe, D. T., and Fazio, G. G.: 1982, Ap. J. (Letters), 257, pp. L77-L81.
Jones, T. J., Hyland, A. R., Robinson, G., Smith, R., and Thomas, J.: 1980, Ap. J. 242, pp. 132-140.
Lada, C. J., Blitz, L., and Elmegreen, B. G.: 1978, "Protostars and Planets," ed. T. Gehrels (Tucson:University of Arizona), pp. 341-367.
Lichten, S. L.: 1982, Ap. J. (Letters) 255, L119-L122.
Linke, R. A.: 1982, "Extragalactic Molecules," ed. L. Blitz and M. Kutner, pp. 87-92.
Manchester, R. N., Whiteoak, J. B., Robinson, B. J., and Rennie, C. J.: 1983, this workshop.
McBreen, B., Fazio, G. G., Stier, M., and Wright, E. L.: 1979, Ap. J. (Letters) 232, pp. L183-L187.
McGee, R. X., and Newton, L. M.: 1981, Mon. Not. Roy. Astr. Soc., 196, pp. 889-905.
Neckel, J.: 1978, Astron. Astrophys 69, pp. 51-56.
Penzias, A. A.: 1979, Ap. J. 228, pp. 430-438.
Ostriker, J. M., and Tinsley, B. M.: 1975, Ap. J. (Letters), 201, pp. L51-L54.
Robinson, B. J., Whiteoak, J. B., McCutcheon, W. H., Manchester, R. N., and Rennie, C. J.: 1983, this workshop.
Rodriguez, L. F., Canto, J., and Moran, J. M.: 1982, Ap. J. 225, pp. 103-110.
Sargent, A. I.: 1979, Ap. J. 233, pp. 163-181.
Sargent, A. I., van Duinen, R. J., Fridlund, C. V. M., Nordh, H. L., Aalders, J. W. G., and Beintema, D.: 1982, in preparation.

Sargent, A. I., van Duinen, R. J., Fridlund, C. V. M., Nordh, H. L., and Aalders, J. W. G.: 1981, Ap. J., 249, pp. 607-621.
Scoville, N. Z., and Hersh, K.: 1979, Ap. J. 229, pp. 578-582.
Tenorio-Tagle, G.: 1982, "Regions of Recent Star Formation," ed. R. S. Roger and P. E. Dewdney (Dordrecht:Reidel), pp. 1-14.
Thum, C., Mezger, P. G., and Pankonin, V.: 1980, Astron. Astrophys. 87, pp. 269-275.
Whiteoak, J. B.: 1983, this workshop.
Wootten, H. A., Loren, R. B., and Snell, R. L.: 1982, Ap. J. 255, pp. 160-174.
Yorke, H. W., and Shustov, B. M.: 1981, Astron. Astrophys. 98, pp. 125-132.
Young, J. S., and Scoville, N. Z.: 1982, Ap. J. 258, pp. 467-489.

CO J=2-1 OBSERVATIONS TOWARD SOUTHERN HII REGIONS

R.N. Martin[1,2], D.T. Emerson[2], K. Ruf[1], T.L. Wilson[1],
P. Zimmermann[1]
1 Max-Planck-Institut für Radioastronomie Bonn, F.R.G.
2 Institut de Radio Astronomie Millimétrique, Grenoble, France.

A spectral line receiver system developed at the Max-Planck-Institut für Radioastronomie in Bonn was installed on the ESO 3.6 m and 1 m telescopes in July 1981. The cooled mixer frontend gave DSB receiver temperatures of 260-600 K at 230 GHz. The spectrometer was a 256 x 1 MHz filterbank. We have observed the CO 2-1 transition towards 42 positions corresponding to the brightest southern HII regions.

INTRODUCTION

While there have been many studies of mm wavelength line emission from the galactic plane, HII regions and dark clouds in the northern hemisphere, it is only within recent years that comparable studies have begun in the southern hemisphere. As a first step toward the study of CO 2-1 (230 GHz) emission from the southern galactic plane, we have surveyed the peaks of the brightest HII regions. Many of these positions are the same as those which have been observed in the CO 1-0

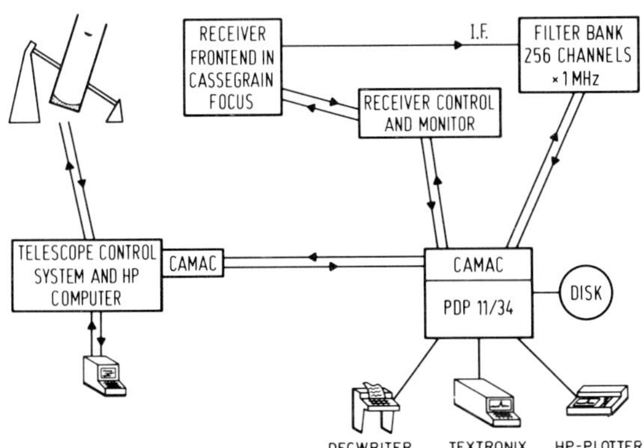

Figure 1: The 1.3 mm wavelength receiver system as it was installed on the ESO telescopes.

W. B. Burton and F. P. Israel (eds.), Surveys of the Southern Galaxy, 217–222.

TABLE 1. SYSTEM PARAMETERS

frontend:	cooled mixer receiver chopper wheel calibration
local oscillator:	doubled klystron, quasi-optically coupled LO injection
receiver noise temperature:	Initially 600 K DSB but optimised to 260 K DSB for the later part of the observations.
backend:	256 x 1 MHz filterbank (1.3 km/s resolution at 230 GHz)
data acquisition and reduction:	PDP11/34 computer
telescope efficiency:	30-40 %
focus:	cassegrain
telescope:	ESO 3.6 m telescope, full width to half power of 1.6 arc min at 230 GHz ESO 1 m telescope, full width to half power of 5.5 arc min at 230 GHz

line by Gillespie et al. (1977) and Whiteoak et al. (1982).

EQUIPMENT

A transportable receiver system has been built at the Max-Planck-Institut für Radioastronomie for spectral line observations at 1.3 mm wavelength. The system consists of a cooled mixer receiver package, a 256 channel x 1 MHz filterbank, and a PDP11/34 computer system to do the data acquisition and reduction. The computer uses a CAMAC interface to control the filterbank and frontend, and to communicate with the telescope drive computer. This system was installed on the ESO 3.6 m and ESO 1 m telescopes at La Silla, Chile during July 1982. Figure 1 is a block diagram of the system. The parameters of the system are summarized in Table 1.

The efficiency and beam shape were measured for both telescopes by continuum scans of the Moon. In addition, scans across Jupiter were made with the 3.6 m telescope. Figure 2 is an example of the beam shape measured on the 1 m telescope using a bright edge of the Moon.

RESULTS

The spectral line data were taken in a position switched mode with on/off pairs of 30-60 seconds each. The data were calibrated against a room temperature chopper every 10-30 minutes. The absolute calibration of all intensities has been arrived at by scaling relative to Orion-KL,

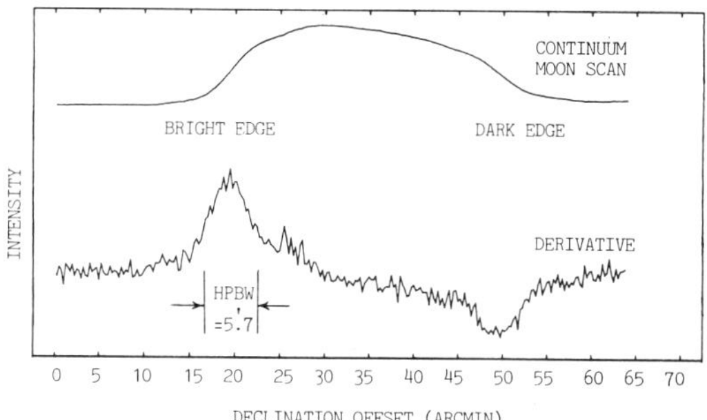

Figure 2: Continuum scan of the Moon and the derivative of the scan as measured with the ESO 1 m telescope. The Moon was approximately 3/4 full during this measurement. The measured beam size is very close to the theoretical diffraction limit of 5.6 arc min. at 220 GHz.

which we assume to be 65 K. Integration times of 2 to 20 minutes were spent on each position. The ultimate sensitivity was limited by baseline problems rather than by receiver noise, in spite of the use of absorbing material at the center of the secondary mirror and around the receiver mount.

We have surveyed 42 positions corresponding to the brightest HII regions in the southern sky. A list of these positions and the intensities we measured on the 3.6 m telescope are given in Table 2. Several of these spectra are shown in Figure 3. The spectrum in Orion-KL shows that the angular resolution of the 3.6 m telescope is still great enough to allow detection of the CO "plateau" emission.

Our relative calibration during the observing period was checked by observing Orion-KL and M17 regularly. The weather was good for the 4 day period during which these spectral line data were obtained. We estimate that our relative calibration is accurate to within 20-30 %, and the absolute calibration is dependent on the value we choose for Orion-KL.

In Figure 4 we have compared the intensity that we observed in the CO J=2-1 transition with that observed in the J=1-0 transition by Gillespie et al. (1977) and Whiteoak et al. (1982) for those positions we have in common. The agreement between the J=2-1 and J=1-0 data is not very good in general. However, the agreement between our J=2-1 intensities and the J=1-0 intensities of either Gillespie et al. (1977) or Whiteoak et al. (1982) is no worse than the agreement between the two J=1-0 data sets. This is not unexpected, however, since all three sets of data are taken with different telescope beam sizes; without a detailed knowledge of the shape of the emission regions such

TABLE 2

Object		RA(1950)	DEC(1950)	T_a^* (K)
209.0-19.4	Orion-KL	05 32 47.0	-05 24 17	65.0
265.1+1.5	RCW36	08 57 38.0	-43 33 24	11.9
267.9-1.1	RCW38	08 57 25.0	-47 19 18	15.5
274.0-1.2	RCW42	09 22 47.0	-51 47 00	<4.5
284.3-0.3	RCW49	10 22 22.0	-57 31 48	<2.5
287.4-0.6	RCW53	10 41 38.0	-59 19 18	17.6
291.3-0.7	RCW57	11 09 47.0	-61 02 36	27.5
298.2-0.3		12 07 22.0	-62 33 06	<2.5
298.9-0.4		12 12 46.0	-62 44 24	<2.5
305.2+0.0	RCW74	13 08 04.0	-62 29 18	10.5
305.3+0.2	RCW74	13 08 23.0	-62 17 42	<2.5
305.4+0.2	RCW74	13 09 21.0	-62 18 54	11.9
316.8-0.1		14 41 31.0	-59 36 54	11.7
320.2+0.8	RCW87	15 01 36.0	-57 19 24	8.8
322.2+0.6	RCW92	15 14 50.0	-56 28 00	17.9
324.2+0.1		15 29 03.0	-55 46 24	<4.5
326.7+0.6		15 40 58.0	-53 57 18	13.5
327.3-0.6	RCW97	15 49 13.0	-54 26 30	20.5
330.9-0.4		16 06 29.0	-51 58 48	12.6
331.5-0.1		16 08 22.0	-51 19 30	9.9
332.2-0.4		16 12 52.0	-51 09 54	5.8
332.8-0.6	RCW106	16 16 25.0	-50 47 30	4.4
333.0-0.4		16 16 52.0	-50 33 00	14.7
333.3-0.4		16 17 47.0	-50 19 12	11.7
333.6-0.2		16 18 26.0	-46 58 54	<4.5
336.5-1.5	RCW108	16 36 20.0	-48 45 36	17.6
337.1-0.2		16 33 02.0	-47 25 18	11.9
345.4-0.9	RCW117	17 06 04.0	-41 32 06	12.5
348.7-1.0	RCW122	17 16 40.0	-38 54 06	17.6
351.1+0.7	RCW127	17 16 37.0	-35 55 12	19.0
351.2+0.5	RCW127	17 17 34.0	-36 00 54	6.2
351.4+0.7	RCW127	17 17 18.0	-35 46 54	21.6
351.6-1.3		17 25 56.0	-36 37 54	16.6
353.1+0.6	W22/RCW131	17 22 18.0	-34 19 54	7.1
353.2+0.9	W22/RCW131	17 21 30.0	-34 08 06	8.1
6.0-1.3	M8	18 01 06.0	-24 28 00	19.0
6.0-1.5	M8	18 01 53.0	-24 28 00	19.0
15.0-0.7	M17	18 17 26.5	-16 14 54	26.1
28.8+3.5	W40	18 28 41.0	-02 08 48	14.6
30.0+0.0		18 43 28.1	-02 39 56	7.6
43.2-0.0	W49	19 07 54.0	09 01 01	13.3
49.5-0.4	W51	19 21 27.0	14 24 30	13.2

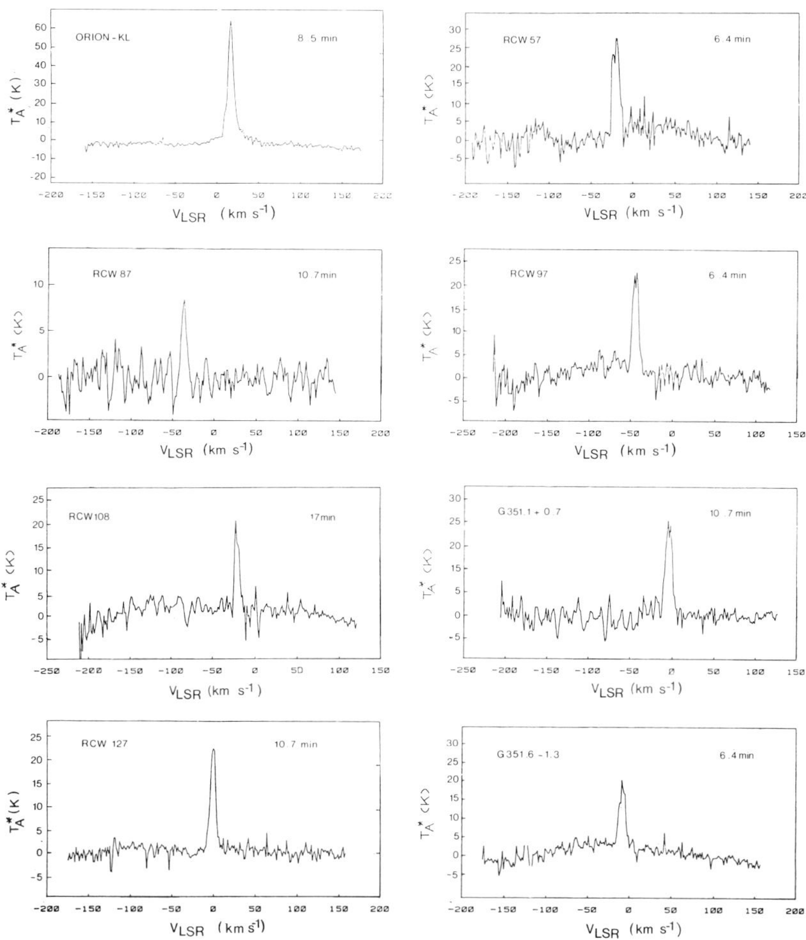

Figure 3: Spectra taken on the 3.6 m telescope. All of the intensities have been scaled to Orion-KL (assuming a value of 65 K). Only linear baselines have been removed from the data. The integration times are given in the top right corner of each spectrum.

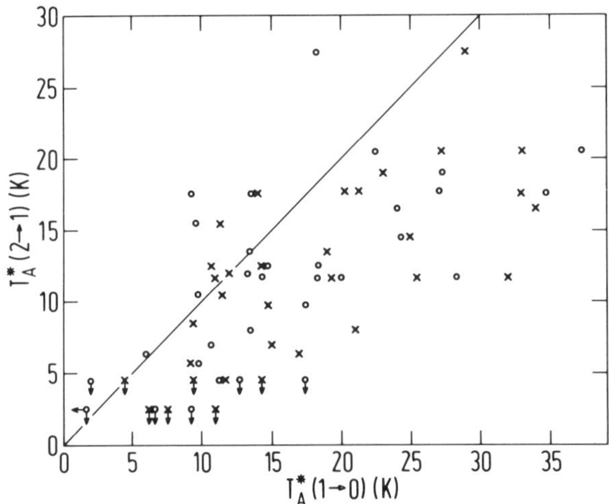

Figure 4: A comparison of the intensities we measure in the J=2-1 transition with those measured in the J=1-0 transition (2.6 mm) by Gillespie et al. (1977) and Whiteoak et al. (1982) at the same positions.

comparisons are difficult. Differences in excitation may also be expected between the J=1-0 and J=2-1 transitions. The results are not inconsistent with the hypothesis that $T_a^*(2-1) \simeq T_a^*(1-0)$ in these optically thick regions. In particular, we obtain $T_a^*(2-1)/T_a^*(1-0) = 0.69 \pm 0.1$ and 0.74 ± 0.1 for the Gillespie et al. and Whiteoak et al. results, respectively.

We thank A. Korn for his assistance during the observing in Chile, the Millimeter Division of the MPIfR for help in building the frontend, and the Electronics Division of the MPIfR for help in building the spectrometer.

REFERENCES

Gillespie, A.R., Huggins, P.J., Sollner, T.C.L.G., Phillips, T.G., Gardner, F.F., Knowles, S.H.: 1977, Astron. Astrophys. 57, 221.

Whiteoak, J.B., Otrucek, R.E., Rennie, C.J., 1982, preprint.

CO IN SOUTHERN SOURCES (*)

Jan Brand
Sterrewacht,Leiden

On behalf of the "Dutch CO-group" (see note at end)

Introduction

Preliminary results are given of CO(J=2→1) observations obtained at the Las Campanas and La Silla Observatories (Chile) using the ESTEC/Utrecht heterodyne (sub-) mm receiver. The data presented here are part of the work of the "Dutch CO-group" which has as its objective to survey the fourth quadrant of the galactic plane and look for CO in discrete southern sources. Use has been made of three telescopes: the 2.5 m at Las Campanas (1980),the ESO 3.6 m at La Silla (1980,1981,1982) and the ESO 1.4 m "CAT" (1981,1982),all stationed in Chile. Beamwidths at 230 GHz are respectively 2.3,1.9 and 5.5 arcmin. The receiver used is described in detail by Lidholm and de Graauw (1979) while specifics of the observing runs are given by de Graauw et al. in this volume. In that same contribution we reported on the galactic survey. This paper is limited to the discrete sources,and two of them in particular.

Results

With the CAT,apart from doing the survey,we looked for CO(J=2→1) emission in ∿50 molecular clouds associated with HII-regions (of which we detected 29; detection usually means $T_A^* > 1$ K (3σ)) and in more than 80 dark clouds,reflection nebulae and HH-type objects (39 detections). Furthermore,three HII-regions and five dark clouds were looked at in more detail. Since at this date we have not yet reduced all data we cannot make a full statistical analysis of the material,and this will be deferred until a later date (two papers are in preparation). It should be remarked however,that none of the HII-regions for which we do not have a detection were observed in CO(J=1→0) by Gillespie et al. (1977). A first look at some POSS-prints suggests that at least part of the non-detections may have been caused by pointing at the ionized gas rather than the obscured parts. For those HII-regions observed in common with Gilles-

(*) *partly based on observations collected at the European Southern Observatory,La Silla (Chile)*

W. B. Burton and F. P. Israel (eds.), Surveys of the Southern Galaxy, 223–228.

pie et al. (1977; beamwidth 4.5 arcmin), the ratio $T_A^*(J=1\rightarrow0)/T_A^*(J=2\rightarrow1)$ is 2.1 (averaged). This factor is accounted for if the J=2→1 data are also corrected for beam efficiency (about 45%-50%), as are the J=1→0 data.

With the 2.5 m at Las Campanas and the ESO 3.6 m some regions were studied in more detail. Below, by way of illustration, we present results on two interesting sources:

(i) G327.30-0.55

G327.30-0.55 is a radio continuum source not associated with an optical HII-region, located close (in projection) to RCW 97. Its near kinematical distance is 3.4 kpc which places it between the Sagittarius and Scutum-Crux arms (as defined by Georgelin and Georgelin, 1976). With the 2.5 m telescope at Las Campanas we mapped a 14 x 14 arcmin grid around this source with a spacing between points of 2 arcmin. Figure 1 summarizes the data; CO temperature and linewidth contours are shown as well as 1415 MHz-continuum contours (Retallack, 1980) for reference. The latter contours show two peaks to be present. While T_A^*(CO) peaks at an elongated area roughly coincident with the secondary radio peak, the largest linewidths (9 kms^{-1}) are reached at the main peak of the radio emission. This whole region seems to be one of star forming activity, as there are compact IR sources (Frogel and Persson, 1974) and OH 1612 MHz emission (Caswell and Haynes, 1975) associated with the main radio peak. Visual

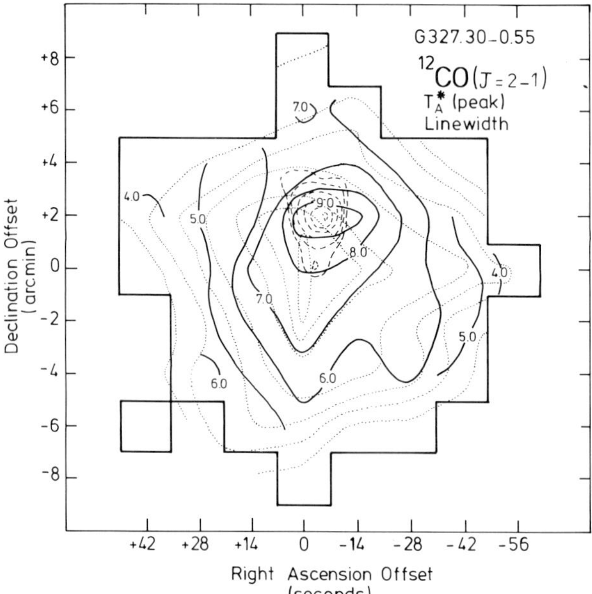

Fig. 1 CO(J=2→1) results

..... T_A^*(CO); contour values 3, 6, 9, 12, 18, 21 and 23 K

——— ΔV(CO); contours labeled in kms^{-1}

----- 1415 MHz contours (Retallack, 1980)

Offset with respect to RA=$15^h49^m16^s$, DEC=$-54^o28'24''$ (1950)

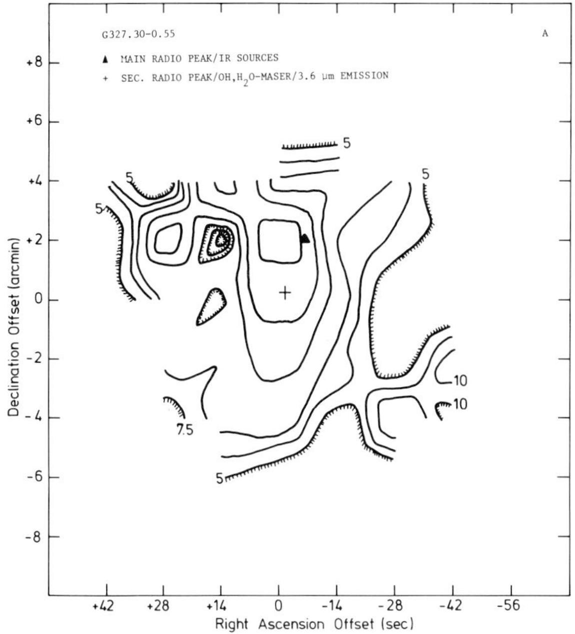

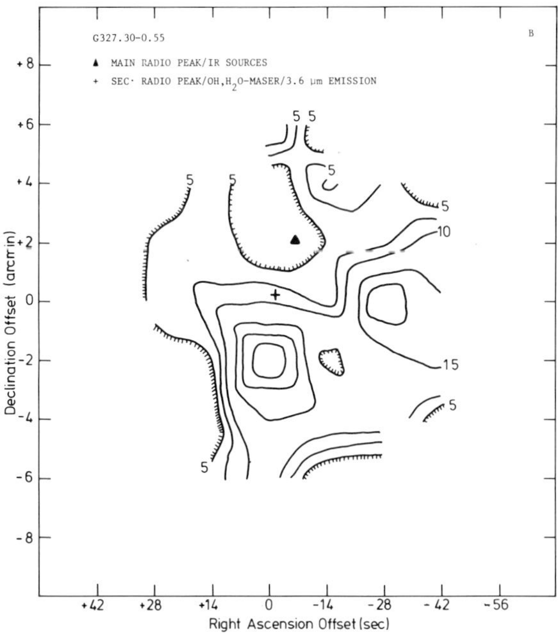

Figure 2 $\int T_A^*(CO)dV$ *A: for blue shoulder B: for red shoulder*
Offset with respect to RA=$15^h49^m16^s$,DEC=$-54^o28'24''$ (1950)

extinction there is at least 20^m (Frogel and Persson,1974; Persson et al.,1976). At the position of the secondary radio peak an OH/H_2O-maser (Caswell et al.,1980; Batchelor et al.,1980) and a 3.6 μm emission source (Epchtein and Lépine,1981) have been found. In the CO data a velocity gradient is present,from -46 kms^{-1} in the SE,to -50 kms^{-1} in the NW. The line profiles obtained are asymmetric,which is illustrated in fig.2a and b. Here we show $\int T_A^* dV$ for the blue and red shoulder of the profiles respectively. The integration is taken from $V_{LSR} \gtrless V(Tpeak) \pm \frac{1}{2}\Delta V$ (where $T_A^* > T_{rms}$). It is clearly seen that the blue and red shoulder emission peak at different positions. Also shown in fig.2a,b are the position of the main radio peak/IR sources (triangle) and that of the secondary radio peak/OH/H_2O-maser/3.6 μm source (cross). Note that the red and blue peaks are located (projected) symmetrically around the latter. This pattern of emission resembles that which is found in other sources that have been identified as being sources exhibiting outflow of gas, from a more or less centrally placed IR source,H_2O-maser or compact HII-region (Rodríguez et al.,1982). Whether G327.30 -0.55 is of this nature cannot yet be established,since mass information is lacking.

(ii) Coalsack

With the ESO 3.6 m telescope we made a 10 point stripscan (plus one additional point) of length ∿50 arcmin through a region in the southern part of the Coalsack,containing Tapia's globules 1,2 and 3 (Tapia,1973). The measured points are indicated in figure 3,shown superimposed on an ESO/SRC print; globule positions are marked with an arrow. This region has been searched for HI absorption by Bowers et al. (1981). The CO results have been condensed in figure 4a and b. First of all we note that there is no significant difference in T_A^* at the globule positions (indicated by the arrows in fig.4a) and off. This may indicate either that CO is so saturated that it makes no difference where one looks,or that the globules add so little extra column density with respect to the surrounding cloud,that no distinction can be made between them and the rest of the cloud (probably both arguments apply here). There is an indication, in the upper panel of fig.4a,that the linewidth at globule 2 is larger than at adjecent points. However,$\int T_A^* dV$ remains roughly constant over the stripscan. In fig.4b we show the position vs. velocity plot. There is a velocity gradient going from globule 3 to globule 1 of 0.6 kms^{-1}. After globule 1 (i.e. to the NE of it),the velocity is constant at -5.6 kms^{-1}. These results agree nicely with what has been found in HI (Bowers et al., 1981). From our T_A^*-values we arrive at a kinetic temperature for this cloud region of 13 K,typical of a dark cloud. This,together with fig.4a, leads to the conclusion that not much seems to be going on in this part of the Coalsack. However, due to undersampling we may have missed small (< 0.4 pc) hot spots.

Note:in the "Dutch CO-group" the following persons collaborate: J.van Amerongen (3),J.van der Biezen (1),J.Brand (2),M.van der Bij (2), T.de Graauw (1),H.Habing (2),F.Israel (1),A.Leene (2),I.Nagtegaal (3), F.Selman (5),H.van de Stadt (3),C.de Vries (2) and J.Wouterloot (4). (1) ESTEC,Noordwijk,(2) Sterrewacht,Leiden,(3) Observatory at Utrecht, (4) ESO,Garching (FRG),(5) Universidad de Chile,Santiago de Chile

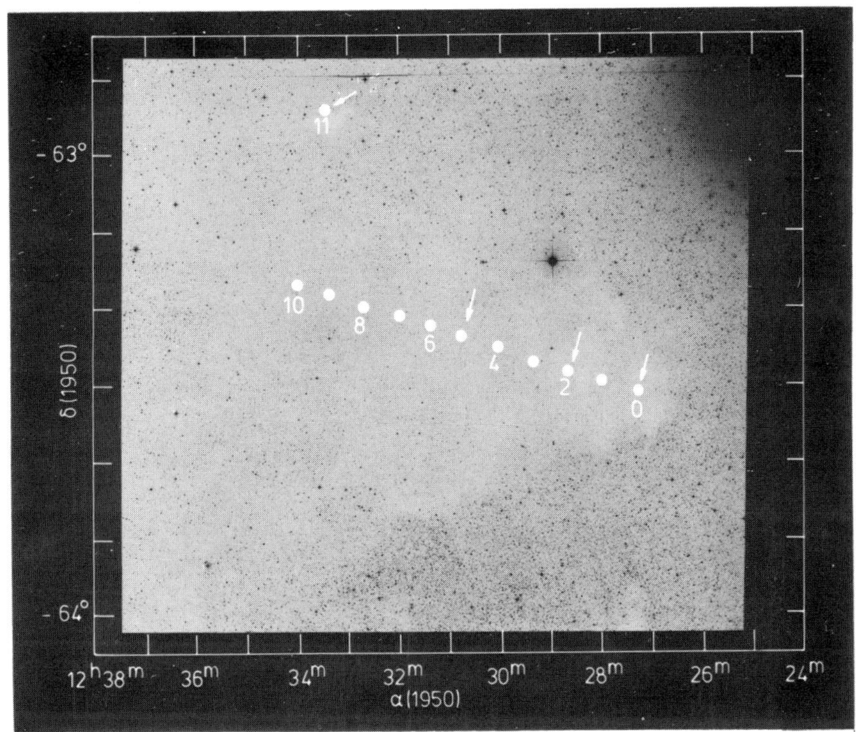

Figure 3 *Part of the Coalsack (from ESO/SRC atlas). Measured points are indicated; arrows mark globule positions.*

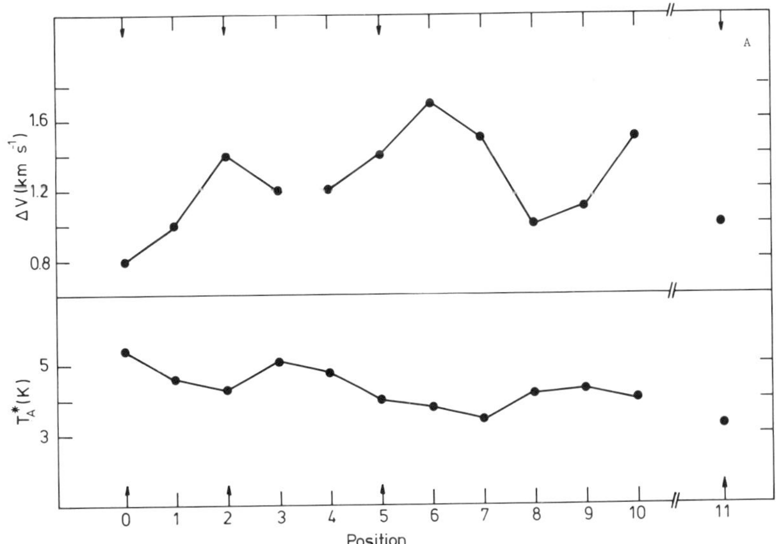

Figure 4 A *Coalsack results. Observed positions are indicated along the abscissa; globule positions are marked by arrows. upper panel: linewidth, lower panel: antenna temperature*

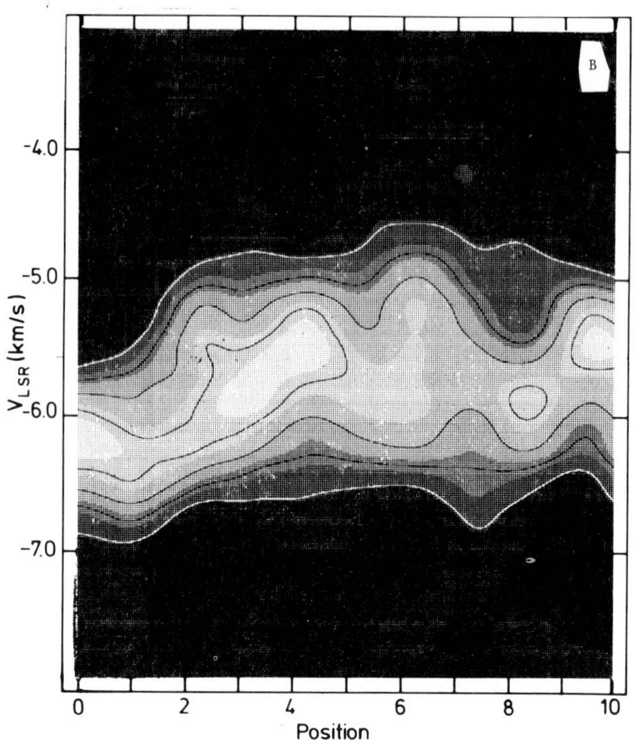

Figure 4 B *Velocity vs. position diagram for the Coalsack. Observed positions are indicated along the abscissa. contour steps 1 K, lowest contour 1 K*

References

Batchelor,R.A.,Caswell,J.L.,Goss,W.M.,Haynes,R.F., Knowles,S.H.,Wellington,K.J.: 1980,Austr. J. Phys. 33,139

Bowers,P.F.,Kerr,F.J.,Hawarden,T.G.: 1981,Astroph. J. 241,183

Caswell,J.L.,Haynes,R.F.: 1975,Mon. Not. R.A.S. 173,649

Caswell,J.L.,Haynes,R.F.,Goss,W.M.: 1980,Austr. J. Phys. 33,639

Epchtein,N.,Lépine,J.R.D.: 1981,Astron. Astroph. 99,210

Frogel,J.A.,Persson,S.E.: 1974,Astroph. J. 192,351

Georgelin,Y.M.,Georgelin,Y.P.: 1976,Astron. Astroph. 49,57

Gillespie,A.R.,Huggins,P.J.,Sollner,T.C.L.G.,Phillips,T.G., Gardner,F.F.,Knowles,S.H.: 1977,Astron. Astroph. 60,221

Lidholm,S.,de Graauw,T.: 1979,in Fourth International Conference on IR and mm-waves and their Applications,Florida,p.App.38 (ed. S. Perkowitz)

Persson,S.E.,Frogel,J.A.,Aaronson,M.: 1976,Astroph. J. 208,753

Retallack,D.S.: Thesis,University of Sydney

Rodríguez,L.F.,Carral,P.,Ho,P.T.P.,Moran,J.M.: 1982 Astroph. J. 260,635

Tapia,S.: 1973,in IAU Symposium 52,Interstellar Dust and Related Topics, p.43 (ed. J.M. Greenberg and H.C. van de Hulst)

STAR FORMATION IN A DUST GLOBULE EMBEDDED IN THE GUM NEBULA

J.A. Graham
Cerro Tololo Inter-American Observatory
Casilla 603
La Serena, Chile

ABSTRACT

New observational data are given for the dark globule ESO 210-6A in which two Herbig-Haro objects were discovered by Schwartz. Images are presented in Hα, [S II] and near-IR light and new velocities as measured from spectrograms taken with the CTIO 4m telescope. Although it has not been possible to locate an obscured source, it seems very probable that in ESO 210-6A we have a good case of bipolar material outflow from a young recently formed star causing shocks which are seen as H-H objects.

ESO 210-6A is a small dust globule seen against and probably embedded within the Gum nebula. Bok (1978) estimates that the total mass is of the order of 25 solar masses. ESO 210-6A is most remarkable, however, for the pair of Herbig-Haro (H-H) objects close to its northern edge. These were discovered by Schwartz (1977) and have been further studied by Dopita (1978). H-H objects have long been known to associate with recently formed stellar objects. However, the field of ESO 210-6A contains only a few emission-line stars characteristic of such a population and Schwartz suggested that ESO 210-6A may be a site of isolated star formation triggered within the cloud by compression from the enveloping, expanding nebula. In this paper, I present some new images of ESO 210-6A taken with narrow band interference filters and, in the near infrared, with a CCD detector. Velocities of various emission knots have also been measured.

Fig. 1. shows the image of ESO 210-6A in Hα light. Details are given in the caption. The two H-H objects, H-H 46 (inside the bright rim) and H-H 47 (outside the bright rim) are clearly seen. They are apparently on the near side of the dark nebula. The two H-H regions are joined by a bridge which also has a bright emission line spectrum. Note that this narrow bridge on passing H-H 47 seems to expand into the surrounding interstellar medium as if a previously operating confinement mechanism no longer applies.

W. B. Burton and F. P. Israel (eds.), Surveys of the Southern Galaxy, 229–232.

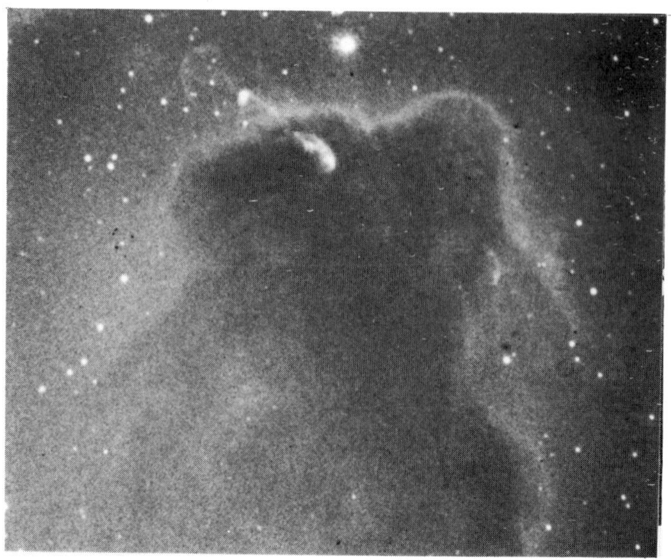

Fig. 1. ESO 210-6A in Hα light. Yale 1m telescope, Carnegie image tube, 17 Å filter, exposure 180 min. North is at top, east to left.

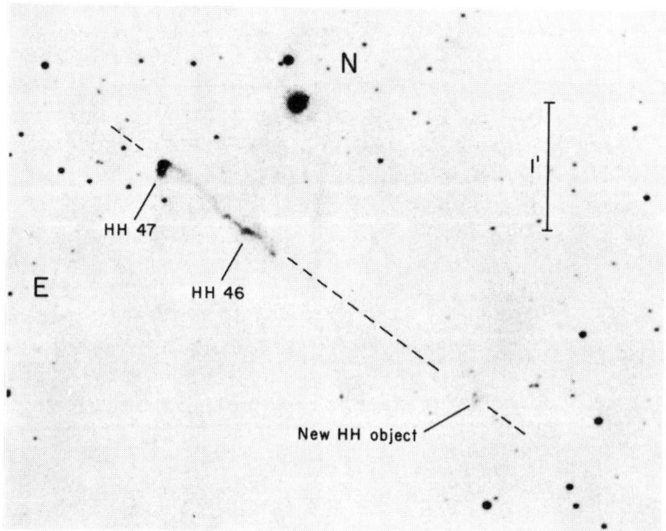

Fig. 2. ESO 210-6A in [S II] light, Yale 1m telescope, Carnegie image tube, 36 Å filter centered at 6727 Å, exposure 180 min. Note the close alignment of the knots of gas which emit strongly in [S II].

Fig. 2 shows in more detail the H-H regions in [S II] light. Prominent in this picture are two new emission regions which lie very close to an axis defined by H-H 46, H-H 47 and their connecting bridge. The new emission regions appear to be on the far side of the dark cloud. Within H-H 46, the [S II] radiation is concentrated into at least 3 knots aligned along this same axis. Neither the bright rim of the globule nor the NE extension of the bridge is especially bright in [S II].

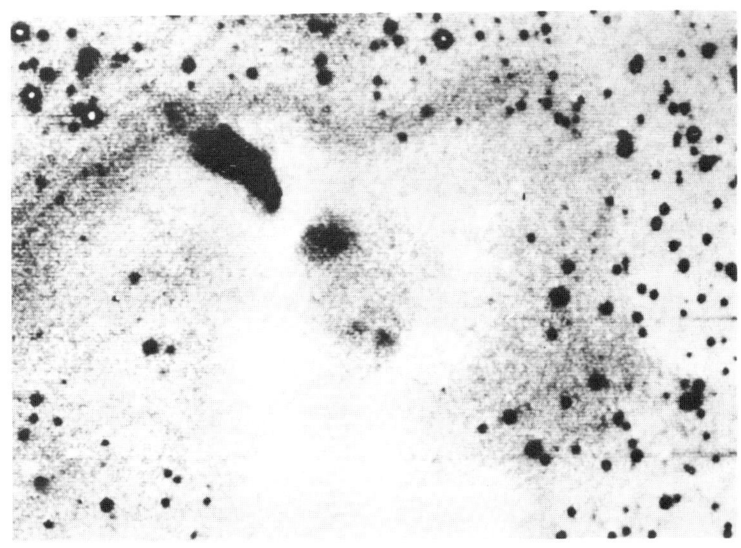

Fig. 3. ESO 210-6A in the near IR. CTIO 4m telescope, CCD camera RG695 filter, exposure 30 min. Note the region of increased IR brightness inside the globule. This is mainly continuum radiation. The brightest part shows weak Hα emission at near-zero velocity.

Fig. 3 shows a negative print of a CCD frame which records essentially near infrared light between 7000 Å and 11000 Å. This frame brings out a number of very red stars. Both H-H 46 and H-H 47 are bright although the linking bridge is weak. An extended, roughly circular area in the center of the globule is illuminated in this image. There is a strong concentration of light on the H-H 46 side of the area. Many very red stars are brought out by this image and comparison with continuum blue images (e.g. the ESO-SRC "J" survey) shows that we are partially but not completely penetrating the dust in the globule at these wavelengths. In fact, the heaviest obscuration appears to be that between the illuminated central region and H-H 46.

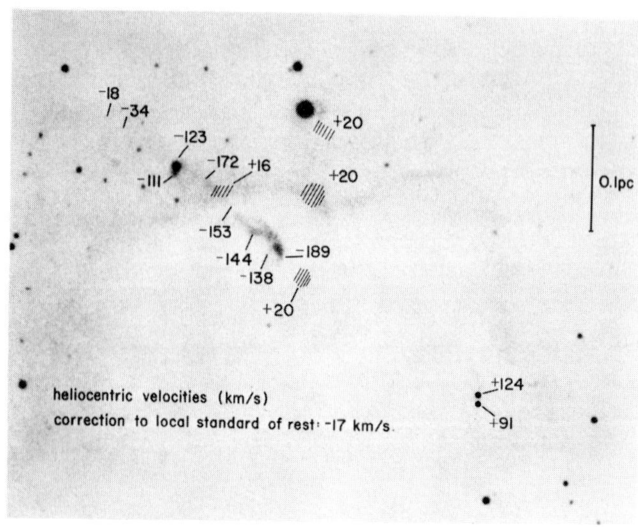

Fig. 4. Heliocentric velocities in ESO 210-6A. In marking the scale a distance of 400 pc is assumed.

Fig. 4 shows a velocity map of the region. The velocities in Fig. 4 are based on spectrograms taken with the 4m telescope at CTIO and its cassegrain spectrograph. Details will be published elsewhere. While the bright rim of the globule and the brightest part of the central area have low velocities close to that of the surrounding Gum nebula (Reynolds 1976), H-H 46 and H-H 47 have high negative velocities and the two new emission knots a high positive velocity.

In ESO 210-6A, there seems to be a good case for the bipolar outflow of gas from a young, recently formed star which is still obscured by surrounding dust. Searches have been made for such a source with the CTIO 4m telescope in the IR at 1.2μm, 1.6μm and 2.2μm by J. Elias without success (Graham and Elias, in preparation). The possibility remains that the source may be coincident with the SW tip of H-H 46 itself but it should be emphasized that we do not see a point radiation source here either in the optical or the IR. The bipolar outflow apparently forms conical cavities on both sides of the source. Along the axes of the cavities, Herbig-Haro objects are formed. The H-H regions cease to be stable once they pass out of the environment of the dust cloud. They interact with the surrounding interstellar medium, slow down and quickly diffuse.

REFERENCES

Bok, B.J.: 1979, P.A.S.P. 90, p. 489.
Dopita, M.A.: 1978, A. & A. 63, p. 237.
Reynolds, R.J.: 1976, Ap. J. 203, p. 151.
Schwartz, R.D.: 1977, Ap. J. 212, p. L25.

CO OBSERVATIONS OF A SAMPLE OF HII REGIONS IN THE SOUTHERN HEMISPHERE

A.R. Gillespie
Max-Planck-Institut für Radioastronomie, Bonn, West Germany

INTRODUCTION

This is a review of both the initial CO, J = 1 - 0, survey of southern HII regions and the subsequent mapping work on regions selected from this survey, carried out by observers from Queen Mary College, London and CSIRO, Sydney. These results are used to show which of the observed HII regions are particularly suitable both for further line work and work at other wavelengths, particularly in the infra-red.

OBSERVATIONS

The observations were made using an InSb mixer mounted at the Coudé focus of the Anglo Australian Telescope. The receiver has a single 2.6 km s^{-1} wide channel which was either frequency stepped to give spectra, or switched between a line and reference frequency whilst scanning the telescope across the source to give maps typically $1^o \times 0.5^o$ with full angular sampling. Full details of the techniques are given in Gillespie et al. (1977 and 1979), called Papers I and II.

SURVEY RESULTS

The survey is given in Paper I and can be summarized as follows: 41 HII regions were observed. 20 of the 28 sources with $\delta < -40^o$ and $T^{Cont}_{5\ GHz} > 5$ K were detected; of the remainder, 3 have upper limits and 5 were not observed. Excluding G 284.3 - 0.3, which has $v_{CO} - v_{H\ 109\alpha} = 16.7$ km s^{-1} there is good velocity agreement between the CO and other lines with $\langle v_{CO} - v_{H\ 109\alpha} \rangle = 0.1 \pm 0.43$ km s^{-1} and $\langle v_{CO} - v_{H_2CO} \rangle = 0.6 \pm 0.47$ km s^{-1}.

CO data has now become available from the CSIRO 4 m telescope (Whiteoak et al. 1982) and there is good agreement between the two sets of data with $\langle T^*_A(AAT)/T^*_A(CSIRO) \rangle = 1.07$. Furthermore, the largest discrepancies can be explained by beam dilution effects or line broadening due to the different instrumentation and also a suspected systematic temperature error for 13 of the sources in Paper I, as stated in that paper.

W. B. Burton and F. P. Israel (eds.), Surveys of the Southern Galaxy, 233–234.

(7 of these being between $350^o < l^{II} < 354^o$).

INDIVIDUAL SOURCES

G 267.9 - 1.1 (RCW 38), G 316.8 - 0.1 and G 348.7 - 1.0 (RCW 122) have been mapped (Paper II). White and Phillips (1982) have shown that the CO 2-1 line in G 267.9 - 1.1 shows self absorption and this is the most interesting of these sources for that reason.

The giant molecular cloud near $l^{II} = 333^o$ and the Carina Nebula are discussed in a separate paper and warrant much further work.

G 305.2 + 0.0, G 305.1 + 0.1 and G 305.4 + 0.2 are in a semicircular ring of CO emission. The diameter of the ring is 16 arcmin and there are several regions of enhanced emission both around the ring and outside it. There is not yet sufficient data for this source to give a definitive explanation for the ring.

In addition to these, attention should be paid to some other unmapped sources: G 284.3 - 0.3 with its large value of $v_{CO} - v_{H\ 109\alpha} = 16.7$ km s^{-1}; G 291.6 - 0.5 has only an upper limit for CO emission although it is one of the most powerful thermal radiosources in the galaxy. It should also be mentioned that virtually all sources mapped showed additional CO peaks near them.

CONCLUSION

The two existing CO J = 1 - 0 surveys in the southern sky have been shown to be in very good agreement. The following sources, in particular, need further work by other observers: the Carina nebula; the area at $l^{II} = 333^o$; the CO ring near G 305.2 + 0.0; G 267.9 - 1.1; G 284.3 - 0.3; G 291.6 - 0.5.

ACKNOWLEDGEMENTS

All the above work was carried out in collaboration with the authors of Papers I and II.

REFERENCES

Gillespie, A.R., Huggins, P.J., Sollner, T.C.L.G., Phillips, T.G., Gardner, F.F., and Knowles, S.H.: 1977, Astron. Astrophys. 60, 221

Gillespie, A.R., White, G.J., and Watt, G.D.: 1979, Mon. Not. R. astr. Soc. 186, 383

White, G.J. and Phillips, J.P.: 1982, in press

Whiteoak, J.B., Otrupcek, R.E., and Rennie, C.J.: 1982, Private Communication

THE GIANT MOLECULAR CLOUDS AT $\ell = 333^o$ AND IN THE CARINA NEBULA

A.R. Gillespie
Max-Planck-Institut für Radioastronomie, Bonn, West Germany
G.J. White
Physics Department Queen Mary College, London, England
G.D. Watt
Mathematics Department, UMIST, Manchester, England

INTRODUCTION

These two clouds have been studied in the J = 1-0 transition of CO and were selected from the list of Gillespie et al. (1977) as suitable for detailed examination because they are two of the most interesting areas in the southern sky. Further details and analysis will be published elswehere as this paper gives only an outline of the results.

Observations were made using the Anglo Australian Telescope and the techniques described fully in Gillespie et al. (1979). Data were collected as: spectra; spatial scans up to 2° long in one frequency channel, 2.6 km s^{-1} wide; and maps made by taking successive scans displaced from each other by about one beamwidth (3!2). Both areas are too large to allow mapping at all velocities but most of the features of the CO clouds are clear from these data and further observations can be based on this work. Some work in the J = 2-1 CO transition has also been done by de Graauw et al. (1981) with higher resolution, but incomplete sampling.

THE GIANT MOLECULAR CLOUD AT $\ell = 333^o$

The extent and velocity of this cloud were first noted by Gillespie et al. (1977). It is located at a distance of 4.2 kpc with CO emission over an area of at least 100 x 20 pc with four major emission peaks along its major axis and a systematic change in their mean CO velocity from V_{lsr} = -55 to -51 km s^{-1}. There are numerous radio continuum peaks in this complex, of which at least six are infrared sources: G333.6-0.2, G333.3-0.4, G333.1-0.4, G333.0-0.4, G332.8-0.6 and G332.7-0.6 (Furniss et al. 1975). The first of these is the strongest IR source, whilst the last two are usually associated with the Hα emission region RCW 106 although the cloud is in a direction of heavy optical obscuration.

Spectra at several positions, scans at seven velocities from -60 to -47 km s^{-1} along the major axis and others at 90° to this line were taken. Fully sampled maps were then made of most of the cloud at velocities of -55, -53.5 and -52 km s^{-1} and the first of these is shown as Figure 1. The

W. B. Burton and F. P. Israel (eds.), Surveys of the Southern Galaxy, 235–238.

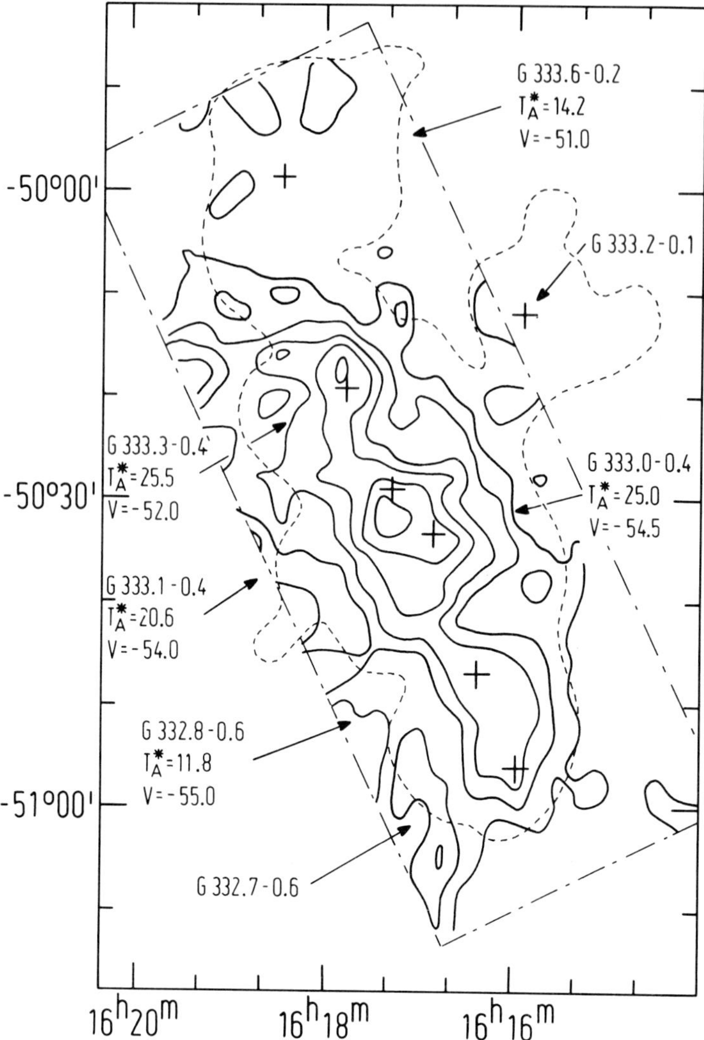

Figure 1. CO emission from the ares around ℓ = 333° at a velocity V_{lsr} = -55 km s^{-1} with a channel width of 2.6 km s^{-1}. The contour interval is 4 K (∿2 σ); the zero and negative contours are suppressed. Crosses mark the positions of radio peaks and the dotted contour shows the extent of the 5 GHz radio emission from Goss and Shaver 1970. CO peak intensities and velocities are given as is the edge of the CO map area.

data show that there is CO emission with T_A > 5 K covering the same area as the 5 GHz radio continuum emission given in Goss and Shaver (1970) except in the direction of G333.2-0.1 which does not lie in the main chain formed by the other radio sources. The H109α line from this source has V_{lsr} = -90.8 km s^{-1} (Wilson et al. 1970) so it is probably a more distant HII region. The strongest CO emission (T_A^* = 25 K) comes from two peaks coincident with G333.0 -0.4 and G333.3-0.4. The brightest IR source,

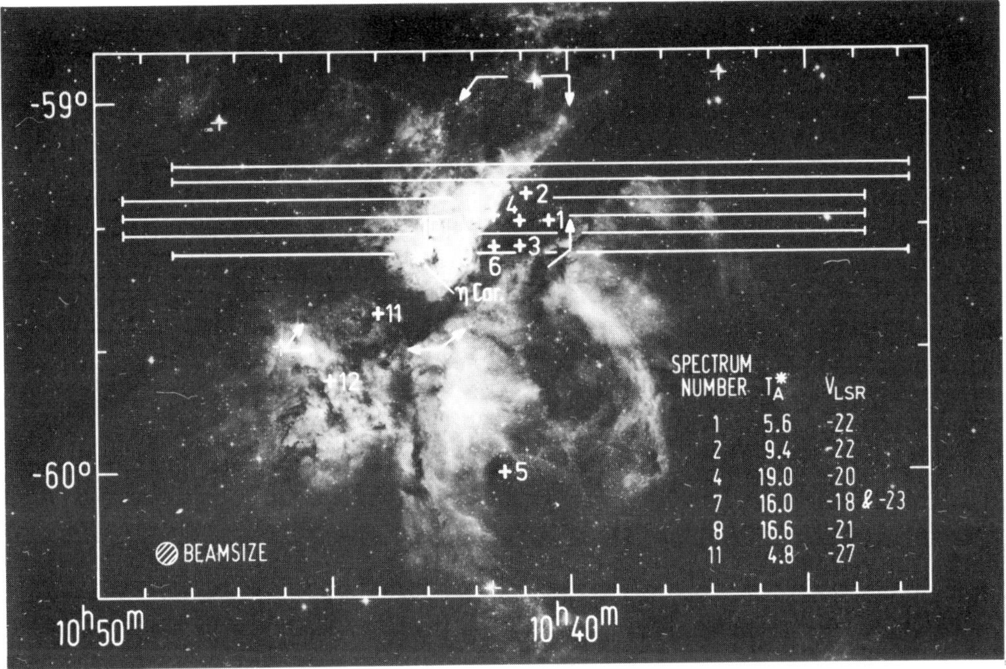

Figure 2. The Carina Nebula in blue light. The 6 constant-declination scans are shown by the horizontal lines; the crosses and numbers refer to spectra. Arrows indicate the approximate boundaries of the map from de Graauw et al. (Photo courtesy of ESO).

G333.6-0.2, contains a very luminous stellar cluster, L = 3.3 x 10^6 $L_\odot$, (Hyland et al. 1980) but it only has T_A^* = 14.5 K, lower than might be expected. There are no serious disagreements with the 2-1 map of de Graauw et al. although their noise level is higher.

The usual picture of sequential star formation such as Elmegreen and Lada (1977) does not fit this cloud well as the CO is in four bright sources, which shows that there are at least this number of active sites separated by distances of about 20 to 30 pc. A mechanism such as triggering by the passage of a density wave through the cloud must be invoked. It could be argued that the whole cloud is a rotating disc seen edge-on, but the long rotation period, a few times 10^8 years, makes this model unlikely.

THE CARINA NEBULA

This prominent optical nebula covers an area of about 2 degrees square, 94 pc by 94 pc, and has been mapped in the radio continuum at a variety of frequencies and several molecular transitions with low angular resolution by several groups (e.g. Goss and Shaver 1970, Gardner et al. 1973, Dickel and Wall 1974). It also contains several young star clusters and two strong infrared sources, Carina I and II (Harvey et al. 1979) and the very unusual star η Carina.

Observations consisted of twelve spectra taken at various positions in the nebula to determine the general extent of the CO emission. At six of these positions (Nos. 3, 5, 6, 9, 10 and 12 in Figure 2) upper limits of $T_A^* < 4$ K were obtained. A series of six right ascension scans were made in the northern part of the nebula at a velocity $V_{lsr} = -18$ km s^{-1}, all 2° long with a 3' separation in declination, centered on $\delta = -59^o17'$. The scans show that there is CO over most of the northern and western part of the nebula and extends over a larger area than that observed by de Graauw et al. (1981). There is broad agreement between the two sets of data in the area in common, showing that the CO is optically thick.

The most northern of the scans show an association between the gas at -18 km s^{-1} and the prominent dust lanes, with more complex structure in the south. There is enhanced CO emission near $\alpha_{1950} = 10^h37^m$, $\delta_{1950} = 59^o22'$ correlating with optical filaments in the west of the nebula. There is also a compact CO source at $\alpha = 10^h42'$, $\delta = -59^o19'$ not seen by de Graauw et al., almost certainly due to their incomplete spatial sampling. Two adjacent spectra (Nos. 7 and 8) give $T_A = 15$ and 16 K; one (9) 3' south of these gives an upper limit of 4 K. This source is only 4' (3 pc) away from the IR source Carina I and must be heated by it. The gas could be either another region of star formation similar to Carina I or a denser remnant of the larger cloud from which Carina I formed.

CONCLUSION

There are very large molecular clouds in both areas which deserve extensive further work. The cloud at $\ell = 333^o$ is difficult to reconcile with the conventional pictures of sequential star formation.

REFERENCES

de Graauw, T., Lidholm, S., Fitton, B., Beckman, J., Israel, F.P., Nieuwenhuizen, H., Vermue, J.: 1981, Astron. Astrophys. 102, 257

Dickel, H.R. and Wall, J.V.: 1974, Astron. Astrophys. 31, 5

Elmegreen, B.G., Lada, C.J.: 1977, Astrophys. J. 214, 251

Furniss, I., Jennings, R.E., Moorwood, A.F.M.: 1975, Astrophys. J. 202, 400

Gardner, F.F., Dickel, H.R. and Whiteoak, J.B.: 1973, Astron. Astrophys. 23, 51

Gillespie, A.R., Huggins, P.J., Sollner, T.C.L.G., Phillips, T.G., Gardner, F.F., Knowles, S.H.: 1977, Astron. Astrophys. 60, 221

Gillespie, A.R., White, G.J., Watt, G.D.: 1979, Mon. Not. R. Astr. Soc. 186, 383

Goss, W.M., Shaver, P.A.: 1970, Austr. J. Phys. Suppl. 14, 1

Harvey, P.M., Hoffmann, W.F., Campbell, M.F.: 1979, Astrophys. J. 227, 114

Hyland, A.R., McGregor, P.J., Robinson, G., Thomas, J.A., Becklin, E.E., Gatley, I., Werner, M.W.: 1980, Astrophys. J. 241, 709

Wilson, T.L., Mezger, P.G., Gardner, F.F., Milne, D.K.: 1970, Astron. Astrophys. 6, 364

A HIGH RESOLUTION HI SURVEY OF M31

E. Brinks
Leiden Observatory, Leiden, The Netherlands

1. INTRODUCTION

HI surveys of our own galaxy have reached a high level of completeness at a relatively high resolution, both spatially and in velocity. But still many questions about the spiral structure of our galaxy, the presence (or not) of a bar, the z-distribution of material, and the detailed structure of its interstellar medium remain unanswered. Even the recent CO surveys although promising and important, have not yet resulted in a major breakthrough, the principal problem being our unfavourable location buried within the system.

The new high resolution 21-cm HI line survey of M31 made with the Westerbork Synthesis Radio Telescope covers the galaxy almost completely at a linear resolution of 80 x 120 pc. This is of the same order as the resolution at the distance of the center of our own galaxy given by a 25-m dish. The M31 survey is consequently comparable to surveys of the Milky Way in wealth of detail as well as in amount of data ($\sim$ 1 Gigabyte), but it has the advantage of being taken from a better perspective. Table 1 gives some general data on M31. Table 2 lists the observational parameters of the survey. The survey is complete except for the extreme northern and southern parts.

Table 1: General data on M31

Hubble type	Sb
Distance (kpc)	690
Holmberg radius (kpc)	21
HI radius (kpc)	30
Position angle	38°
Inclination	74°
Systemic vel. (km s^{-1})	-310

Table 2: Observational parameters

Mean HPBW	24" x 36"
System temperature (K)	90
R.m.s. noise at field centre (K)	2
Flux equivalent of 1 K (mJy)	1.39
Interferometer spacings (m)	0(18)1458
Velocity range (km s^{-1})	-618.1(4.1)-16.2
Velocity resolution (km s^{-1})	8.2

W. B. Burton and F. P. Israel (eds.), Surveys of the Southern Galaxy, 239–242.

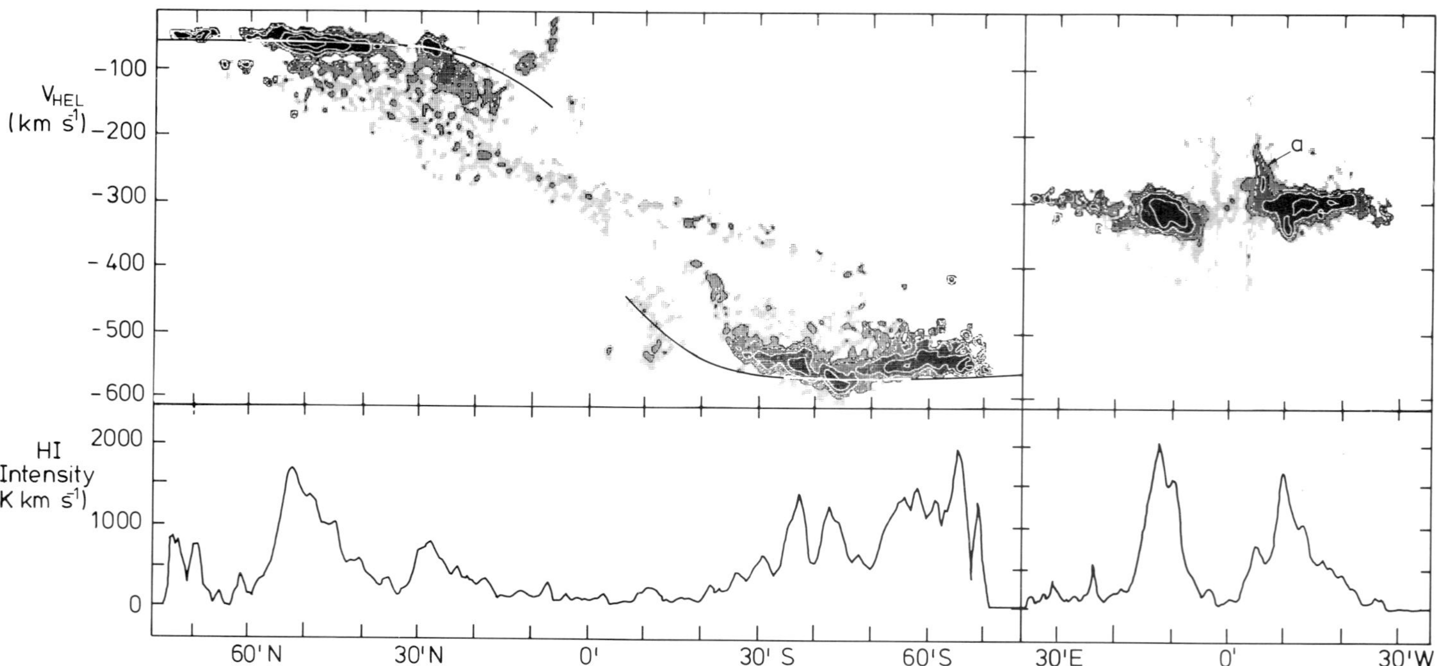

Figure 1: upper left: Position-velocity map along the major axis. Contour levels are at 2.5, 5, 10, and 25 K. The thick line is based on the mean rotation curve used by Bajaja and Shane (1982).
upper right: Position-velocity map along the minor axis, same contour levels as above.
lower left: HI intensity profile along the major axis. Due to the limited coverage of the galaxy, the noise increases rapidly at the ends of the profile.
lower right: HI intensity profile along the minor axis.

2. OBSERVATIONS AND DATA REDUCTION

The observations were made in 1978 and 1979 using the WSRT in its 1.5 km configuration and equipped with the new digital line back-end. In total five fields were used to cover M31 out to a 15 kpc radius measured along the major axis. In the observations which cover the northern part, HI from the Milky Way with velocities between about 0 and -100 km s^{-1} enters the survey. The amount of such contamination is quite appreciable because of the low galactic latitude of M31 ($b \simeq -21^{o}$). This foreground HI also affects the calibration. This was accounted for during data reduction.

The continuum background in the line observations was removed by subtracting a separate 21-cm broadband continuum survey (Walterbos, Brinks, and Shane, in preparation). The low spatial frequencies, 0-m and 18-m, which are absent in the interferometer data were Fourier filtered from the survey of M31 by Cram, Roberts, and Whitehurst (1980) using the 100-m Effelsberg telescope, except at those velocities, where we have galactic foreground HI. The latter maps were corrected by extensive use of the CLEAN technique (Högbom, 1974).

The final step in the reduction was the combination of the five survey areas to one big field. At each velocity interval the maps of the five fields were interpolated onto one common grid and combined, correcting them at the same time for the primary beam attenuation. This resulted in a set of 147 channel maps, each covering M31 at a 4.1 km s^{-1} interval. A few sample maps are presented by Shostak and Brinks (this volume). In each channel map the points which contain no emission above a certain level were set to zero to suppress the noise, when combining the maps. This level was set at two times the rms noise after smoothing to four times the original beam size. Allowance is made for continuity in successive channels and for the fact that the noise in neighbouring channels is correlated.

3. RESULTS

The maps produced this way were added to give a map of the total HI surface density at full resolution, shown in the fold-out reference map published elsewhere in these proceedings. The noise varies across the map and increases rapidly at the extreme northern and southern ends. Only HI within a rectangular area of 140' x 50' is shown. As can be seen from the figure the HI is concentrated in a ring at about 10 kpc. This main ring coincides with the ring of radio continuum emission, of HII-regions, of OB-associations, etc. The HI shows much fine scale structure. There is also a strong correlation of HI with dust.

To improve the signal to noise ratio for weak extended features, the maps were also smoothed to twice the original beam size. Figure 1 shows a set of position-velocity maps along the principal axes of M31 based on these maps (upper panel) and two intensity profiles along the same lines through the total surface density map, also at a lower resolution (lower panel). The intensity profiles show very well the asymmetry between the northern and southern halves of M31 and between the eastern and western sides. Several arm crossings can be seen along the major axes, but on the minor axes, especially on the far, i.e. East side of M31, the arms are blended. The arm separation is of order 4 kpc along the major axis, but due to the unfavourable inclination no firm statements can as yet be made about spiral structure, although it is possible to trace over several kiloparsecs arm segments or near circular arm-like structures.

The major axis position-velocity map gives a first order approximation of the rotation curve of the system. The thick line is a rotation curve based on previous HI data. As in our own galaxy, the kinematics is quite complex. Three velocity systems can be seen, one being the expected disk in differential rotation, which roughly follows the thick line. Just as in our own galaxy the disk in M31 is warped. This warped component shows up in the figure as HI running diagonally across the map along an almost straight line.

The third component is restricted to the inner 4-5 kpc, where velocities as high as 200 km s^{-1} with respect to the systemic velocity are found. The minor axis position-velocity map shows that there is a steep velocity gradient on both sides of the nucleus. Figure 1 and similar maps support the idea that the high velocities are due to rotation rather than produced by infall of material or ejection from the nucleus. The latter two hypotheses are unlikely because the central radio source in M31 is intrinsically an order of magnitude weaker than the one in our own galaxy and because it shows no signs of recent activity. It has been demonstrated that the isophotes of the bulge show a gradual change in ellipticity and position angle (Matsumoto _et al._, 1977; Lindblad, 1956). This suggests a model in which the gas flows along elliptical stream lines, perhaps even tilted out of the plane, much like the distribution proposed for the central region of our own galaxy.

Feature "a", which is indicated in the minor axis plot, lies at about 5 kpc and coincides with the prominent dust arm North-East of the nucleus. This arm, which has its counterpart also measured in CO (Stark, 1979), was already discussed by Bajaja and Shane (1982), who point out the resemblance between this arm and the galactic 3-kpc arm.

REFERENCES

Bajaja, E., Shane, W.W.: 1982, _Astron. Astrophys. Suppl. Ser._ _49_, 745
Cram, T.R., Roberts, M.S., Whitehurst, R.N.: 1980, _Astron. Astrophys. Suppl. Ser._ _40_, 215
Högbom, J.A.: 1974, _Astron. Astrophys. Suppl. Ser._ _15_, 417
Lindblad, B.: 1956, _Stockholm Obs. Ann._ _19_, No. 2
Matsumoto, T., Murakami, H., Hamajima, K.: 1977, _P.A.S. Japan_ _29_, 583
Stark, A.A.: 1979, Ph.D. Thesis, Princeton University
Walterbos, R.A.M., Brinks, E., Shane, W.W.: in preparation

THE PRODUCTION OF A 16-MM FILM OF M31

G.S. Shostak
Kapteyn Astronomical Institute
Groningen, The Netherlands

and

E. Brinks
Leiden Observatory
Leiden, The Netherlands

1. INTRODUCTION

The new high resolution HI survey of M31 (Brinks, this volume) was effected at a resolution of $\Delta\alpha \times \Delta\delta \times \Delta V = 24'' \times 36'' \times 8.2$ km s^{-1}. These measures comprise a data cube consisting of 147 channel maps, separated in velocity by 4.1 km s^{-1}. Each of the maps is 1024 x 1024 pixels in size. Interpretation of this very large data set is not only extremely time consuming, but also difficult because the information is so widely distributed over the coordinate space. It is especially hard to see continuity along the velocity axis or along a single coordinate axis. Simply presenting the data is a formidable task.

For these reasons we chose to make a record of the data set on 16-mm film. Although facilities were available to quickly make video recordings of the maps, we preferred a film record as it is more transportable than video, being of a standardized format.

2. HARDWARE DESCRIPTION

All data reduction was performed by Brinks at Leiden Observatory. The processed data were then brought to Groningen where they were displayed and photographed using the GIPSY image processing system (Shostak and Allen, 1979). This facility consists of an International Imaging Systems Model 70 image processor connected to a PDP 11/70 host computer. Westerbork data were available as intensity maps of size 512 x 512 or 256 x 256 pixels, written in the form of 32-bit real values on magnetic tape. For convenience of access, it was first necessary to transfer these maps to disk. Several PDP RPO 5 (88 Mbyte) disk packs were used for this purpose. Maps on disk could be loaded into one of the Model 70's 512 x 512 x 8-bit deep display memories, which could then be viewed on a video monitor after digital-to-analogue conversion. The analogue output signal could also be recorded on a video disk recorder, which had the capacity to store up to 300 monochrome images. The quality of the recorded images was virtually indistinguishable from the originals. Playback speed of the

W. B. Burton and F. P. Israel (eds.), Surveys of the Southern Galaxy, 243–246.

video disk was selectable and ranged from one frame per four seconds up to standard video rate (25 frames per second).

One of the fundamental decisions which had to be made concerning the production of the film was whether to shoot frame by frame or in "real time", using the video disk in playback mode. The former method (similar to conventional animation techniques) allowed exposure times of one second, thus forestalling the appearance of a hum bar in frame. Shooting real time was obviously faster (especially since many of the sequences in the film were repetitive), but required synchronization of video disk and film camera. (Note that the disk is not synchronized to the power line frequency). Further any jitter in the video disk would degrade the image quality.

For reasons of speed, it was decided to first load map sequences onto the video disk, and then shoot in real time upon playback. Although most sequences were photographed at a playback rate of 12.5 maps per second, the video display rate was always 25 frames per second. (In this case each map was simply displayed twice.) A once-per-frame synchronization pulse was output from the disk and used to drive a synchronous motor mounted on a 16-mm Bolex film camera. The camera thus also ran at approximately 25 frames per second. The relative phase between the video image and the opening of the camera shutter could be mechanically adjusted. A piece of translucent paper was introduced into the film gate and observed while this phase was varied until the hum bar was no longer visible. In this way the camera was once-and-for-all synched to the video disk.

Since exposure times were short (0.02 sec), a reasonably sensitive emulsion was demanded. We used Eastman type 7250 high speed video news film, daylight type, which has an ASA of 250. This is a reversal color stock; however most of the sequences of the film were photographed from a high resolution black and white monitor.

3. DISPLAY CONSIDERATIONS

All maps had to be numerically scaled to the 8-bit range of the display memories. Since the dynamic range of the data was approximately 20 dB, this presented no problem. However, because of the transfer characteristics of the monitors, the range of displayed brightnesses was not 256 to 1, but rather 256^{γ} to 1, where $\gamma \simeq 2.4$. Most reversal films have a density range over the linear part of their characteristic curve of 2 to 3, or 20 to 30 dB brightness. Since we intend to make a print of the original film for projection purposes, a procedure which normally increases contrast, we tried to keep the brightness range of the images below 20 dB. To this end, special software was used to reduce image contrast by effectively taking the $(1/\gamma)$-th root of display memory values.

Other software was constructed to document images with crosses, text, etc. Further, some maps were displayed in false color, where the

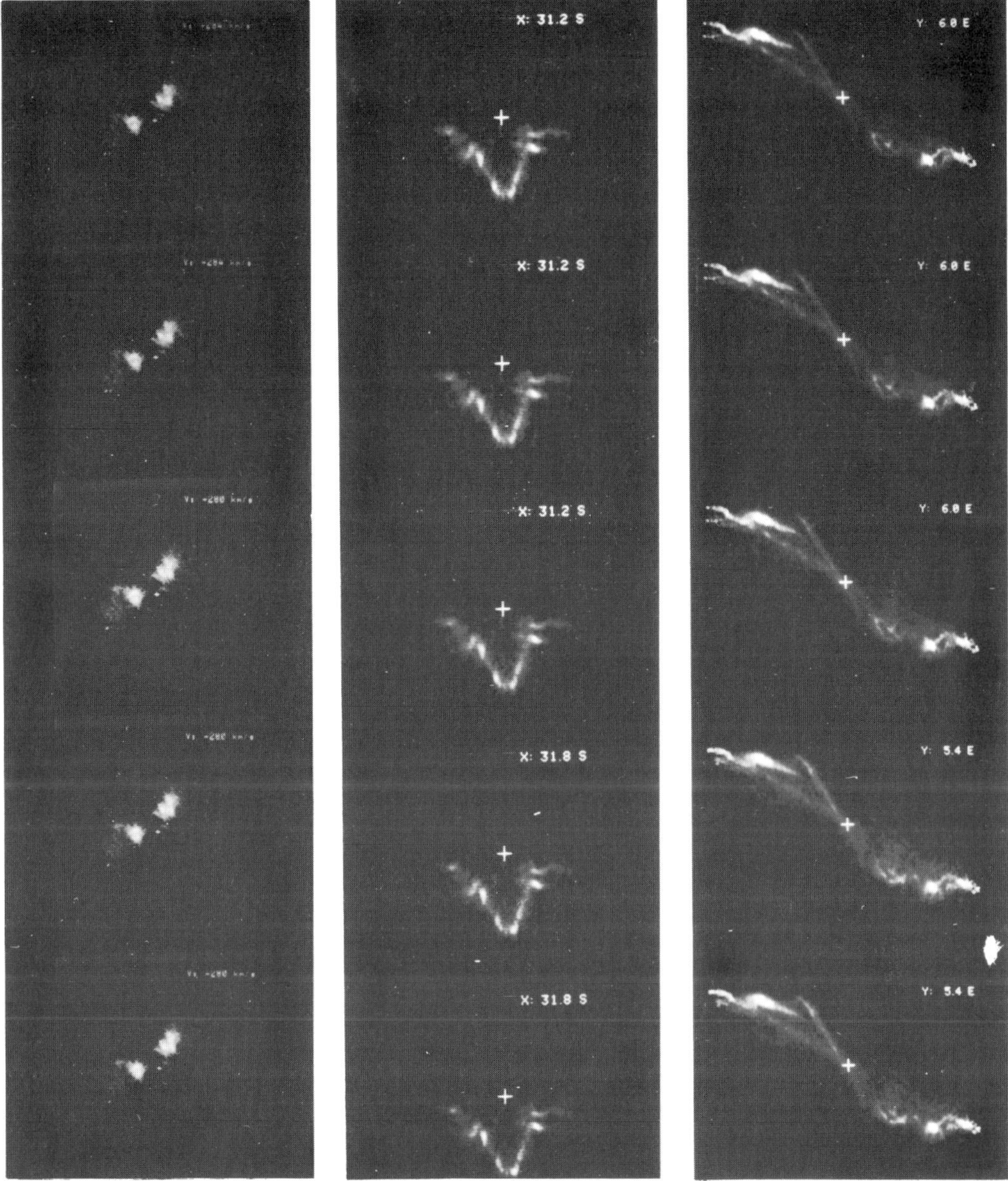

Figure 1: (<u>left</u>): Sample frames from the sequence of channel maps. Field of view: $2^{\circ}.15 \times 2^{\circ}.15$. The velocity is indicated in the upper right hand corner; the nucleus is indicated by a + sign. (<u>middle</u>): Position-velocity maps perpendicular to the major axis through channel maps smoothed to 48" x 72". The width is 76'.8; East is to the left. Ordinate: velocity ranging from -618.1 (bottom) to -12.1 km s^{-1}. The + sign indicates the major axis crossing at -317.2 km s^{-1}. X gives the distance from the minor axis in arcmin. (<u>right</u>): As above, but now parallel to the major axis. The width is 153'; North is to the left. The + sign indicates the minor axis crossing at -317.2 km s^{-1}. Y gives the distance from the major axis in arcmin.

hue was determined by intensity values in the same or another map. Titles describing the image sequences were made using conventional cine techniques.

The final printed film shows the full dynamic range of the data and has a resolution comparable to the original images.

4. FILMED SEQUENCES

The film primarily consists of three sequences, each of which shows the data cube along one of its principal axes. A sequence is repeated four to five times at a relatively high frame rate and then three times at a lower rate. The recordings are in black and white as we feel it is easier to translate gray scale levels into intensity levels, and because it appears less "noisy".

The first sequence shows the channel maps at full resolution, with a velocity range from -618.1 to -16.2 km s^{-1} in steps of 4.1 km s^{-1}, i.e. at half the velocity resolution. The second series consists of position velocity maps, made along a line perpendicular to the major axis. The position velocity maps were made after the channel maps had been smoothed to twice the original spatial resolution in order to enhance the extended low level emission. The film shows this sequence of position velocity maps, each separated by 0.6 arcmin, starting about 70 arcmin north of the nucleus. The last sequence comprises position velocity maps made parallel to the major axis, also at 0.6 arcmin intervals, running from east to west of the nucleus.

In the figure is shown a set of sample frames for each of the three sequences. In addition to the sequences described here, the film also contains static shots of the mean velocity field of M31 and the total HI surface brightness maps. Some of these are in color.

The duration of the film is approximately 8 minutes. Copies may be obtained at cost (about 250 Dutch guilders) from E. Brinks.

REFERENCE

Shostak, G.S., Allen, R.J.: 1979, in "ESO Workshop on Two Dimensional Photometry" (ed. P. Crane and K. Kjar; Noordwijkerhout)

HI STRUCTURES IN M31

E. Brinks and E. Bajaja*
Leiden Observatory, Leiden, The Netherlands
*On leave from the Instituto Argentino de Radioastronomia, Argentine.

1. INTRODUCTION

Recent high resolution HI observations of M31 made with the Westerbork Synthesis Radio Telescope shed new light on the structure of the interstellar medium in a galaxy supposedly similar to the Milky Way. Many filaments, arcs, and shell-like structures are found delineating regions which are devoid of neutral hydrogen. These regions, which we shall in the rest of this paper refer to as holes, are very similar to the supershells in our own Galaxy discovered by Heiles (1976, 1979) and the supergiant shells in the Large Magellanic Cloud described by Meaburn (1980).

At present we are preparing a list of some 250 holes found in our HI survey. The analysis is now concentrated on checking if there is a correlation of holes with OB-associations, HII-regions, etc. Heiles (1979) found a weak correlation of supershells with OB-associations, but his search was of course hampered by interstellar extinction. A large hole in M31 has been discussed already by Brinks (1981). This hole coincides with the bright OB-association NGC 206 (OB78 of van den Bergh, 1964), and shares many characteristics with LMC 1, 4, and 5 from Meaburn (1980).

2. A CATALOGUE OF HI HOLES

The maps which are used to search for holes in the HI distribution are those described by Brinks (this volume). The observations were done at a resolution of $\Delta\alpha \times \Delta\delta \times \Delta V = 24'' \times 36'' \times 8.2$ km s^{-1}. The angular resolution corresponds to a linear resolution of 80 x 120 pc at the assumed distance of M31 of 690 kpc. In total 147 channel maps, separated by 4.1 km s^{-1}, cover the whole velocity range of HI over most of the optical image of M31. A natural way to look for holes in the HI distribution would seem to be an examination of the total surface-density map. In fact, this map is rich in all kinds of features: filaments, shells, large and small holes, unresolved bright spots (see the fold-out reference

W. B. Burton and F. P. Israel (eds.), Surveys of the Southern Galaxy, 247–251.

map elsewhere in these proceedings). But in this map all velocity components along the line of sight are simply added, so a hole which is present in one velocity structure in M31, for instance the differentially rotating disk, looses contrast if HI from the warp is added. In consequence, holes should be also searched for in single channel maps. Even in such maps one should realize that a spatially localized hole might exert no influence in regions on the map where the velocity distance gradient is small and where therefore long stretches of path contribute to a short range of velocities.

The observations were Hanning tapered, which introduces a correlation between neighbouring channels. The final maps have a varying noise level with the noise increasing at the extreme northern and southern ends of the galaxy. To make the search for holes uniform in terms of the signal-to-noise ratio the maps were corrected to give channel maps with a constant noise level. They were loaded five at a time into the Leiden Image Processing System for visual inspection and displayed in sequence as a mini film loop. Recognizing holes in this way is necessarily subjective and depends on the way the HI structures are perceived by the human eye and the readiness of the brain to "see" structures such as holes and shells built out of noise. To minimize any personal bias, an independent list of objects was compiled by each of us. To make the search as objective as possible, objects were included only if they fulfilled the following conditions: the hole must

- be present in at least three successive maps.
- have its center stationary in consecutive channel maps.
- have a good contrast.

The degree to which these conditions are met determines a quality factor Q, ranging from 1 (poorly defined object) to 5. The first condition is a consequence of the fact that neighbouring channels are correlated. The second and very important condition makes a first selection of holes which are caused by absence of neutral hydrogen and pseudo-holes created by kinematical effects. The third one is hard to quantify and depends also on the brightness and contrast setting of the monitor. Of great help was the option on the image processing system to make intensity profiles through the maps. Coordinates of the holes were obtained with an accuracy of one pixel, i.e. about 10".

The catalogue has some 250 entries, which can be binned into roughly three classes: spatially unresolved holes, which are not or barely resolved in velocity, resolved small holes (50-150 pc), and large holes (> 150 pc), generally showing shell-like brighter edges. They are distributed amongst the bins as 30%, 30% and 40%. Analysis is at the moment concentrated on objects with Q = 4 or 5, which form a subset of 65 objects. Two of them are discussed in more detail below.

3. EXAMPLES

Figure 1 shows two channel maps, one at -210.0 and the other at

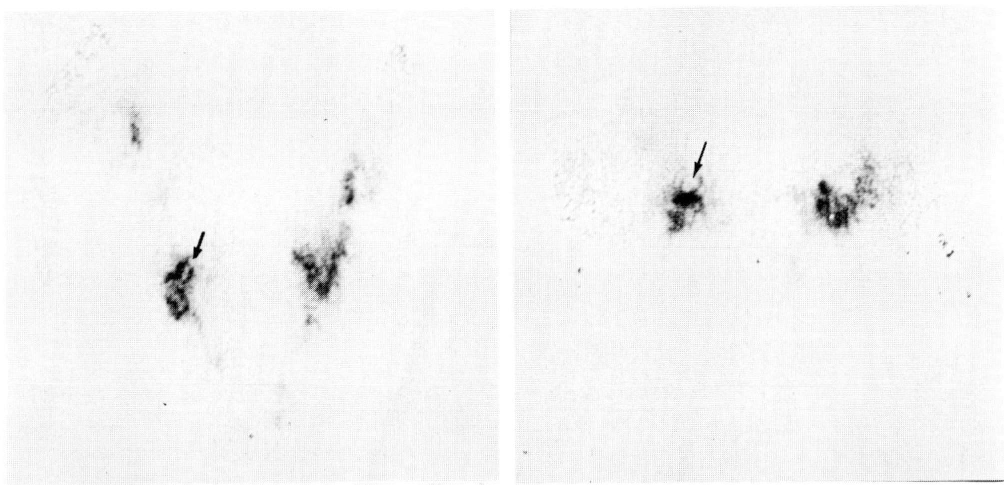

Figure 1. Channel maps at -210.0 km s^{-1} (left) and at -271.8 km s^{-1} showing the holes considered here, which are indicated by arrows. The + indicates the position of the nucleus of M31. Scale: 1 cm = 10'.

-271.8 km s^{-1}. On each map, many features are present. Two prominent holes, called A and B, are indicated by arrows. Their X-Y positions (with X measured along the major axis and Y along the minor axis) are for A : X = 25.8N, Y = 9.9E and for B : X = 13.1N, Y = 10.4E. Hole A measures ∿ 300 pc and sits on side of a large HII region of moderate emission measure as listed by Pellet et al. (1978) and lies between two OB-associations named OB41 and OB42 by van den Bergh (1964). At the position of B, which is about 400 pc in size, we find OB36 and close to the hole we find two HII regions, P526 and P532, from the catalogue of Pellet et al. Figure 2 shows for each hole a set of position-velocity maps, one parallel to the major axis, the other perpendicular to the major axis through the center of the hole, which is indicated by arrows. The lower panel of Figure 2 gives the spectrum measured at the position of the center of each hole. In these graphes, the arrow is placed at the velocity of the channel maps of Figure 1.

The black dots in the position-velocity maps are Hα-velocities from Deharveng and Pellet (1975). We took from their data the velocities measured at positions which fall within a 1' wide strip centered at the line along which the profiles were made. The Hα velocities are in excellent agreement with the HI data and in case A, where we are dealing with a double valued HI profile, there is a tendency for the Hα data to follow the HI component with the highest velocity. The hole in the channel map just shows up in the lowest velocity component and is visible as a gap. This is a clear case of a hole where HI is missing. In case B the HI is not absent, but seems pushed to a higher velocity. So in a range of channel maps around -271.8 km s^{-1} no HI is present at that position, but at higher velocities there is an excess of HI. The velocity pertur-

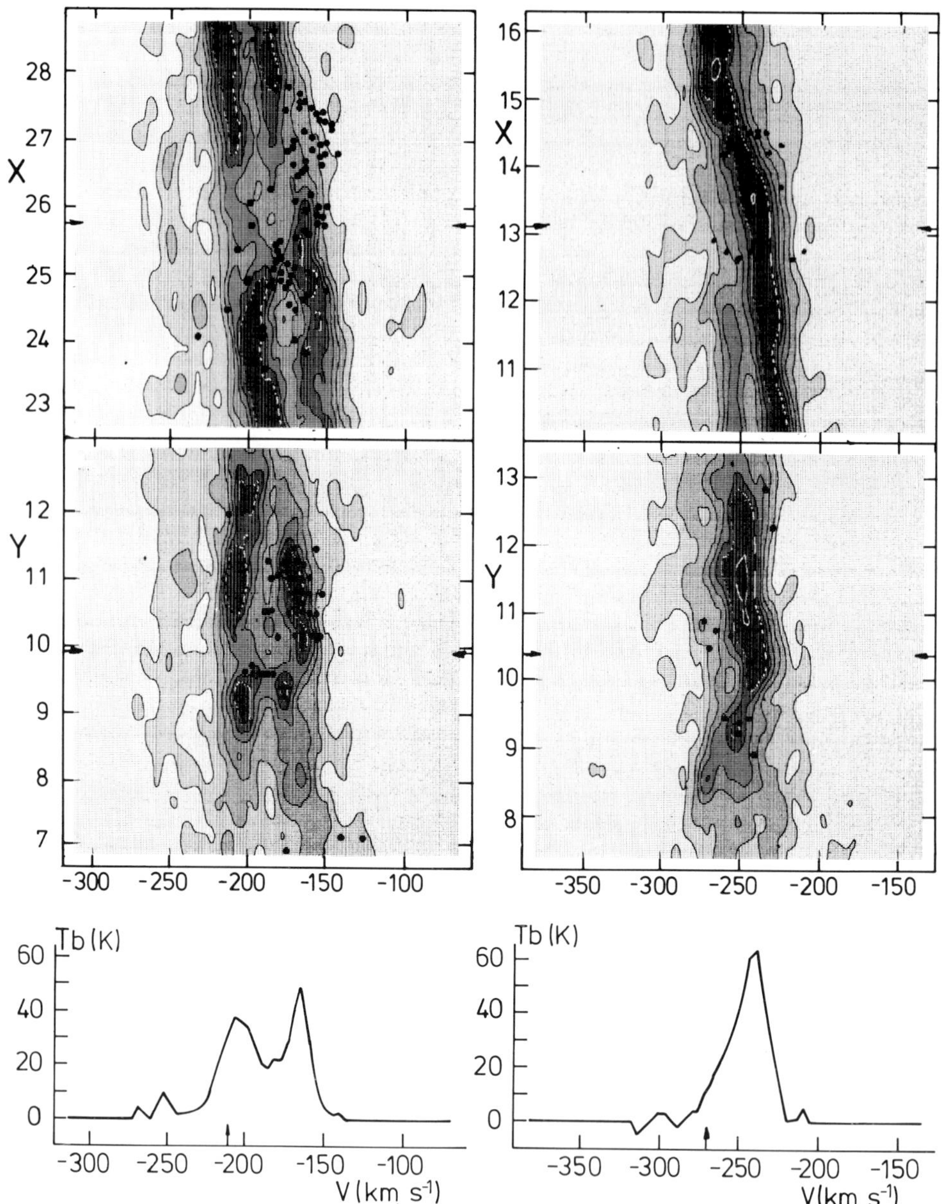

Figure 2. Position-velocity maps through hole A at 25.8N, 9.9E (left) and hole B at 13.1N, 10.4E (cf. Figure 1) and velocity profiles at these positions. The top panel shows the position velocity map through the center of the hole and parallel to the major axis of the galaxy (X-axis), the other map is made parallel to the Y-axis. Contour levels are at 2.5, 10, 20, ... K. The arrows indicate the location of the holes. The black dots correspond to the Hα velocities of emission regions measured by Deharveng and Pellet (1975). The arrows on the velocity profiles (bottom panel) indicate the velocities of the channel maps of Figure 1.

bation amounts to $\sim$ 10 km s^{-1}, which gives 7 x 10^{51} erg as an estimate for the kinetic energy involved. In case A, of the order of 0.5 x 10^6 $M_\odot$ of HI is missing and only a mild indication for a velocity perturbation is present.

We can not yet on the basis of these two examples draw any firm conclusions, although they seem to confirm the picture in which the holes are correlated with regions of recent star formation. Due to the combined effects of stellar winds and supernova explosions it seems possible to create holes in the HI distribution which are filled with hot (T $\sim$ 10^6 K) gas. A number of models have been proposed, Elmegreen (1982), Tenorio-Tagle et al. (1982), Bruhweiler et al. (1980), some involving other mechanisms like infall of extragalactic clouds (Tenorio-Tagle, 1980). Further analysis of the data at hand should result in tighter constraints for these models.

REFERENCES

van den Bergh, S.: 1964, Astrophys. J. Suppl. 9, 65.
Brinks, E.: 1981, Astron. Astrophys. 95, L1.
Bruhweiler, F.C., Gull, T.R., Kafatos, M., Sofia, S.: 1980, Astrophys. J. Lett. 238, L27.
Deharveng, J.M., Pellet, A.: 1975, Astron. Astrophys. Suppl. 19, 351.
Elmegreen, B.G., Chiang, W.H.: 1982, Astrophys. J. 253, 666.
Heiles, C.: 1976, Astrophys. J. Lett. 208, L137.
Heiles, C.: 1979, Astrophys. J. 229, 533.
Meaburn, J.: 1980, Monthly Notices Roy. Astron. Soc. 192, 365.
Pellet, A., Astier, N., Viale, A., Courtès, G., Maucherat, A., Monnet, G., Simien, F.: 1978, Astron. Astrophys. Suppl. 31, 439.
Tenorio-Tagle, G.: 1980, Astron. Astrophys. 88, 61.
Tenorio-Tagle, G., Beltrametti, M., Bodenheimer, P., Yorke, H.W.: 1982, Astron. Astrophys. 112, 104.

MOLECULAR CLOUDS AND STAR FORMATION IN SPIRAL GALAXIES

Judith S. Young
Five College Radio Astronomy Observatory, U. of Massachusetts

ABSTRACT

A large observational program investigating the 2.6 mm CO line in spiral galaxies is being conducted by myself and Nick Scoville using the 14 m telescope of the Five College Radio Astronomy Observatory (HPBW = 50"). Thus far we have observed 46 galaxies of types Sa, Sb, Sc and Irr, detected 31, and mapped 16. Our major findings are:

(1) In several relatively face-on Sc galaxies (IC 342, NGC 6946 and M51) the radial distribution of molecular gas out to 10 kpc follows the exponential blue luminosity profile of the disk within each galaxy (Young and Scoville 1982a, Scoville and Young 1983).

(2) From a comparison of the CO and B luminosities of the central 5 kpc in a sample of Sc galaxies both in the field and in the Virgo cluster, we find that the blue luminosity is proportional to the first power of the CO content (Young and Scoville 1982b). We interpret this to mean that the star formation rate per H_2 in Sc galaxies (indicated by the B luminosity) is constant.

(3) We have found molecular rings in two Sb galaxies, NGC 7331 and NGC 2841, with peaks at radii of 4-5 kpc (Young and Scoville 1982c). The central holes in the CO distributions are possibly related to the presence of large nuclear bulges in these galaxies.

(4) Molecular rings like the one in the Milky Way at radii 4 to 8 kpc were not seen in the Sc galaxies IC 342, NGC 6946 and M51. However, the CO radial distribution in NGC 253 shows a central peak and a small ring, not unlike what we would observe for our own galaxy viewed at the same distance and inclination as NGC 253.

1. INTRODUCTION

Studies of the molecular gas contents of galaxies of all types and luminosities are essential for understanding the evolution of galaxies, since giant molecular clouds are the sites of future star formation. Surveys of CO emission from our galaxy indicate that this gas shows strong maxima within 400 pc of the galactic center and in a ring between

W. B. Burton and F. P. Israel (eds.), Surveys of the Southern Galaxy, 253–264.

radii 4 and 8 kpc (Scoville and Solomon 1975; Burton et al. 1975). Observations of the radial distributions of molecular clouds in external galaxies shed light on the structure of our own galaxy as well as star formation in other galaxies. The aims of our CO observations of other galaxies, made with the 14 m telescope of the Five College Radio Astronomy Observatory (HPBW = 50"), are to determine the radial distributions of molecular gas in galaxies of all types, and the dependence of the distributions on Hubble type and luminosity. In addition, for the more face-on and nearby galaxies, it is possible to determine the relative confinement of molecular clouds to spiral arms.

2. SC RADIAL DISTRIBUTIONS

In IC 342, NGC 6946 and M51 (types Sc and Scd) we have mapped the CO radial distributions out to radii of 10 kpc at 1-2 kpc resolution (Young and Scoville 1982a, Paper I; Scoville and Young 1983, Paper II). Although there are azimuthal variations in the CO intensity at a particular radius, the mean distribution in each galaxy decreases smoothly with radius. We have compared the CO distributions with the optical light profiles in IC 342, NGC 6946 and M51 and find that the radial distribution of molecular gas out to 10 kpc follows the exponential blue luminosity profile of the disk within each galaxy, as shown in Figure 1 for NGC 6946. Although the B luminosity may not itself be primarily from the youngest stars, it is an indicator of the recent star formation rate for several reasons. First, the B-V colors are relatively constant with radius in these Sc galaxies (Ables 1971; Penston 1973; Schweizer 1976). Second, as we have shown for M51 (Paper II), in galaxies for which Hα light distribution has been measured the Hα and B luminosity profiles have similiar exponential scale lengths. Assuming the abundance of molecular gas is proportional to the CO emission (see Paper I, Appendix), the close correspondence between the CO emission and blue light indicates that the star formation rate per H_2 is constant within a particular galaxy.

We have derived an empirical relationship for determining H_2 column densities (N_{H_2}) from CO intensities (I_{CO}) based on visual extinction and CO observations of both dark clouds and giant molecular clouds in our own galaxy. In the Appendix of Paper I we showed that these samples are consistent with $N_{H_2}/I_{CO} = 4 \pm 2 \times 10^{20}$ H_2 $cm^{-2}/(K\ km\ s^{-1})$. Using this conversion in the external galaxies we find that H_2 dominates HI by as much as a factor of 100 in the interiors of the high luminosity Sc galaxies. The H_2 and HI distributions diverge at the centers of these galaxies as shown in Figure 2 for M51; the H_2 exhibits an exponential increase while the HI profile is flat with a central hole. In the Sc galaxies the H_2 masses within ~ 10 kpc are comparable to those of HI interior to 25 kpc.

Schmidt (1959) made the original suggestion that the star formation rate is proportional to the second power of the gas density, based on the distribution and scale height of neutral hydrogen in our galaxy.

Since the scale height of the young stars (50 pc) is less than that of the HI (120 pc), Schmidt argued the denser regions must preferentially form stars. This conclusion was also supported by the flat radial distribution of HI in the Milky Way. However, in the inner part of our galaxy where the bulk of the ISM is in molecular form (Sanders 1981), the CO scale height is similar to that for the young stars. Thus, when the molecular and atomic components of the ISM are considered, the data in our own galaxy and in the external galaxies are consistent with star formation proportional to the first and not second power of the gas density.

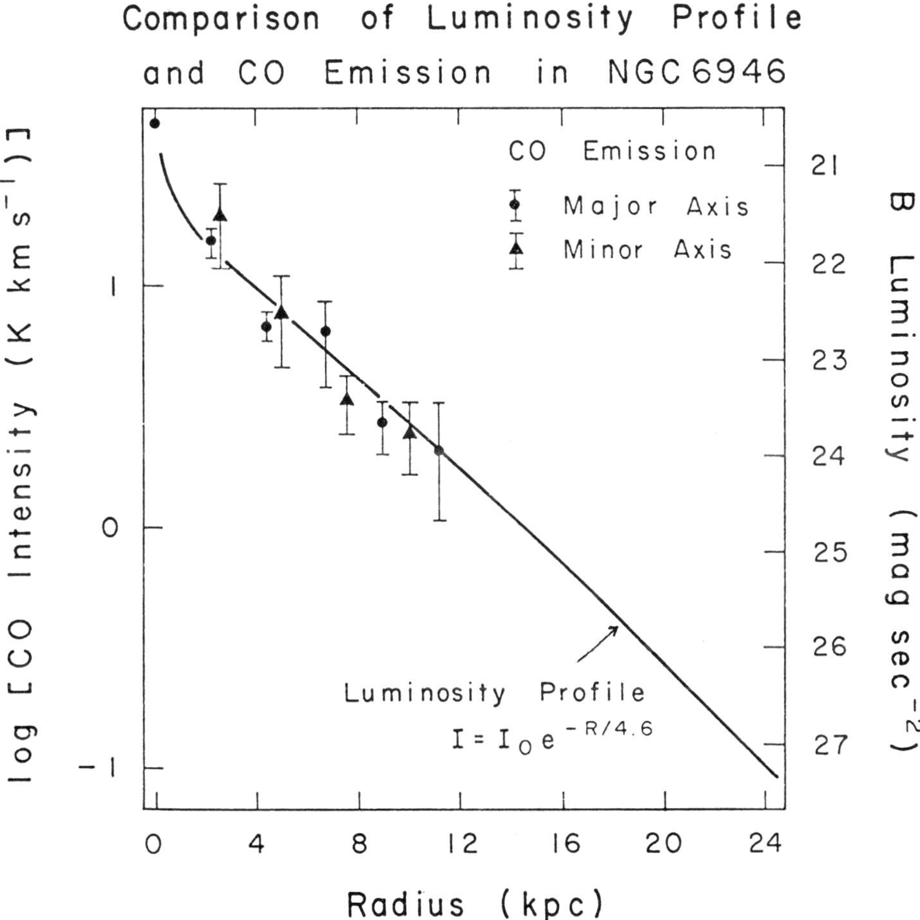

Figure 1. Comparison of CO emission with the exponential B luminosity profile of Ables (1971) for NGC 6946. Points plotted are the mean CO intensities at each radius, with bars indicating the spread in observed intensities; uncertainties in the individual intensities are a small fraction of the dispersion at each radius. The close agreement between the radial dependences of the CO emission and the luminosity profile is striking. The exponential nature of the luminosity profile is evident on this semi-log plot as a straight line.

In order to determine whether or not the molecular emission correlates with spiral features in M51, we have compared those positions which show the largest deviations from the exponential CO distribution with optical photographs of the galaxy. The two positions which show the largest deviations below the exponential (a factor of 2) are between spiral arms in the outer part of the galaxy. However, optical features do not always correlate with CO enhancements and interarm regions do not always have corresponding CO deficiencies. (Although the angular resolution and continuity of the CO observations are sufficient to separate

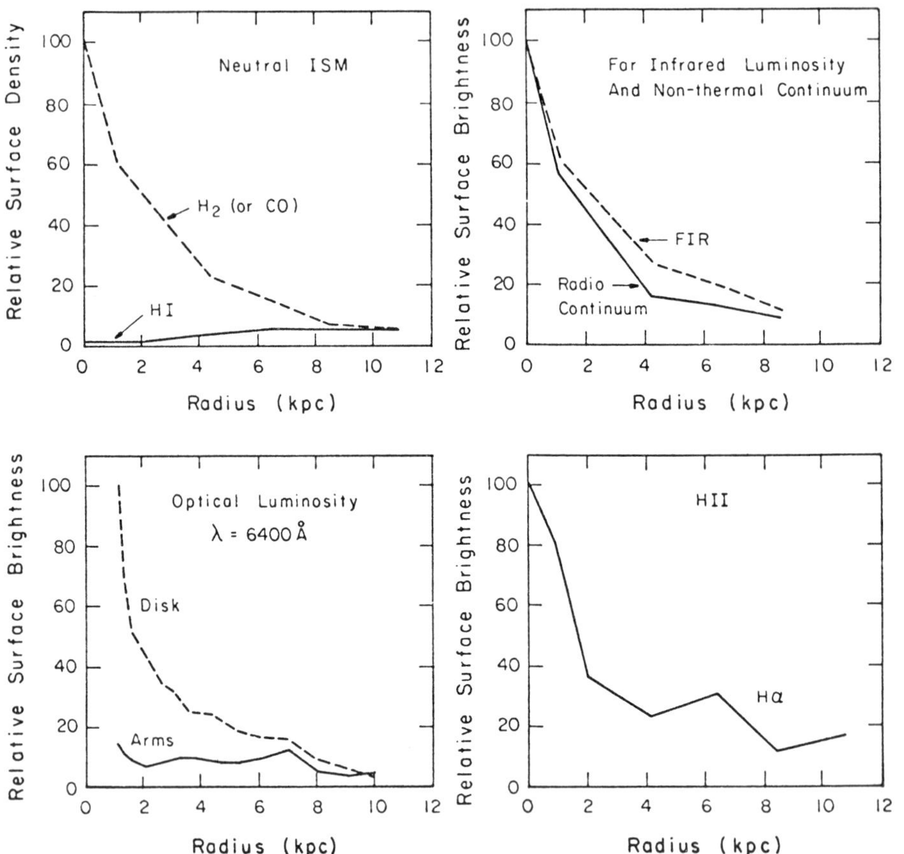

Figure 2. Comparison of mean radial distributions in M51. H_2 values are derived assuming a linear relationship between CO flux and H_2 mass. HI (Weliachew and Gottesman 1973) $\ll$ H_2 over the interior of this and other high luminosity galaxies. Also shown are the far infrared (FIR) emission (λ ≃ 170 m; Smith 1982), the optical disk light (Schweizer 1976), and the 21 cm continuum (Mathewson, van der Kruit and Brouw 1972). Similar exponential fall-offs are seen for the molecular gas, optical luminosity, FIR flux and nonthermal radio emission.

arm from interarm regions in M51, the observed positions were not chosen specifically to maximize the arm/interarm contrasts.) In M51 as well as the other high luminosity galaxies, the dominant feature in the CO radial distribution is the exponential character; global spiral structure is not evident in the molecular data at the present resolution.

3. COMPARISON OF SC GALAXIES

The correlation of CO intensity with blue luminosity within a particular Sc galaxy led us to investigate a larger sample of Sc's covering a wide range of size, mass, and total luminosity both in the field (Young and Scoville 1982b) and in the Virgo cluster (Young and Brady 1982). The spectra observed in the central 50" of 4 galaxies in the Virgo cluster (corresponding to 5 kpc at a distance of 20 Mpc) indicate that high luminosity galaxies have more CO than low luminosity galaxies (see Figure 3).

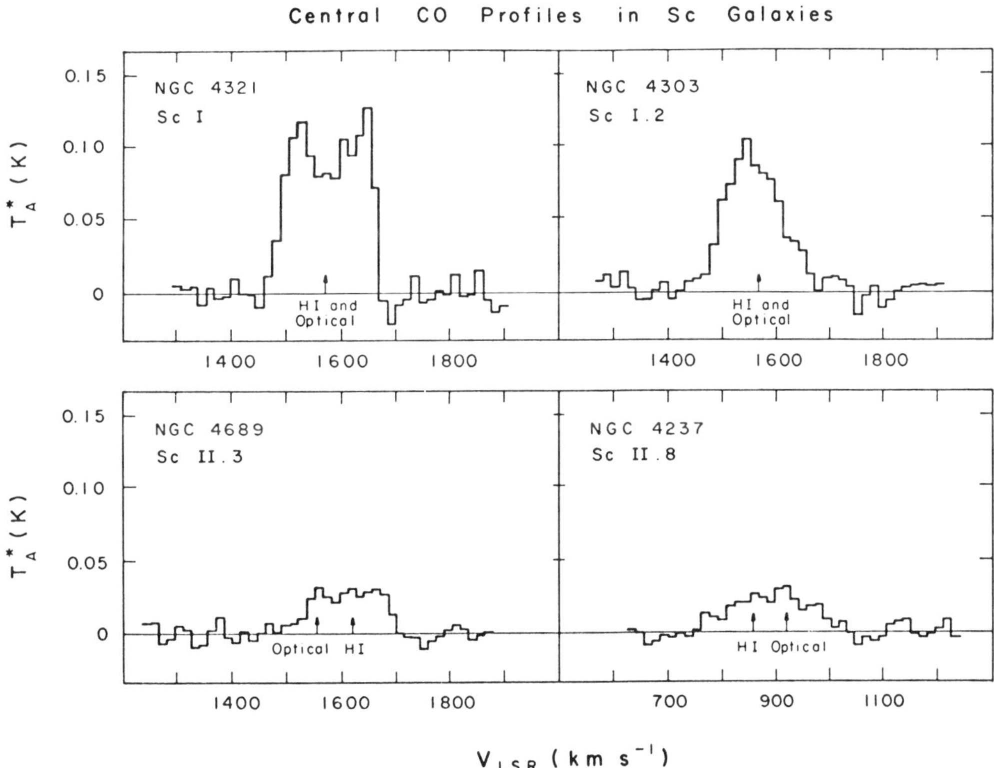

Figure 3. Central CO profiles in 4 Sc galaxies in the Virgo cluster (at 20 Mpc, 50" corresponds to 5 kpc). The high luminosity galaxies (NGC 4321 and NGC 4303) have more CO than regions of the same size in low luminosity galaxies.

We have compared the CO and blue luminosities of the central 5 kpc for 14 galaxies, as shown in Figure 4, and an approximately linear correlation is revealed between these quantities. Assuming the CO emission is proportional to the abundance of molecular gas, this correlation implies that low luminosity regions have little H_2 while high luminosity regions have large amounts. If the B luminosity is an indicator of the recent star formation rate, these results suggest that the star formation rate per H_2 in Sc galaxies is constant.

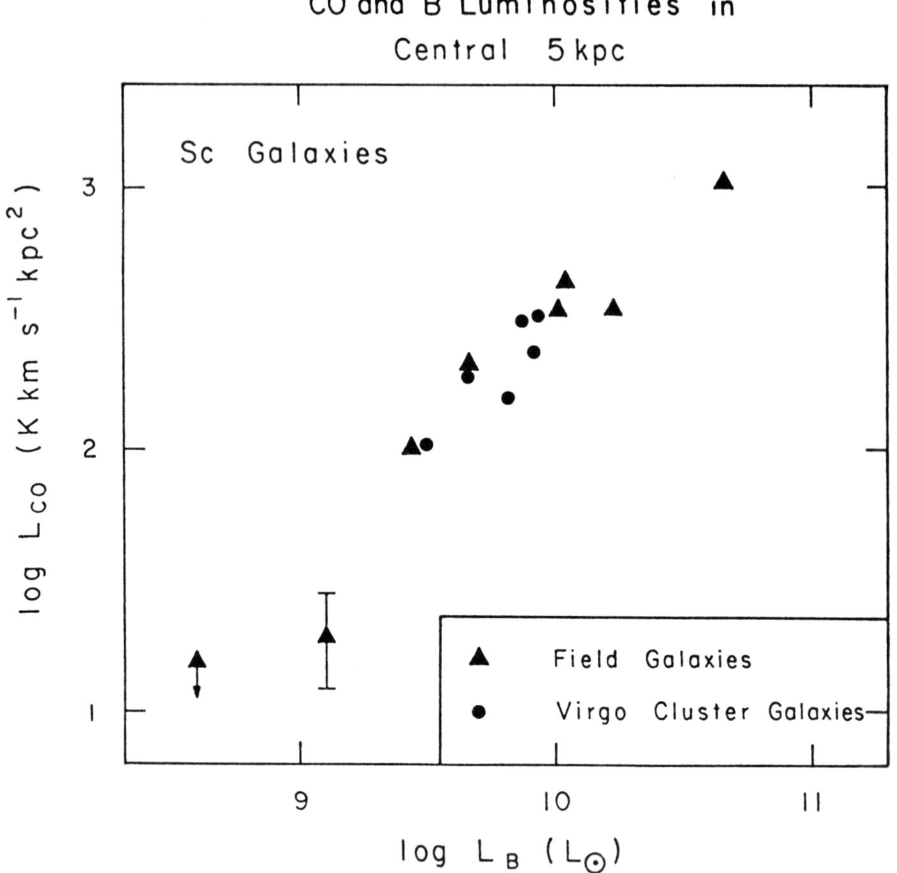

Figure 4. Comparison of CO and blue luminosities in regions 5 kpc in diameter in 13 Sc galaxies and NGC 1068. Dots represent the Virgo cluster objects and X's mark the field galaxies. Over two orders of magnitude in luminosity a linear correlation is evident: $L_B = 4 \times 10^7 L_{CO}^{1.0}$. We interpret this to indicate that the star formation rate per nucleon is constant from one Sc galaxy to another. In order of decreasing CO luminosity, the Virgo cluster galaxies are NGC 4254, NGC 4321, NGC 4303, NGC 4654, NGC 4651 and NGC 4535; the other galaxies plotted are NGC 1068, NGC 6946, NGC 5194, NGC 5236, NGC 5457, IC 342, NGC 2403, and NGC 598.

Within this sample the molecular masses out to a fixed radius, $R \leq 2.5$ kpc, range from $< 6 \times 10^7$ $M_\odot$ for M33 to $\sim 2 \times 10^9$ $M_\odot$ for NGC 6946. However, the H_2 mass to blue luminosity ratio is relatively constant, with $M_{H_2}/L_B = 0.17 \pm 0.08$ $M_\odot/L_\odot$ over two orders of magnitude in L_B. In contrast, these galaxies all have similar amounts of HI in the central $R < 2.5$ kpc, so that H_2/HI ratio varies from < 0.4 in M33 to ~ 32 in M51. However, the total ISM mass (H_2 + HI) to B luminosity ratio is also relatively constant in these galaxies. These results are summarized in Table 1 for the field galaxy sample.

The observations of Rubin et al. (1980) for Sc galaxies indicate that high luminosity galaxies have steeper rotation curves and higher values of V_{max} than low luminosity galaxies. Thus, at all radii high luminosity galaxies have more total mass (from the rotation curve) and more molecular gas (from these CO observations) than low luminosity galaxies; that is, they have higher mass densities. However, these Sc galaxies all have similar HI distributions with surface densities of $\sim 10^{21}$ atoms cm^{-2} over much of the disks. In the high luminosity galaxies the excess ISM above this surface density is in molecular form, while in low luminosity galaxies the H_2 and HI surface densities are comparable. Since our CO observations indicate that both high and low luminosity galaxies have similar star formation rates per ISM nucleon, more stars have formed in the high luminosity galaxies--making them have high luminosities--because more molecular gas is present.

Table 1 -- Masses of H_2 and HI

Galaxy	Central R < 2.5 kpc[a]			$M(H_2)/M(HI)$	$M(H_2)/L_B$	$M(HI)/L_B$
	L_B ($L_\odot$)	$M(H_2)$[a] ($M_\odot$)	$M(HI)$[a] ($M_\odot$)		($M_\odot/L_\odot$)	($M_\odot/L_\odot$)
N1068	4.6×10^{10}	4.2×10^9	5×10^6	840	0.091	1×10^{-4}
N5236	1.7×10^{10}	1.9×10^9	9.5×10^7	20	0.11	5.6×10^{-3}
N6946	1.1×10^{10}	2.4×10^9	1.2×10^8	20	0.22	1.1×10^{-2}
N5194	1.0×10^{10}	2.0×10^9	6.3×10^7	32	0.20	6.3×10^{-3}
N4321	7.4×10^9	1.5×10^9			0.20	
I342	4.5×10^9	1.2×10^9	6.8×10^7	18	0.27	1.5×10^{-2}
N5457	2.8×10^9	5.6×10^8	9.1×10^7	6.2	0.20	3.2×10^{-2}
N2403	1.3×10^9	$<6.0\times10^7$	1.3×10^8	<0.46	<0.046	1.0×10^{-1}
N598	3.9×10^8	$<5.5\times10^7$	1.3×10^8	<0.42	<0.14	3.3×10^{-1}

[a] B luminosities and HI surface densities are determined from observations reported in the literature, and H_2 masses are derived from the CO observations (see Young and Scoville 1982b).

4. MOLECULAR RINGS

No molecular rings like the one in the Milky Way at radii 4 to 8 kpc were seen in the late type galaxies IC 342, NGC 6946 and M51. In Paper I we suggested that the difference in the radial distributions is the absence of gas at R ~ 1 to 4 kpc in the Milky Way. Recently, however, we discovered molecular rings in two Sb galaxies, NGC 7331 and NGC 2841, with peaks at radii of ~ 4 kpc (Young and Scoville 1982c). Figure 5 shows the CO radial distributions in these galaxies, where the integrated intensities toward the nuclei are low relative to the disk. The presence of the central CO holes in these Sb galaxies provides clues to the origin of such a distribution.

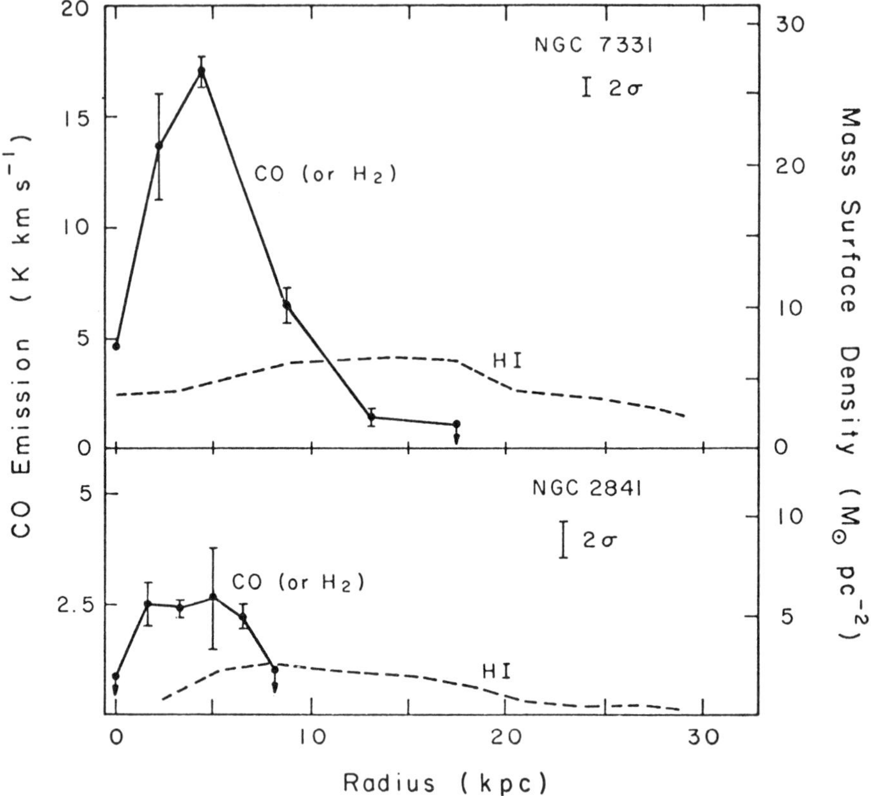

Figure 5. CO radial distributions in NGC 7331 and NGC 2841 (left-hand axis) and mass surface densities (right-hand axis). The solid dots represent the average CO emission at each radius, with bars indicating the dispersion in intensities observed. The magnitude of the 2σ uncertainty on each individual measurement is also indicated. In both galaxies the CO radial distributions exhibit molecular rings which peak around 4-5 kpc. Also plotted are the HI distributions (Bosma 1978); NGC 7331 has higher surface densities in both H_2 and HI than NGC 2841.

The observed CO rotational velocities in NGC 7331 are shown in Figure 6. The previously measured Hα and HI rotation curves for these galaxies (Rubin et al. 1965; Bosma 1978) reach their maximum velocities at the same radii as the peaks in the molecular rings, so that we can rule out the possibility that the hole in the distribution arises from dynamical action at the ILR. Instead, the CO distributions in NGC 7331, NGC 2841 and possibly the Milky Way are related to the nuclear bulges. In the external galaxies, the central CO holes are coincident with the size of the bulge components measured by Boroson (1981). Rather than being in molecular clouds the gas which was present during the formation of the galaxy may have been depleted in forming stars in the nuclear bulge.

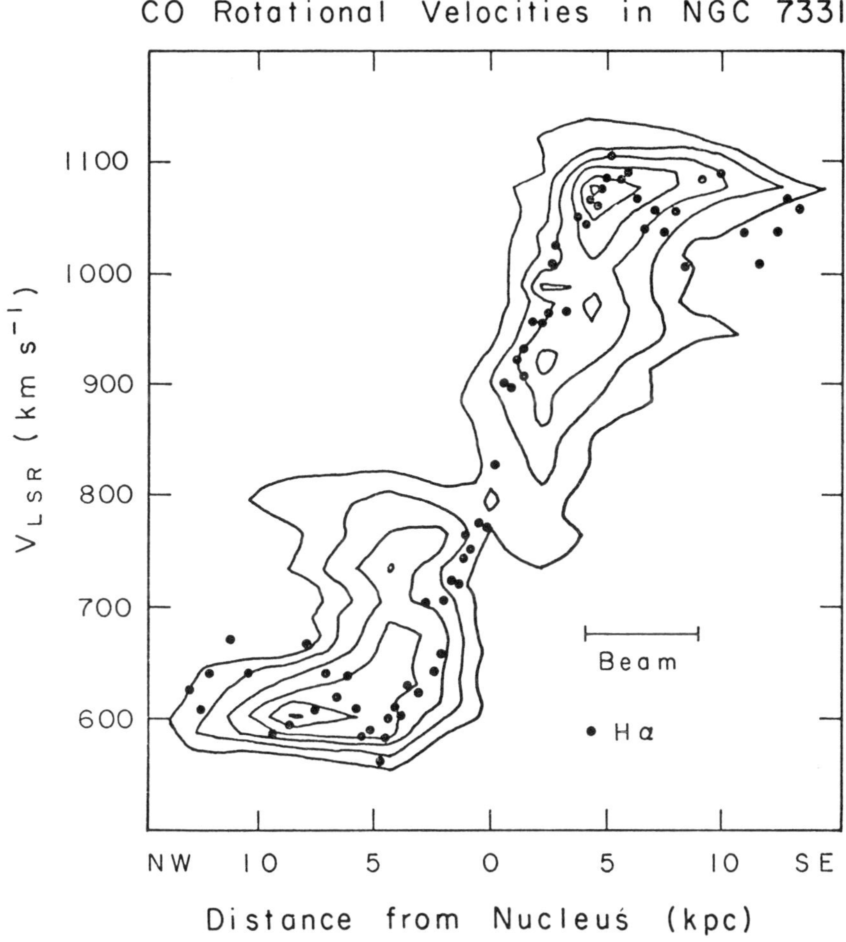

Figure 6. Observed CO rotational velocities along the major axis of NGC 7331; the Hα emission (Rubin et al. 1965) shows similar velocities (solid dots). The peak in the rotation curve coincides in radius with the peak in the molecular emission from which we infer that the hole in the gas distribution is probably related to the nuclear bulge.

A galaxy whose CO radial distribution may resemble that in the Milky Way is NGC 253. This galaxy shows strong CO emission in the center, a sharp fall off out to a radius of 2' along the major axis, and then an increase of a factor of 2 as shown in Figure 7. The CO distribution in the Milky Way, viewed at the same resolution and inclination as NGC 253, shows a similar decrease in intensity from the nucleus out to 2 kpc radius, followed by a factor of 4 increase at the peak of the molecular ring. Thus, the CO morphologies in these two galaxies may be similar.

Additional similarities between NGC 253 and the Milky Way are apparent. Detailed photometric and kinematic observations of NGC 253 (Pence 1978) indicate a velocity structure in the central 200" which resembles simple gas flow near a bar. This is an important conclusion in view of the fact that it has been suggested that our galaxy has a bar in the center and that such a feature could be related to the unusal CO morphology.

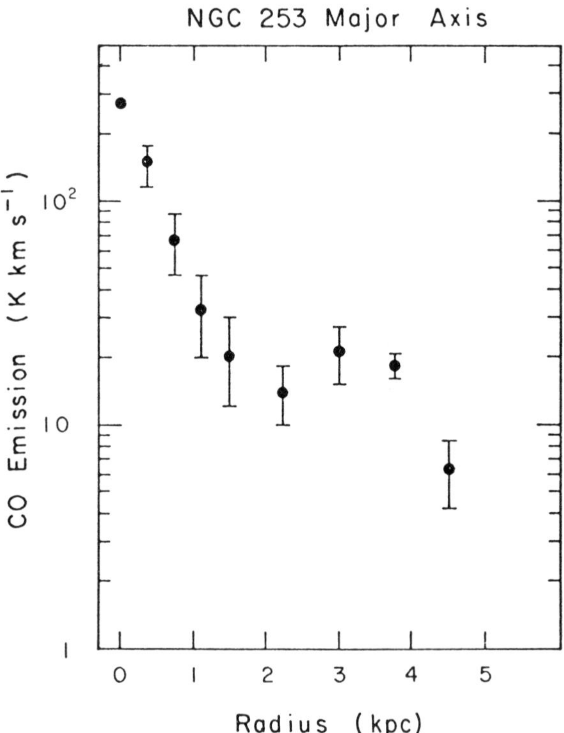

Figure 7. CO radial distribution in NGC 253, showing a hole at ~ 2' radius followed by an increase at 3' and then the continued decrease in intensity with radius. (At a distance of 3.4 Mpc 1' = 1 kpc.) The Milky Way CO distribution, viewed at the same inclination and resolution as NGC 253, shows a hole at 2 kpc with the peak in the ring at 5 kpc. The distribution of CO in Milky Way is the most similar to that in NGC 253 of the galaxies I have thus far observed.

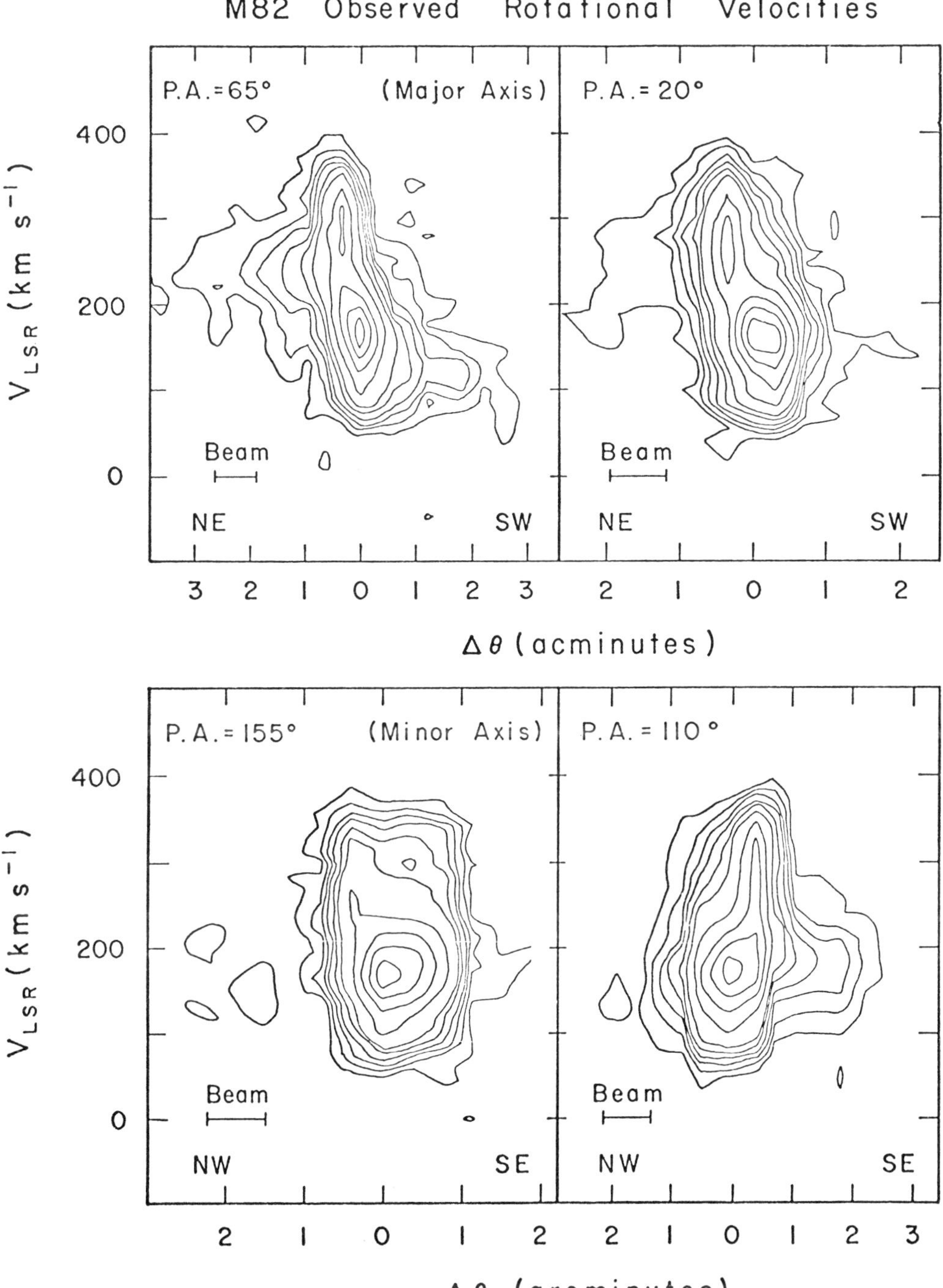

Figure 8. Observed CO velocities along 4 axes in M82. On the NE side of the major axis the CO emission does not exhibit rotation, but occurs at the systemic velocity of the galaxy.

5. M82

We have observed the CO radial distribution at 70 positions in the irregular galaxy M82 and found several anomalous features (Young and Scoville 1982d). First, CO emission was detected on the minor axis 2' from the nucleus in the vicinity of the extensive system of optical filaments associated with this galaxy. Second, the kinematics of the molecular gas along the major axis of M82 show peculiarities in the NE in the sense that no rotation is evident > 2' out in the disk (see Figure 8). These unusual features are probably related to the evolutionary history of this galaxy.

REFERENCES

Ables, H.D.: 1971, Publ. U.S. Naval Obs. Sec. Ser., Vol XX, Part IV, Washington, D.C.
Boroson, T.: 1981, Ap.J. Supp., 46, 177.
Bosma, A.: 1978, Ph.D. dissertation, University of Groningen.
Burton, W.B., Gordon, M.A., Bania, T.M. and Lockman, F.J.: 1975, Astrophys. J., 202, 30.
Kormendy, J. and Norman, C.A.: 1979, Astrophys. J., 233, 539.
Mathewson, D.S., Kruit, P.C. van der, Brouw, W.N.: 1972, Astron. Astrophys., 17, 468.
Pence, W.D.: 1978, Ph.D. dissertation, University of Texas.
Penston, M.V.: 1973, Monthly Not. Royal Ast. Soc., 162, 359.
Rubin, V.C., Burbidge, E.M., Burbidge, G.R., Crampin, D.M., and Prendergast, K.H.: 1965, Astrophys. J., 141, 759.
Rubin, V.C., Ford, W.K. Jr. and Thonnard, N.: 1980, Astrophys. J., 238, 471.
Sanders, D.B.: 1981, Ph.D. dissertation, State Univ. of N.Y.
Schmidt, M.: 1959, Astrophys. J., 129, 243.
Schweizer, F.: 1976, Astrophys. J. Suppl., 31, 313.
Scoville, N.Z. and Solomon, P.M.: 1975, Astrophys. J. (Letters), 199, L105.
Scoville, N.Z. and Young, J.S.: 1983, Astrophys. J., in press (Paper II).
Smith, J.: 1982, Astrophys. J., in press.
Weliachew, L. and Gottesman, S.T. 1973, Astron. Astrophys., 24, 59.
Young, J.S. and Brady, E. 1982, in preparation.
Young, J.S. and Scoville, N.Z.: 1982a, Astrophys. J., 258, 467 (Paper I).
________ 1982b, Astrophys. J. (Letters), 260, L11.
________ 1982c, Astrophys. J. (Letters), 260, L41.
________ 1982d, in preparation.

COLUMBIA UNIVERSITY SOUTHERN HEMISPHERE MILLIMETER-WAVE SURVEY TELESCOPE

Richard S. Cohen
Department of Astronomy, Columbia University
New York City

After seven years of use the Columbia millimeter-wave telescope has demonstrated the utility of an instrument in the one meter class dedicated to CO observations. It has almost finished a task that would be totaly impractical at the resolution of a 10 meter telescope, namely a complete northern hemisphere inventory of molecular matter both in the distant galactic plane and in the nearby clouds in the local arm and Gould's belt. This has been possible because the 8 minute beamwidth of our telescope is well matched to such projects, allowing, for example, the first quadrant of the distant galactic plane to be surveyed in a few hundred days of observation.

A few years ago we became convinced that there were several reasons why a duplicate of the Columbia telescope would be an ideal instrument for the Southern Hemisphere. First, since we knew there would be difficult operational problems in the Southern Hemisphere, we thought that it would be important to minimize the technical problems by building a fairly simple telescope of proven design. A larger instrument would more logically follow once the broad outlines of the southern molecular sky were known. Equally important, completing the survey of the galaxy in the southern sky with an instrument identical to the one used in the north would bypass the difficult problems encountered in reconciling the classical northern and southern 21-cm

Table 1

Specifications of Telescope

Antenna	1.2 meter Cassegrain
Surface Accuracy	20 μm
Beamwidth	8 arc minutes (FWHM)
Beam Efficiency	75% (calculated)
Bandwidth	128 MHz
Spectral Resolution	0.5 MHz
Receiver Noise	385 K (SSB)

W. B. Burton and F. P. Israel (eds.), Surveys of the Southern Galaxy, 265–269.

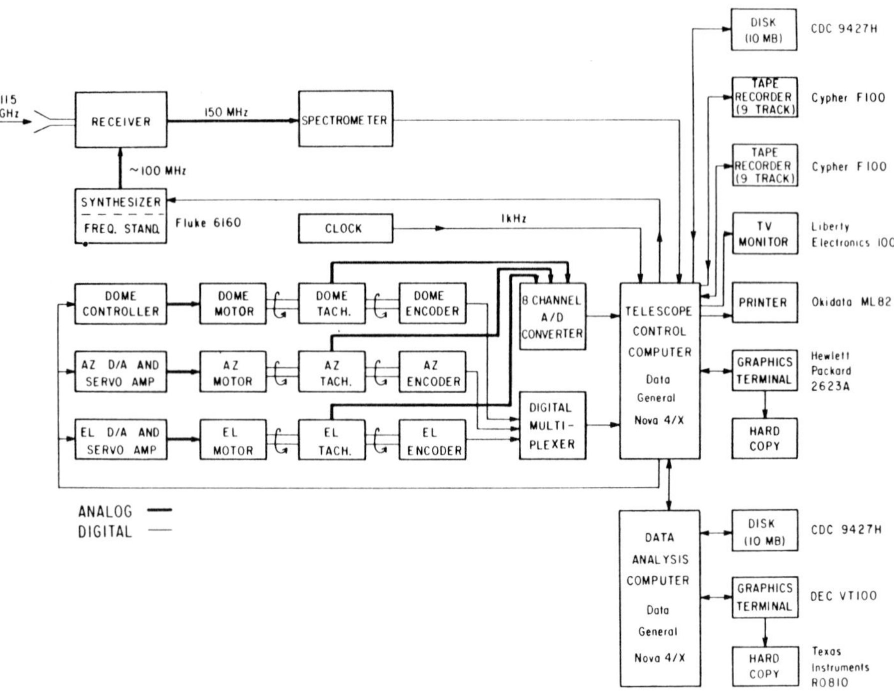

Figure 1. Block diagram of the telescope.

surveys made with different beams, sensitivities, calibrations, etc.

Preliminary work on this new telescope began in 1980, and in April 1981 major construction started under a three year grant from the National Science Foundation. During 1982 the telescope was completely assembled for testing on the Columbia University campus in New York and astronomical observations were made. The telescope is now (November 1982) being shipped to South America and by mid-1983 around-the-clock observations should be underway at the Cerro Tololo Interamerican Observatory near La Serena, Chile.

Like its twin in New York, the new telescope is a 1.2 meter Cassegrain. (See Table 1 for the major specifications.) The primary is made from a single aluminium casting numerically milled to better than 12 μm. Gravitational deformations, calculated using the NASTRAN computer code, add another 8 μm of error, resulting in a total surface accuracy of $\lambda/100$ at the 2.6 mm CO wavelength -- essentially perfect by conventional radio astronomy standards. Because of the small size of the antenna, we have none of the usual problems of antenna alignment. The secondary can be positioned in a few minutes using a precision mechanical jig that simply holds the secondary in place with respect to

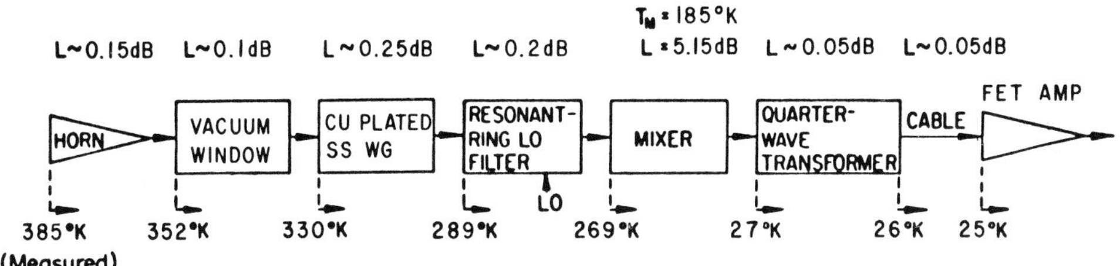

Figure 2. Block diagram of the receiver showing the noise temperature at various points along the signal path.

the primary while the adjusting bolts are tightened. Besides focusing the feed horn, this is the only mechanical adjustment of the antenna required. The optics were checked in New York by a range test using a transmitter located in the Mt. Sinai hospital, a tall building about 2 km (4 D^2/λ) across Central Park from the Columbia campus. The main beam was measured to be about 10% wider than that predicted by a simple scalar diffraction model, well within the error of this calculation. The sidelobes were, as calculated, down by more than 18 dB.

For the kind of large-scale CO surveys planned, it is usually necessary when position switching to use a reference position five or even ten degrees from the observed point. The antenna mount was therefore designed with a very fast drive using direct drive torque motors: it can switch up to five degrees in less than one second. (See Fig. 1 for a block diagram of the telescope.) The pointing was checked by sighting on stars using a small optical telescope mounted on the dish and collimated with the radio axis. Using about forty stars spread over the entire sky, we determined that the peak error is less than a tenth of a beamwidth.

For the receiver we chose a conservative design based on a Schottky barrier diode cooled to 77 K by liquid nitrogen in a standard Infrared Labs Dewar. By cooling 80% of the way from room temperature to the 20 K typical of refrigerator-cooled systems, we estimated that we would get a receiver almost as good while avoiding possible tracking problems caused by mounting a mechanical refregerator on our lightweight telescope drive. This scheme also condiderably speeds the receiver development, because, though it takes several hours to cool a mixer to 20 K with a refrigerator, cooling to 77 K can be accomplished in a matter of minutes using an open bucket of liquid nitrogen.

The rest of the receiver is quite standard. Inside the Dewar (Fig. 2) there are a resonant ring filter for injecting the local oscillator signal, a Schottky barrier diode (made by R. Mattauch) mounted in reduced height waveguide, an impedance-matching transformer, and a three stage 30 dB gain FET amplifier built from the design of S. Weinreb. The output of the Dewar is down-converted to the 150 MHz second IF using standard commercial components.

This receiver has fully met our expectations. It has a single sideband noise temperature of 385 K measured at the feed horn. Furthermore, this performance is entirely reproducible: we have several mixer blocks with essentially identical characteristics. Work is continuing on the mixers. We are now testing several more types of diodes as well as modifications to the mixer block and we expect that the system temperature can probably be lowered by at least another 50 K.

The spectrometer is a 256 channel, 500 kHz resolution filter bank based on the NRAO design. Its resolution (1.3 km sec^{-1} at the 2.6 mm CO line) and bandwidth (330 km sec^{-1}) are well suited to the galactic survey. For cold cloud observations, however, more resolution is desirable, so a second 256 channel filter bank with 100 kHz resolution, is now being built.

The computer system is an updated copy of the New York system. The Data General Nova 1200 has been replaced with its current version, the Nova 4/X. For data storage we have a 10 megabyte CDC 9427H disk drive and two ten-inch tape drives. Using a Hewlett-Packard 26223A graphics terminal, we can display scans and can do simple data reduction such as summing scans, removing baselines, and fitting gaussians. For straightforward cloud maps these on-line programs are frequently all that is needed. For more complex tasks, such as producing contour maps, we have generally used the large IBM computer at the Institute for Space Studies. We can continue to do this with the Southern telescope by bringing tapes to New York for processing. We will also be developing software to run on a second Nova computer system that will serve both for data analysis and as a spare for the telescope control computer.

In addition to the pointing and range tests already described, we made actual astronomical observations. Calibration checks on several sources agreed with the New York instrument to within 10%. Our longest single observation, 10 hours on W51, had an extremely flat baseline and showed no sign of a deviation from a $1/\sqrt{t}$ decrease in the noise. We also repeated a small section of the Columbia galactic survey and found complete agreement with the original spectra. A typical spectrum, shown in Figure 3, although made in the poor atmospheric conditions typical of New York in September, required no baseline removal except for subtraction of a constant tc compensate for the difference in sky emission at the observation and reference positions.

The initial observing program in the south is fairly obvious. It will be largely a mirror image of what was done in New York: a high resolution survey of the distant inner arms of the Milky Way, detailed maps of several individual objects and, finally, a wide angle survey of the entire local arm. We will start with the galactic survey. When combined with the New York data it will yield a single map with uniform coverage of a 2° wide latitude strip from ℓ = 300° through the galactic

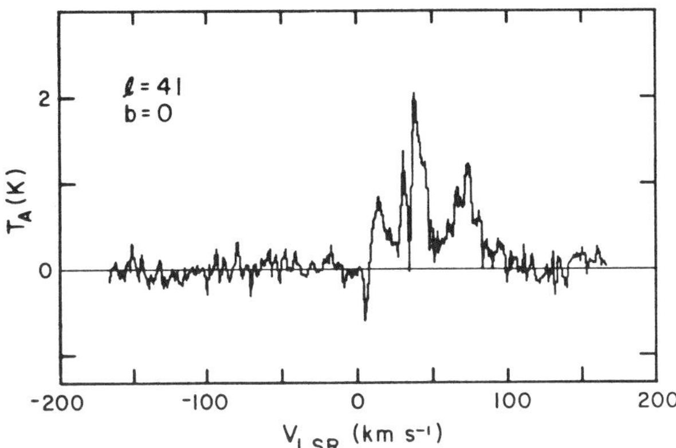

Figure 3. A sample galactic survey spectrum made in 14 minutes of integration.

center to ℓ = 90°. Simultaneously we will map the Magellanic Clouds, which conveniently transit during the six hours that the fourth galactic quadrant is down. Other obvious objects to be covered in the first year include the Coal Sack and the Carina cloud complex. We will also observe the galactic center. Although considerable effort has been expended on the center already, observations are very slow from the latitudes of existing telescopes, and the maps are still incomplete. At Cerro Tololo, on the other hand, the center transits 2° degrees from the zenith, and we will be able to make a wide-angle map of the entire galactic center region.

Given the almost perfect atmospheric conditions at Cerro Tololo at 2.6 mm over almost the entire year and the high sensitivity of our receiver, we expect observations to proceed rapidly. For example, when the effects of source elevation are included, the southern galactic survey should take an order of magnitude less time than the northern survey, reducing the original two years to only a few months. By mid-1984, we expect to have completed a substantial fraction of the work outlined here.

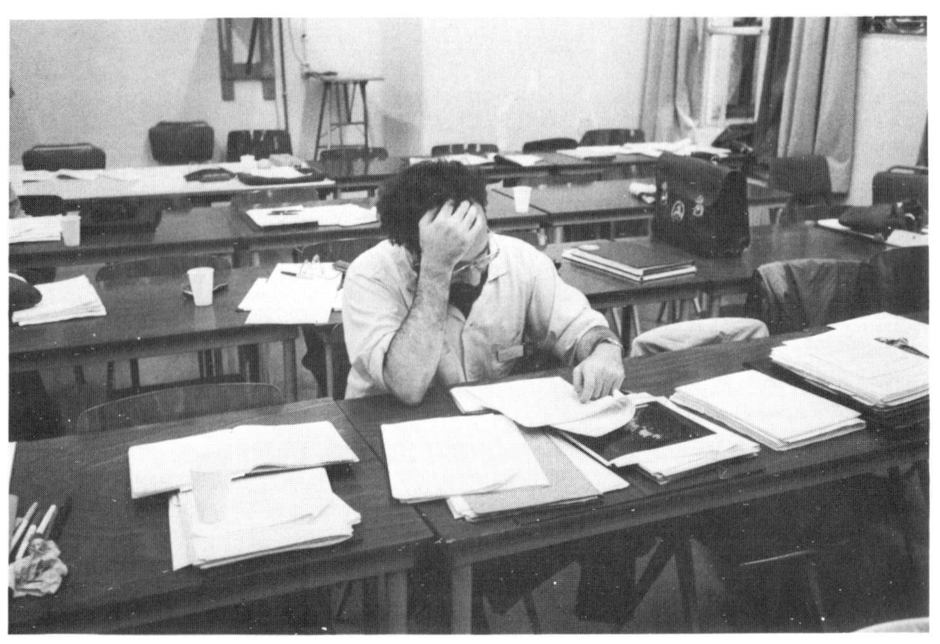

HIGH ENERGY SATELLITE SURVEYS

K. Bennett
Space Science Department of the European Space Agency
European Space Research and Technology Centre
Noordwijk, The Netherlands

ABSTRACT

During the last decade significant advances have been made in the fields of X-ray and gamma-ray astronomy.

A short review of the prominent satellite missions is presented, outlining the experimental techniques employed and illustrated by the major discoveries.

1. INTRODUCTION

Restricting the topic of this presentation to surveys carried out in X- and gamma-rays very effectively limits the number of relevant missions about which I can talk! Rather than provide a history of the science (which is now so well established as to merit numerous text books on the subject), I will briefly present the milestone missions, the detection techniques and major findings.

2. X-RAY ASTRONOMY

2.1 UHURU - The First X-ray Mission

The UHURU satellite (Giacconi et al. 1971) was launched by NASA in 1970. The experiment payload consisted of 840 cm^2 collimated proportional counters operating in the range 2-6 keV. The detectors viewed perpendicular to the axis of the spinning satellite which was in a low background equatorial orbit. This mission lasted 2 years, the results of which are displayed in Figure 1. This shows the X-ray sky to contain several hundred sources above the threshold sensitivity. Prior to UHURU, the X-ray sources amounted to a few dozen observed during rocket flights.

The UHURU sources covered a dynamic range of 10^5 in intensity, the weakest being about 10^{-3} of the Crab (1μJy). The sources were found to exhibit both regular and irregular time variability which provided

W. B. Burton and F. P. Israel (eds.), Surveys of the Southern Galaxy, 271–280.

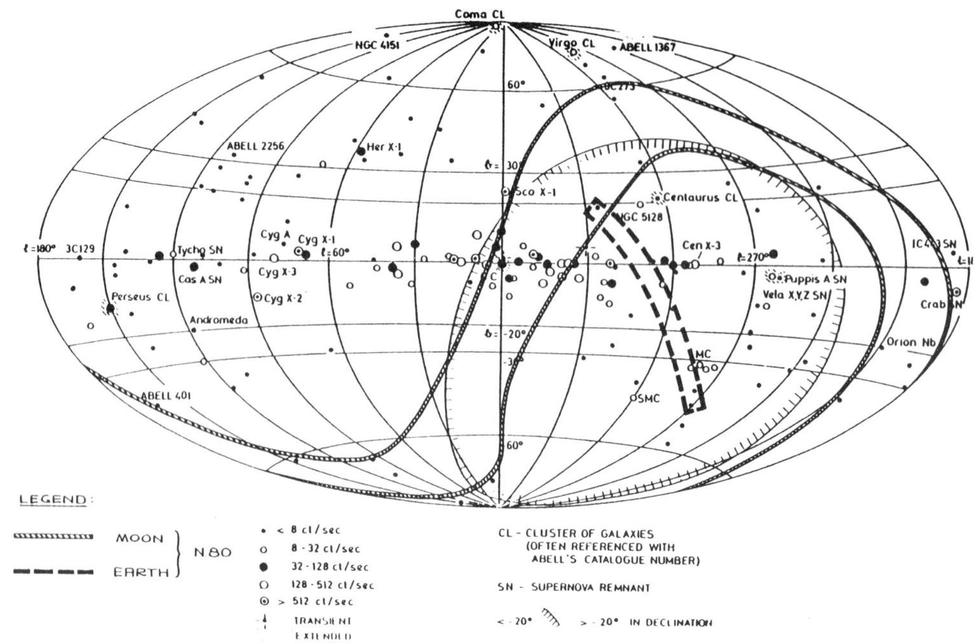

Figure 1: The X-ray sky as seen by UHURU.

evidence for X-ray binary systems, neutron stars and, possibly as for Cyg X-1, the black holes. The sources divide naturally into 2 populations, galactic and extragalactic, although even the nearest galaxies remained unresolved by UHURU.

2.2 The Einstein Observatory - The Next Logical Step

The next significant advance in X-ray astronomy resulted from the successful application of a new technique, namely, X-ray imaging optics, aboard the HEAO-B (Einstein) observatory launched in 1978 (Giacconi et al. 1979a). The principle of X-ray imaging is illustrated in Figure 2. The telescope depicted is a so-called Wolter 1 system comprising a paraboloidal mirror (1st reflection) followed by a hyperboloidal mirror (2nd reflection). In such a system, X-rays impinging on the highly polished, gold coated surface at grazing incidence are reflected onto a detector in the focal plane of the system. In the Einstein telescope, focal length 340 m, 4 mirror pairs are nested to increase the collecting area and the image is formed at any one of a cluster of detectors mounted on a turret in the focal plane of the telescope. This technique is effective in the energy range 0.15-3 keV. The lower limit is determined by the focal plane detector efficiency and the higher limit by the fall in reflectivity. The large collecting area together with low background due to the small size of the detector combined to improve the sensitivity to about 100 times that of the UHURU mission (10^{-5} Crab) and permit source location to ~ arc sec.

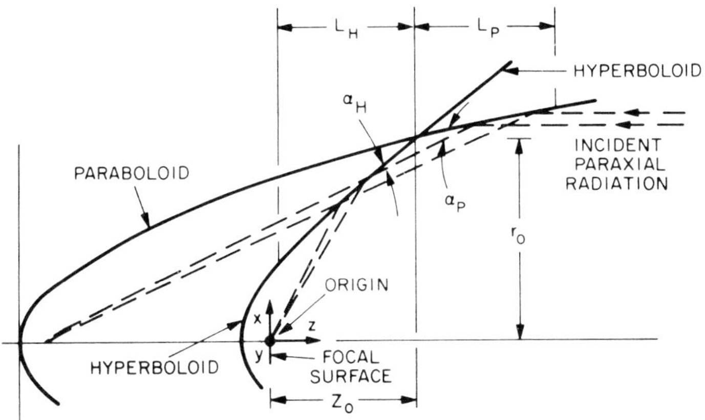

Figure 2: The principle of X-ray imaging optics.

As an example of the resolving power of Einstein, the Andromeda galaxy unresolved by and at the sensitivity threshold of UHURU, was observed using the High Resolution Image Detector (HRI) to comprise > 70 sources! (van Speybroeck et al. 1979).

Apart from the obvious task of studying the UHURU sources in detail, major discoveries relating to deep extragalactic surveys have revealed quasars of red shifts > 2.5, as well as details of galactic cluster emission, AGN and Seyferts. These deep surveys have used the HRI which has a field of view of only 25 arc min. The IPC, with an FOV of 75' has yielded superb images of, among other things, supernova remnants. Nevertheless, despite the fact that the known sources have, with Einstein increased a hundredfold, less than a hundredth of the sky has been surveyed.

2.3 The EXOSAT Observatory - By Public Demand

The European EXOSAT satellite presently awaiting launch will extend the work done by the Einstein observatory in terms of energy range and spectroscopic capability (Taylor et al. 1981).

The payload, shown in Figure 3, comprises 2 mirror systems offering similar source sensitivity and angular resolution to Einstein, but using gratings can offer spectral capability down to .04 keV, while the imaging proportional counters have significantly better energy resolution than Einstein.

Complementary instruments comprising large area (2000 cm^2) proportional counters with field-of-view collimated to 45 arc min, provide for sensitive measurements of time variability and spectroscopy in the energy range 1-50 keV (so-called ME detector). A unique feature of this mission is the ability to occult sources using the moon (as illustrated in Figure 4). In Figure 1 the swath encloses those UHURU sources which can be observed to occult by EXOSAT.

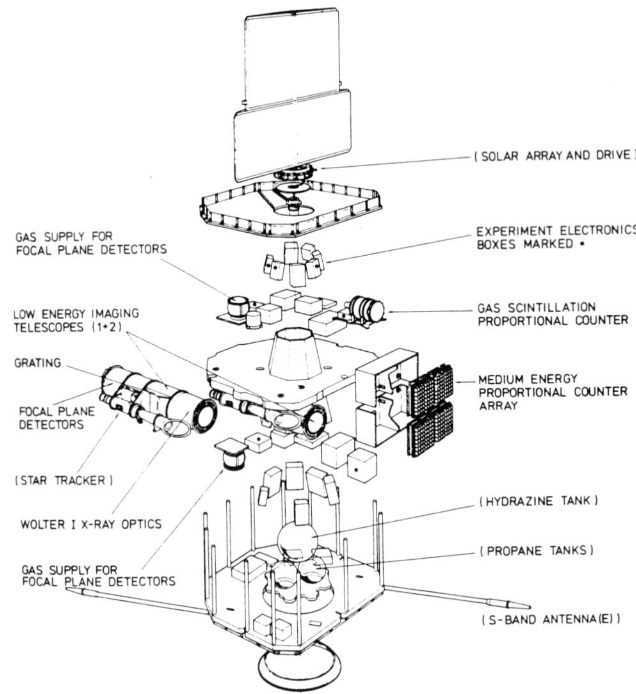

Figure 3: Exploded view of EXOSAT showing payload elements.

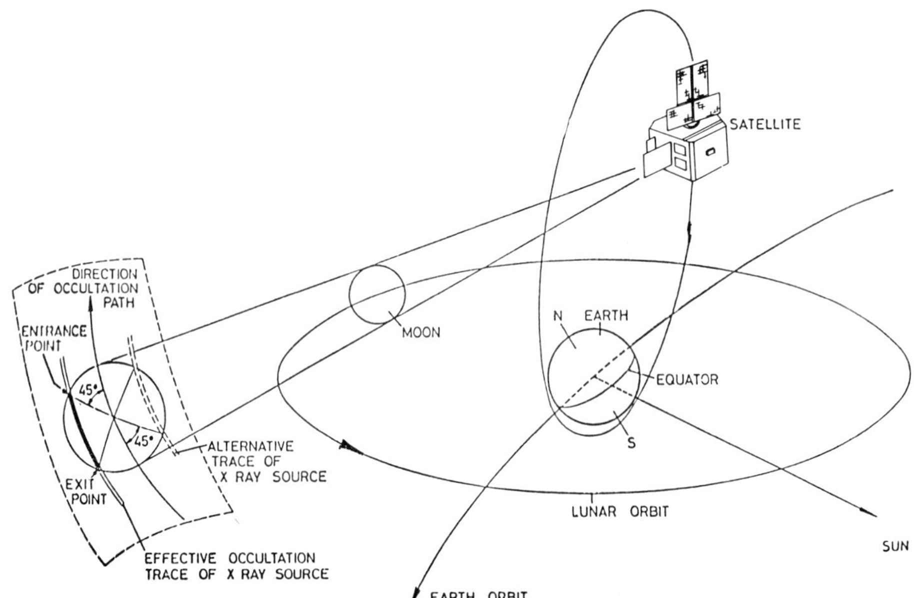

Figure 4: The method of lunar occultation used by EXOSAT.

Using the ME proportional counter array this technique offers position determination to a few arc sec in the 1-50 keV energy range as well as the means to study possible extended features.

More detailed spectroscopy in the ME energy range than possible with the proportional counters (resolution ~ 20% at 6 keV) is achieved using a novel gas scintillation proportional counter (resolution ~ 10% at 6 keV), which has a similar field-of-view to the ME detectors (Peacock et al. 1981).

Given the limited coverage of Einstein and the larger energy range and improved spectroscopic capability of the EXOSAT payload, there is little doubt that this mission will be as fruitful as Einstein.

2.4 ROSAT - A True Survey Instrument

While both EXOSAT and Einstein allow detailed studies of known (and postulated) sources of X-rays, neither can be said to be a survey instrument. It is clear, as in the case of UHURU, that a survey instrument can mark a 'break-point' in the evolution of an observational science such as X-ray astronomy. The next break-point is likely to come with the launch of ROSAT (Trümper, 1982), shown in Figure 5, built by Max Planck Institute, Garching, West Germany. This mission promises to carry out a complete sky survey in the energy range 0.1-2 keV, within the first six months of its orbital lifetime with a sensitivity competitive with that of Einstein and EXOSAT. ROSAT exploits imaging optics using a set of nested Wolter I-type mirrors with a focal length of 240 cm and offers a collecting area 3 times greater than Einstein.

This mission is further enhanced by a secondary instrument to conduct a sky survey in the range 50-250 Å. The instrument, called the Soft X-ray Wide Field Camera will be provided by a consortium of British groups. It is at these low energies that observations of relatively nearby stars will be made through the 'haze' of the interstellar medium.

The EXUV waveband is an almost virgin territory and likely to yield an abundance of sources, colder than classical X-ray sources, being white dwards visible only in this energy range. Furthermore, spectral features relating to the interstellar medium itself will be studied against the light of X-ray emitters, as well as the diffuse soft X-ray emission reported by Kraushaar (1973).

3. HIGH ENERGY GAMMA-RAY ASTRONOMY (30 MeV-10 GeV) - PHOTONS ARE BECOMING SCARCE!

As opposed to X-ray astronomy, gamma-ray astronomy began with the goal of surveying the emission from the galaxy. Early, albeit, optimistic predictions of gamma ray emission from cosmic-ray interstellar medium interactions led to the birth of gamma-ray astronomy pursued with balloon observations. While yielding a detection of the Crab

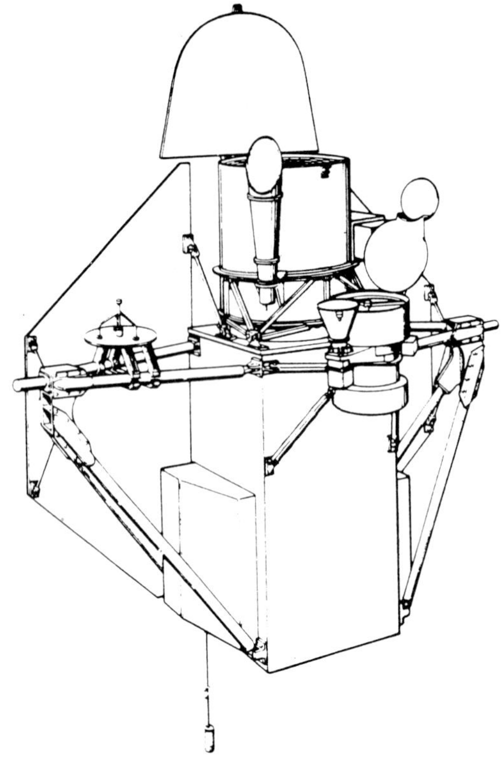

Figure 5: The ROSAT experiment and spacecraft concept.

(e.g. Browning et al. 1971) and a hint of galactic emission from the disc (Sood et al. 1969), it was left to satellite missions to make the major impact. It was OSO III in 1968 (Kraushaar et al. 1972) which yielded the first data to show the gamma-ray emission of the galaxy as a function of longitude. The scintillation telescope used in OSO III had only limited angular resolution (24° FWHM), so despite providing an observational milestone, the interpretation of the data was ambiguous.

It was not until the launch of SAS II (Fichtel et al. 1975) in 1972 and COS-B (Scarsi et al. 1977) in 1975 that the observational situation improved significantly. The detection techniques of these two missions was essentially the same, so I will use the example of COS-B shown in Figure 6 by way of explanation: At energies > 30 MeV gamma-rays interact with matter by the pair production process. In COS-B, this interaction takes place within a spark chamber (SC), which is triggered by a 3-element scintillation counter telescope (B1, B2, C). A plastic scintillator dome surrounds these two units to detect and discriminate against charged particles which would otherwise trigger the system. Beneath the telescope is an energy calorimeter consisting of a Caesium Iodide scintillator which absorbs the electron-positron pair originating from the gamma-ray.

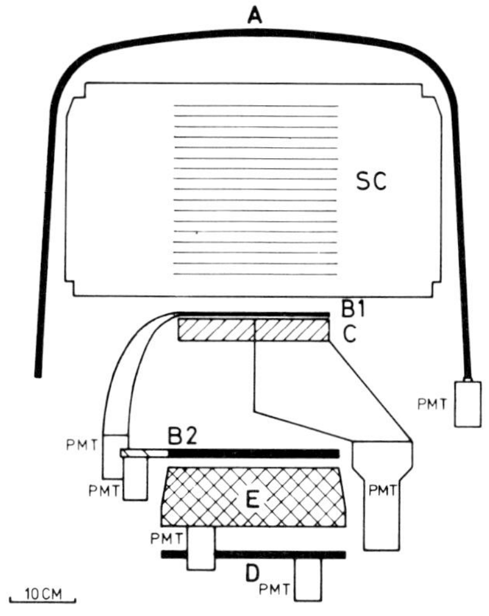

Figure 6: Sectional view of the COS-B experiment.

The detection parameters of COS-B are illustrated in Figure 7, and these are such that about 1-2 months are necessary to yield sufficient detection sensitivity. The field-of-view is roughly 30° FWHM, so within the mission lifetime of almost 7 years, the galactic plane has been surveyed in great detail. A malfunction on SAS-II unfortunately limited

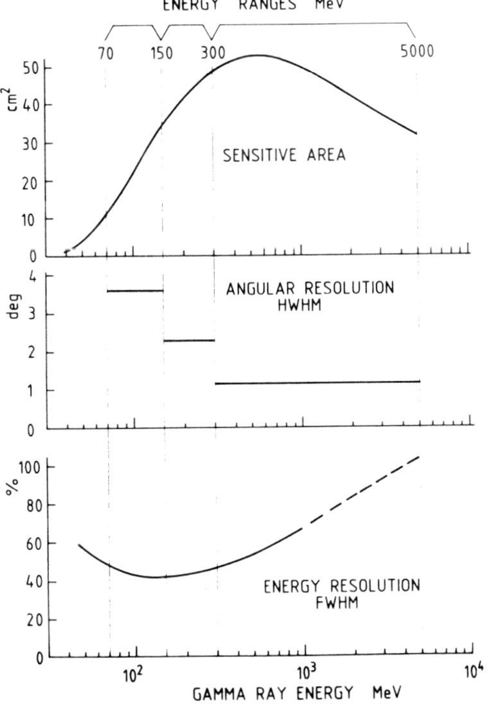

Figure 7: Detection parameters of COS-B.

its operational lifetime to 7 months, nevertheless, a significant contribution to gamma-ray astronomy was made, namely:

- extensive survey of the galactic disc
- discovery of the Vela gamma-ray pulsar
- discovery of the gamma-ray γ-195 source in the anticentre
- measurement of the diffuse cosmic gamma-ray background.

With the benefit of extended mission lifetime (almost 7 years), COS-B has been able to confirm and, in many cases, extend the observations of SAS-II. A complete galactic survey, essentially in agreement with the SAS-II results, was achieved and is presented at this conference by W. Hermsen (these proceedings).

As regards pulsars, searches in the data using contemporary radio parameters, has failed to yield more than the two seen in the SAS-II data, namely PSR 0531+21 (Crab) and PSR 0833-45 (Vela). Repeated observations have, however, permitted COS-B to examine the evolution in the light curves throughout the lifetime of the mission. In Figure 8 the observed light curves are shown for several epochs. Ignoring the absolute intensities (the histograms show only counts, not fluxes) it is evident that the Crab first to second pulse intensity is variable.

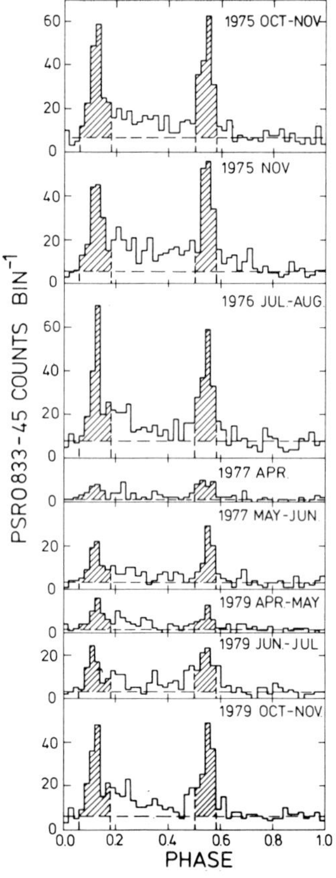

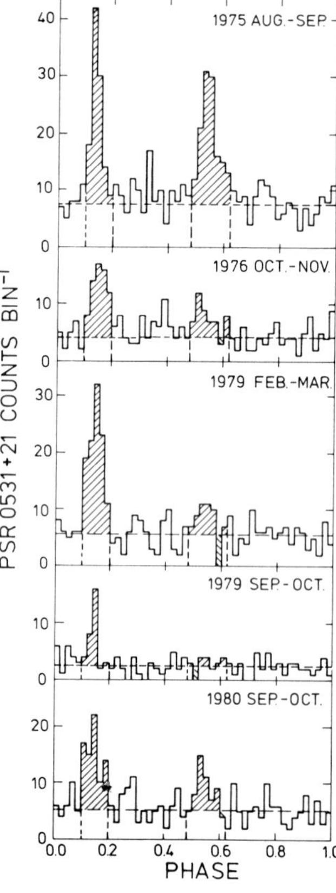

Figure 8:
COS-B gamma-ray light curves of PSR0833-45 (Vela) and PSR0531+21 (Crab) at the epochs indicated.

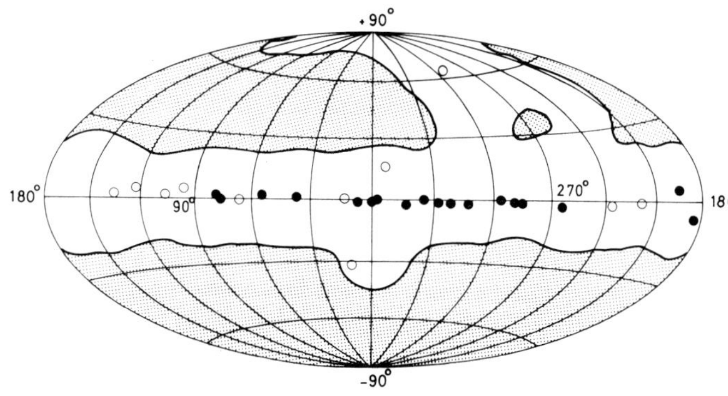

Figure 9:
The gamma-ray sources seen by COS-B.

COS-B has observed some 29 regions of localised emission, including those observed by SAS-II. These are distributed as shown in Figure 9. Until now only the 2 pulsars have been identified as galactic sources while CG291+65 has been identified as the quasar 3C273 - the first extragalactic gamma-ray source at these energies. This remains so despite innumerable searches for counter-parts within the COS-B error box as summarised by Bignami and Hermsen (1983).

4.2 Future Gamma Ray Surveys

It is evident that both SAS-II and COS-B have made dramatic achievements in galactic gamma-ray astronomy and gamma-rays have proved to be a most important tracer of the interstellar medium. At lower gamma-ray energies (< 30 MeV) survey data are still meagre.

The next major gamma-ray survey mission will be the NASA Gamma Ray Observatory (GRO) to be launched in 1988. GRO carries 4 major instruments complementary in energy range and detection techniques as indicated in Table 1.

Table 1: GRO Payload

Experiment	Technique	Energy Range	Angular Resolution	Source Sensitivity	Location Accuracy
EGRET	Spark chambers	10 MeV-10 GeV	~ 1 deg.	0.01 Crab	5'
COMPTEL	Compton telescope	1-30 MeV	~ 1 deg.	0.01 Crab	5'
OSSE	Scint. telescope	.1-10 MeV	~ 7 deg.	.005 Crab	10'
BATSE	Scintillators	50-600keV	4 π FOV	Gamma bursts	~ deg.

While it is clear that a prime task of the EGRET instrument will be to further the survey and point source work of COS-B and SAS-II, COMPTEL and OSSE embark on a exploration of an, as yet, poorly known energy range, but which is destined to show up localised and diffuse sources - for instance the interstellar medium, bombarded by low energy cosmic rays will emit many gamma ray lines and high energy electrons will emit bremsstrahlung radiation by their interaction with the ISM.

ACKNOWLEDGEMENT

I am grateful for the assistance of Dr. Mark Sims in the preparation of this presentation.

REFERENCES

Bignami, G.F. and Hermsen, W.: 1983, Ann. Rev. of Astr. & Astrophys. 21.
Browning, R. et al., 1971, Nature Phys. Sci. 232, 99-101.
Fichtel, C.E. et al., 1975, Astrophys. J. 198, 163-182.
Giacconi, R. et al., 1971, Astrophys. J. 165, 127.
Giacconi, R. et al., 1979, Astrophys. J. 230, 540.
Giacconi, R. et al., 1981, Space Sci. Rev. 30, 3-32.
Kraushaar, W.L., 1973, Proc. Symposium "Gamma Ray Astrophysics", NASA SP-339.
Kraushaar, W.L., 1972, Astrophys. J. 177, 341.
Peacock, A. et al., 1981, Space Sci. Rev. 30 525-534.
Scarsi, L. et al., 1977, Proc. 12th ESLAB Symp., ESA SP-124, 3-11.
Sood, R.K., 1969, Nature 222.
Taylor, B.G. et al., 1981, Space Sci. Rev. 30, 479-494.
Trümper, J., MPE Preprint No. 10.
van Speybroeck, L. et al., 1979, Astrophys. J. 234, 145-147.

High Energy Satellite Survey

X-rays

4th UHURU Catalogue: Foreman, W. et al., 1979, Ap. J. Suppl. 39.
Ariel 5 Catalogue ($|b|<10$): Warwick, R.S. et al., 1981, MNRAS 197, 865.
Ariel 5 Catalogue ($|b|>10$): McHardy, I.M. et al., 1981, MNRAS 197, 893.
General: Ammuel, P.R. et al., 1979, Ap. J. Suppl. 41, 327.
Diffuse soft X-ray background (southern hemisphere): Sanders et al., 1977, Ap. J. 217, L87.
Diffuse soft X-ray background: McCammon, D., 1979, Proc. 16th ICRC OG S-3.

Gamma-Rays

2 CG catalogue of COS-B gamma rays sources: Swanenburg, B.N. et al., 1981, Ap. J. 243, L69.
Large scale distribution of gamma radiation observed by COS-B. Mayer-Hasselwander, H.A. et al., 1982, Astron. and Astrophys. 164.

THE HIPPARCOS MISSION - ASTROMETRY FROM SPACE

M.A.C. Perryman
Space Science Department of ESA,
ESTEC, Noordwijk, The Netherlands

ABSTRACT

A summary of the principal scientific objectives of the ESA space astrometry mission HIPPARCOS is presented, along with a concise description of the payload and its operations.

1. AN OVERVIEW OF THE SCIENTIFIC OBJECTIVES

The scientific goals of the space-astrometry mission HIPPARCOS are the accurate measurements of the positions, proper motions and trigonometric parallaxes of some 100 000 selected stars, mostly brighter than magnitude B = 10 mag and with a limiting magnitude of about B = 13 mag. After the nominal satellite lifetime of 2.5 years, the accuracy expected for stars brighter than magnitude B = 9 mag is 0.002 arcsec in the parallaxes, and 0.002 arcsec in each component of the positions and annual proper motions.

The mission will provide a uniform whole-sky stellar catalogue suitable for detailed astrometric and astrophysical studies. Compared with existing stellar catalogues, the HIPPARCOS catalogue will offer (in addition to the significant improvement on the errors of these quantities) absolute rather than relative parallaxes, a relatively dense reference network, and homogeneous sky coverage. In particular, it will ensure a firm connection between the northern and southern celestial hemispheres.

The large number of stars contained in the HIPPARCOS observing programme (some 2.5 stars per square degree) will fulfil two fundamental and interdependent roles:
- irrespective of their detailed astrophysical importance, the precise positions will result in a basic reference frame against which all celestial objects, detected in different wavebands, can be identified;

W. B. Burton and F. P. Israel (eds.), Surveys of the Southern Galaxy, 281–284.

- at the same time, the changes in position of celestial objects relative to this reference frame will be measured. It is the changes of position of stars within the reference frame that allows the measurement of stellar distances, luminosities, etc. (via the parallaxes) and which permits the study of the dynamics of the solar system and of our galaxy (via the proper motions).

Some of the most spectacular advances to be expected from the mission are likely to come from the fivefold increase in precision of measurements of trigonometric parallax compared with typical earth-based observations, and from the very much larger number of stars that will be measurable. Presently, for all but late K and M dwarfs, calibration of the Hertzsprung-Russell diagram generally involves statistical assumptions about the kinematic behaviour of stars or, in the case of main-sequence fitting, assumptions about the relative luminosity of stars with the same spectral type in different clusters, which may be significantly affected by chemical composition. It is clearly desirable that direct calibration by means of trigonometric parallaxes should be carried out for as much of the Hertzsprung-Russell diagram as possible. The increase by a factor of about five in the precision with which absolute parallaxes will be measurable by HIPPARCOS extends the effective horizon for individual significant parallaxes to about 100 pc.

Further details of the scientific objectives of the HIPPARCOS mission may be found in the Proceedings of the Colloquium on the Scientific Aspects of the HIPPARCOS Space Astrometry Mission, held in Strasbourg earlier in the year (Perryman & Guyenne 1982). Within this volume, articles by Fricke (1982), Wielen (1982) and Blaauw (1982) provide more detailed discussions of the present situation concerning reference systems, proper motions and parallaxes respectively, and of the contributions in these areas to be expected as result of the HIPPARCOS mission.

In parallel with the main astrometric experiment, a further experiment, TYCHO, will result in supplementary astrometric and photometric data whilst generating attitude information necessary for operation of the primary detection system. A preliminary investigation into the results expected from the TYCHO experiment suggests that, in addition to the high-precision positions, proper motions and parallaxes obtained for the 100 000 pre-selected programme stars, positions for all stars down to magnitudes B = 11-12 mag (some 400 000 or more stars) will be derived with an accuracy of some 0.05 arcsec. Although an order of magnitude poorer than the results for the pre-selected programme stars, this accuracy is considerably better than that routinely obtainable from the ground. The positions will be directly tied into the HIPPARCOS reference grid, providing a dense stellar network of some 10 or more stars per square degree. Two-colour photometric information will be determined for each star at several different epochs throughout the 2.5 year mission lifetime. Høg, Jaschek & Lindegren (1982) and Jaschek (1982) provide further details on the goals and motivations for TYCHO.

2. PAYLOAD DESCRIPTION

The basic principle of observation is to scan the entire sky continuously and systematically with a telescope capable of accurately measuring the angles between widely separated stars. By making the observations from space, the effects of the Earth's rotation, of latitude variations, of refraction, and of instrumental flexure due to the Earth's gravity will be eliminated.

The angles are measured by superimposing in the focal plane of a single telescope two fields of view, approximately 70° apart on the sky, each field containing one of the stars in a pair. The angle between the two fields of view, referred to as the basic angle, is maintained by a complex mirror in front of the telescope, splitting its entrance pupil into two parts and directing the light from one star into one half of the entrance pupil and that from the other star into the other half.

The attitude of the satellite about its centre of gravity is controlled in such a way as to scan the whole celestial sphere in a regular movement. As a result of the scanning motion, the star images move across the focal surface of the telescope. Measurements are made by means of a system of grids situated at the focal surface, composed of alternately opaque and transparent bands normal to the apparent direction of motion of the stars, which modulate the incoming light. Behind these grids, the field of view of an image dissector tube follows the star paths, converting the modulated light into a sequence of photon counts from which the phase of the entire pulse train corresponding to a star can be derived. The apparent angle between two stars in the combined field of view is obtained from the phase difference between the corresponding pulse trains compared by interlacing measurements of different stars.

In order to direct the instantaneous field of view of the image dissector tube to the correct position and follow a star image correctly across the field of view, the celestial co-ordinates of the star and the triaxial attitude of the instrument must be known in real time to within a few seconds of arc. Star co-ordinates are uplinked from the ground, and the attitude determination is performed on board, using periodic observations with a star mapper, and continuous sampling of the on-board gyroscope signals.

The basic principle of observation is one-dimensional, i.e. to measure instantaneous relative positions of stars along the scanning great circles. Since all five astrometric parameters of a particular star combine to produce a certain displacement at the instant of observation, each star has to be observed at at least five epochs, epochs at which the astrometric parameters combine in distinctly different ways.

The photon counts generated by the image dissector tube are transmitted to the ground and are processed to yield the times at which the star images crossed the different slits, and hence the relative positions of the stars in the combined field of view. The true angle between two stars belonging to different fields of view is obtained by adding the value of the basic angle to the apparent angle in the combined field of view.

The grid system, located at the focal surface of the telescope, comprises two sets of transparent slits manufactured on the curved surface of a supporting glass plate: (a) the main modulating grid, which permits the accurate determination of the one-dimensional coordinates of the programme stars as they move across the field of view, as described above; and (b) the star mapper grids, used for the determination of the three-axis attitude of the telescope and for the TYCHO experiment.

Based on detailed payload models and on simulations, the final expected precision on the astrometric parameters as a function of star magnitude and ecliptic coordinates can be derived. Table 1 shows a model of the possible stellar distribution of the final HIPPARCOS Catalogue, with the resulting mean sky errors for each of the astrometric errors tabulated.

Table 1. Expected number of stars in the HIPPARCOS Catalogue as a function of magnitude, and the expected astrometric accuracy.

B	Number of stars	Completeness (%)	Mean errors (milli-arcsec) λ, β	μ_λ, μ_β	$\tilde{\omega}$
0 - 6	3 000	100	1.0	1.4	1.4
6 - 7	5 400	100	1.0	1.4	1.4
7 - 8	14 800	100	1.1	1.5	1.5
8 - 9	40 800	100	1.2	1.6	1.7
9 -10	16 000	15	1.3	1.9	1.9
10-11	12 000	5	1.7	2.3	2.4
11-12	6 000	0.8	2.2	3.0	3.1
12-13	2 000	0.1	3.0	4.3	4.3

Further details of the payload design and of the satellite operation, and of the scientific organisation of the project are given by Kovalevsky (1982) and Perryman (1982) respectively. The final astrometric catalogue may be available by about 1991.

REFERENCES

Blaauw, A., 1982. ESA SP-177 pp 97-100.
Fricke, W., 1982. ESA SP-177 pp 43-47.
Høg, E., Jaschek, C. & Lindegren, L., 1982. ESA SP-177 pp 21-25.
Jaschek, C., 1982. ESA SP-177. pp 133-135.
Kovalevsky, J., 1982. ESA SP-177 pp 15-20.
Perryman, M.A.C., 1982. ESA SP-177 pp 31-38.
Perryman, M.A.C. & Guyenne, T.D. (eds). 'Scientific Aspects of the HIPPARCOS Space Astrometry Mission', Proceedings of the Strasbourg Colloquium, ESA SP-177, 1982.

CLOSING SUMMARY

F. J. Kerr
University of Maryland

I have been asked to make a few summarizing remarks. This has been an interesting workshop, with a small enough attendance to make good discussions possible.

Returning to Leiden is always interesting for people involved in galactic structure studies. Many of us know the city and the old Sterrewacht well from former visits, and it has been especially moving to hold the meetings in the "real" Sterrewacht and in the very room where Henk van de Hulst made his famous prediction in 1944 about the detectability of the 21-cm hydrogen line. Those of us with Australian connections are very conscious of the fact that Australia was "New Holland" for over 150 years in the early discovery period. The name New Zealand survived in world geography, but New Holland did not.

Probably the most important of the material presented to this workshop was the new CO work in the south. A new method of investigating the spiral structure and other aspects of the interstellar material is now available for the whole galactic circle. The CO survey by the CSIRO group is the most elegant so far, and adds a great deal of new information. The Dutch and Queen Mary College groups have also obtained some interesting results, while the Columbia group is nearly ready to begin a large survey from Chile.

Most of the meeting was concerned with radio astronomical work. More attention might have been given to the optical and infrared spectral ranges, but there was probably not enough time for this to be done. A selection always has to be made.

We have heard several suggestions about the spiral pattern for the whole Galaxy or for just the outer parts, based on either a two-armed or four-armed structure. I feel there is a tendency to look too hard for a very regular pattern. We should remember that in external galaxies, the spiral structure is never perfectly regular, but shows a greater or lesser degree of irregularity. I prefer an approach in which each major spiral feature is looked at separately,

W. B. Burton and F. P. Israel (eds.), Surveys of the Southern Galaxy, 285–286.

instead of being necessarily a part of a single "grand design". Historically the extreme example of a solution with an improbable type of regularity was presented by a Soviet astronomer in 1962, based on a study of 21-cm observations. This model consisted of a 12-armed spiral, in which essentially every major hydrogen complex has led to a distinct spiral arm.

There has been a bias in some of the CO surveys so far towards looking at HII regions. We need more searches which are quite general and objective. Another need is for the determination of the outer-region rotation curve for the southern side of the Galaxy, as has been done so well for the northern side.

We had some controversies about "clouds" and the extent to which they define spiral arms. Much of the argument seems to arise from the fact that people talk about different things, as each group's particular technique tends to control the size of object which is regarded as a "cloud".

Discussions of symmetry and asymmetry were frequently heard at the workshop, arising from the increased interest in comparisons between south and north. Such discussions arose in studies of CO, HI, and gamma rays.

Astrophysics and observations are intertwined in derivations of the molecular hydrogen abundance from CO-line observations. It is clear that the various steps in this conversion cannot yet be carried out in a manner which is beyond controversy. One point which may not have received sufficient attention is the possibility that the conversion may be different in the inner and outer parts of the Galaxy, due to metal abundance differences.

Several papers at the workshop pointed to the importance of studies of other galaxies for the understanding of our own Galaxy. Extragalactic and Galactic observations must continue to go hand-in-hand.

Space observations are beginning to contribute to the subjects discussed at the workshop, and will continue to do so to an increasing extent. It is worth noting that in studies from space there is no longer any disparity in observations of the southern and northern hemispheres.

In closing, I want to thank the organizers and our hosts for all their contributions, especially for establishing an atmosphere that was just right for a successful workshop.

INDEX OF SOUTHERN SURVEY LISTINGS

INDEX OF SURVEYS DESCRIBED IN THIS VOLUME

AUTHOR INDEX

SUBJECT INDEX

OBJECT INDEX - GALACTIC OBJECTS

OBJECT INDEX - EXTRAGALACTIC OBJECTS

A1.

HI EMISSION FROM THE GALACTIC EQUATOR (ℓ,v)

DATA: F.J. Kerr, P.F. Bowers, and M. Kerr, 1983, in preparation.
G. Westerhout and H.H. Wendlandt, 1982, Astronomy and Astrophysics Supplement Series, Vol. 49, 143.
W.B. Burton and H.S. Liszt, 1983, Astronomy and Astrophysics Supplement Series, Vol. 52, 63.

MAP PREPARATION: H.S. Liszt, National Radio Astronomy Observatory, Charlottesville.

The map shows the longitude, velocity distribution in the galactic equator of emission from the hyperfine line at λ21 cm of neutral atomic hydrogen. The velocities are expressed with respect to the local standard of rest.

The data along the southern galactic equator at $\ell < 350^\circ$ were gathered with the CSIRO 64-m telescope in Parkes with a HPBW of 14'. The data at $\ell > 10^\circ$ were observed with the 9' HPBW of the Effelsberg 100-m telescope of the Max Planck Institute for Radio Astronomy. Both of these sets of data involved almost complete sampling in longitude. The data at $350^\circ \leq \ell \leq 10^\circ$ were sampled at half-degree longitude intervals using the NRAO 140-foot telescope in Green Bank. The intensities of all three sets of data are in comparable brightness temperature units based on observations of the IAU standard fields.

The longitude and velocity scales of this map are the same as those of the CO ℓ,v map.

A2.

CO EMISSION FROM THE GALACTIC EQUATOR (ℓ,v)

DATA: B.J. Robinson, W.H. McCutheon, R.N. Manchester, and J.B. Whiteoak, this volume.
R.N. Manchester, J.B. Whiteoak, B.J. Robinson, R.E. Outrupcek, and C.J. Rennie, this volume.
D.B. Sanders and P.M. Solomon, this volume.

MAP PREPARATION: H.S. Liszt, National Radio Astronomy Observatory, Charlottesville.

The map shows the longitude, velocity distribution in the galactic equator of emission from the J=1-0 line at λ2.6 mm of the molecule $^{12}C^{16}O$. The velocities are expressed with respect to the local standard of rest.

The data along the southern equator between $\ell=294^{\circ}$ and 358° were gathered by Manchester, Robinson, and collaborators with the 4-m Cassegrain telescope of the CSIRO Division of Radiophysics. The effective angular resolution of 8' arc is determined by the set of nine observations on a 3' grid centered on each target position. After smoothing the velocity resolution of the southern data is 1.6 km s^{-1}.

The data along the equator between $\ell=358^{\circ}$ and $\ell=90^{\circ}$ were gathered by Sanders and Solomon with the 14-m telescope of the Five College Radio Astronomy Observatory. The sampling interval of the northern data is 3' arc. Away from the galactic core, the velocity resolution is 2 km s^{-1}. Both the northern and the southern data sets represent intensity units of T_A^*.

The longitude and velocity scales of this map are the same as those of the HI ℓ,v map.

A3.

408 MHz RADIO CONTINUUM FLUX AT $|b| < 40^{\circ}$

DATA: C.G.T. Haslam, C.J. Salter, H. Stoffel, and W.E. Wilson, 1982, Astronomy and Astrophysics Supplement Series, Vol. 47, 1-143.

MAP PREPARATION: C.G.T. Haslam, Max Planck Institut für Radioastronomie, Bonn.

The map shows the radio continuum brightness at a frequency of 408 MHz (λ73 cm). The atlas of the all-sky survey published by Haslam et al. contains data from four surveys, made with the Jodrell Bank Mk I (anticenter region), the Bonn 100-m ($-8^{\circ} < \delta < 48^{\circ}$), the Parkes 64-m (southern sky), and the Jodrell Bank Mk I A (north celestial pole region) telescopes. The respective references for the four component surveys are Haslam, Quiqley, and Salter, 1970, MNRAS 147, 405; Haslam, Wilson, Graham, and Hunt, 1974, A&A Suppl. 13, 359; Haslam, Wilson, Cooke, Cleary, Graham, Wielebinski, and Day, 1975, Proc. ASA 2 (6), 331; and Haslam, Klein, Salter, Stoffel, Wilson, Cleary, Cooke, and Thomasson, 1981, A&A 100, 209. The angular resolution of the original data base is $0\overset{\circ}{.}85$. For the map in galactic coordinates produced here, the data have been smoothed and placed on a 2° rectangular grid. The contours represent brightness temperatures labelled in Kelvins.

A4.

HIGH-ENERGY GAMMA-RAY EMISSION (ℓ,b)

DATA: COS-B Caravane Collaboration: H.A. Mayer-Hasselwander, K. Bennett, G.F. Bignami, R. Buccheri, P.A. Caraveo, W. Hermsen, G. Kanback, F. Lebrun, G.G. Lichte, J.L. Masnon, J.A. Paul, K. Pinkau, B. Sacco, L. Scarsi, B.N. Swanenburg, and R.D. Wills, 1982, Astronomy and Astrophysics, Vol. 105, 164.

MAP PREPARATION: J.B.G.M. Bloemen, Cosmic Ray Working Group and Sterrewacht, Leiden.

The maps show the distribution on the plane of the sky of gamma-ray emission as measured by the COS-B satellite. Three of the maps show the photon-intensity rate in units of ph $cm^{-2}s^{-1}sr^{-1}$. The fourth map gives an on-axis count rate for the total observed energy range. Although the photon intensity is a directly interpretable physical unit, the on-axis-rate (on-axis counts $s^{-1}sr^{-1}$) is, as discussed by Mayer-Hasselwander et al., a more useful unit for showing the spatial properties of the gamma-ray emission as observed over the full energy range of the COS-B experiment.

Map 1. Gamma-ray emission over the energy range 70-150 MeV. Contours are drawn at multiples of 5×10^{-5} ph $cm^{-2}s^{-1}sr^{-1}$. The black/white division occurs at 2.5×10^{-4} ph $cm^{-2}s^{-1}sr^{-1}$.

Map 2. Gamma-ray emission over the energy range 150-300 MeV. Contours are drawn at multiples of 3×10^{-5} ph $cm^{-2}s^{-1}sr^{-1}$. The black/white division occurs at 1.5×10^{-4} ph $cm^{-2}s^{-1}sr^{-1}$.

Map 3. Gamma-ray emission over the energy range 300-5000 MeV. Contours are drawn at multiples of 4×10^{-5} ph $cm^{-2}s^{-1}sr^{-1}$. The black/white division occurs at 1.6×10^{-4} ph $cm^{-2}s^{-1}sr^{-1}$.

Map 4. Gamma-ray emission over the energy range 70-5000 MeV. Contours are drawn at multiples of 3×10^{-3} on-axis counts $s^{-1}sr^{-1}$. The black/white division occurs at 2.1×10^{-2} on-axis counts $s^{-1}sr^{-1}$.

A5.

INTEGRATED HI EMISSION (ℓ,b)

DATA: H.F. Weaver and D.R.W. Williams, 1973, Astronomy and Astrophysics Supplement Series, Vol. 8, 1.

C. Heiles and H.J. Habing, 1974, Astronomy and Astrophysics Supplement Series, Vol. 14, 1.

C. Heiles and M.N. Cleary, 1979, Australian Journal of Physics Supplement, Vol. 47, 1.

A.W. Strong, P.A. Riley, J.L. Osborne, and J.D. Murray, 1982, Monthly Notices of the Royal Astronomical Society, Vol. 201, 495.

MAP PREPARATION: J.B.G.M. Bloemen, Cosmic Ray Working Group and Sterrewacht, Leiden.

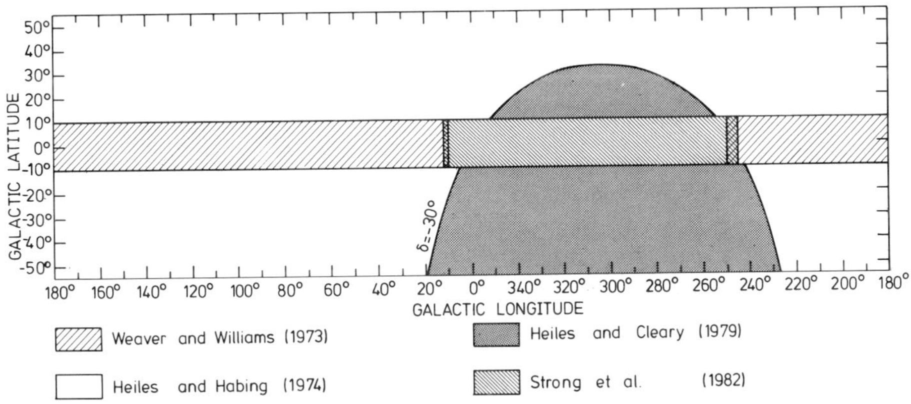

The line-of-sight integrals $\int T_b dv$ of the brightness temperature of the HI 21-cm line were calculated from the four surveys, containing in total some 190,000 spectra, whose sky coverage is given by the above figure. The procedure for combining the surveys involved projection onto a uniform grid with spacings $0\overset{\circ}{.}5$ in longitude and $0\overset{\circ}{.}25$ in latitude. The effective resolution of the maps in set by the grid spacings used by the individual surveys. Unobserved points were provided by interpolation. The antenna temperature scales were converted to a uniform brightness temperature scale using the conversion factors suggested by the authors of the various surveys based on observations of the IAU standard fields. Overlapping parts of the surveys were averaged.

A lower limit to the HI column density in the velocity intervals represented on each map is provided by the integral $1.823 \times 10^{18} \int T_b dv$ H atom cm^{-1}. At intermediate and high latitudes this estimate is probably reliable; within a few degrees of the galactic equator optical depth effects may be severe.

Map 1. $\int T_b dv$ calculated over the velocity range given by the different surveys. Contours are drawn at levels of 240, 300, 380, 460, 540, 680, 800, 1000, 1200, 1500, 2000, 2500, 3500, 4000, 5000, ..., 10 000, and 12000 K km s^{-1}. The black/white division for intense regions near the galactic equator occurs at 3000 K km s^{-1}; the white regions at high latitudes represent integrals less than 100 K km s^{-1}.

Map 2. $\int T_b dv$ calculated over the velocity range $|v| \leq 20$ km s^{-1}, representing gas near the solar circle distance R_0. Except near the equator, this gas is predominantly local. Contours are drawn at levels of 150, 200, ..., 400, 500, ..., 800, 950, 1100, 1350, 1700, 1900, 2200, ..., 3500, 3900, 4300 K km s^{-1}. The white regions represent integrals less than 80 K km s^{-1}.

Map 3. $\int T_b dv$ calculated over velocities $v > +20$ km s^{-1} at $\ell < 180^o$ and over $v < -20$ km s^{-1} at $\ell > 180^o$. This velocity selection allows depiction of gas predominantly interior to the solar circle. Contours are drawn at levels of 80, 120, 180, 260, 380, 500, 700, 900, 1100, 1500, 2560, 3000, 3500, 4000, 5000, 6000, 7000, 8000 K km s^{-1}. The black/white division for intense regions near the galactic equator occurs at 2000 K km s^{-1}; the white regions at high latitudes represent integrals less than 30 K km s^{-1}.

Map 4. $\int T_b dv$ calculated over velocity $v < -20$ km s^{-1} at $\ell < 180^o$ and over $v > +20$ km s^{-1} at $\ell > 180^o$. This velocity selection allows depiction of gas predominantly exterior to the solar circle. Contours are drawn at levels of 80, 120, 180, 260, 380, 500, 700, 900, 1100, 1500, 2500, 3000, 3500, 4000, 5000, 6000 K km s^{-1}. The black/white division for intense regions near the galactic equator occurs at 200 K km s^{-1}; the white regions at high latitudes represent integrals less than 30 K km s^{-1}.

A6.

HI SURFACE DENSITY IN M31

DATA: E. Brinks and W.W. Shane, 1983, submitted to Astronomy and Astrophysics Supplement Series.

MAP PREPARATION: E. Brinks, Sterrewacht, Leiden.

The map of integrated HI emission from M31 was constructed from observations made with the Westerbork Synthesis Radio Telescope. The noise varies across the map according to the locations of the primary beam field centers; the noise increases rapidly as the extreme northern and southern ends of the 140' x 50' area of the map are reached. The mean HPBW of 24" x 36" corresponds to a linear resolution of 80 x 120 pc. This resolution is approximately the same as that available in the surveys of our own galaxy made with a 25-m dish.

ASTROPHYSICS AND SPACE SCIENCE LIBRARY

Edited by

1. C. de Jager (ed.), *The Solar Spectrum, Proceedings of the Symposium held at the University of Utrecht, 26–31 August, 1963.* 1965, XIV + 417 pp.
2. J. Orthner and H. Maseland (eds.), *Introduction to Solar Terrestrial Relations, Proceedings of the Summer School in Space Physics held in Alpbach, Austria, July 15–August 10, 1963 and Organized by the European Preparatory Commission for Space Research.* 1965, IX + 506 pp.
3. C. C. Chang and S. S. Huang (eds.), *Proceedings of the Plasma Space Science Symposium, held at the Catholic University of America, Washington, D.C., June 11–14, 1963.* 1965, IX + 377 pp.
4. Zdeněk Kopal, *An Introduction to the Study of the Moon.* 1966, XII + 464 pp.
5. B. M. McCormac (ed.), *Radiation Trapped in the Earth's Magnetic Field. Proceedings of the Advanced Study Institude, held at the Chr. Michelsen Institute, Bergen, Norway, August 16–September 3, 1965.* 1966, XII + 901 pp.
6. A. B. Underhill, *The Early Type Stars.* 1966, XII + 282 pp.
7. Jean Kovalevsky, *Introduction to Celestial Mechanics.* 1967, VIII + 427 pp.
8. Zdeněk Kopal and Constantine L. Goudas (eds.), *Measure of the Moon. Proceedings of the 2nd International Conference on Selenodesy and Lunar Topography, held in the University of Manchester, England, May 30–June 4, 1966.* 1967, XVIII + 479 pp.
8. J. G. Emming (ed.), *Electromagnetic Radiation in Space. Proceedings of the 3rd ESRO Summer School in Space Physics, held in Alpbach, Austria, from 19 July to 13 August, 1965.* 1968, VIII + 307 pp.
10. R. L. Carovillano, John F. McClay, and Henry R. Radoski (eds.), *Physics of the Magnetosphere, Based upon the Proceedings of the Conference held at Boston College, June 19–28, 1967.* 1968, X + 686 pp.
11. Syun-Ichi Akasofu, *Polar and Magnetospheric Substorms.* 1968, XVIII + 280 pp.
12. Peter M. Millman (ed.), *Meteorite Research. Proceedings of a Symposium on Meteorite Research, held in Vienna, Austria, 7–13 August, 1968.* 1969, XV + 941 pp.
13. Margherita Hack (ed.), *Mass Loss from Stars. Proceedings of the 2nd Trieste Colloquium on Astrophysics, 12–17 September, 1968.* 1969, XII + 345 pp.
14. N. D'Angelo (ed.), *Low-Frequency Waves and Irregularities in the Ionosphere. Proceedings of the 2nd ESRIN-ESLAB Symposium, held in Frascati, Italy, 23–27 September, 1968.* 1969, VII + 218 pp.
15. G. A. Partel (ed.), *Space Engineering. Proceedings of the 2nd International Conference on Space Engineering, held at the Fondazione Giorgio Cini, Isola di San Giorgio, Venice, Italy, May 7–10, 1969.* 1970, XI + 728 pp.
16. S. Fred Singer (ed.), *Manned Laboratories in Space. Second International Orbital Laboratory Symposium.* 1969, XIII + 133 pp.
17. B. M. McCormac (ed.), *Particles and Fields in the Magnetosphere. Symposium Organized by the Summer Advanced Study Institute, held at the University of California, Santa Barbara, Calif., August 4–15, 1969.* 1970, XI + 450 pp.
18. Jean-Claude Pecker, *Experimental Astronomy.* 1970, X + 105 pp.
19. V. Manno and D. E. Page (eds.), *Intercorrelated Satellite Observations related to Solar Events. Proceedings of the 3rd ESLAB/ESRIN Symposium held in Noordwijk, The Netherlands, September 16–19, 1969.* 1970, XVI + 627 pp.
20. L. Mansinha, D. E. Smylie, and A. E. Beck, *Earthquake Displacement Fields and the Rotation of the Earth, A NATO Advanced Study Institute Conference Organized by the Department of Geophysics, University of Western Ontario, London, Canada, June 22–28, 1969.* 1970, XI + 308 pp.
21. Jean-Claude Pecker, *Space Observatories.* 1970, XI + 120 pp.
22. L. N. Mavridis (ed.), *Structure and Evolution of the Galaxy. Proceedings of the NATO Advanced Study Institute, held in Athens, September 8–19, 1969.* 109 1971, VII + 312 pp.

23. A. Muller (ed.), *The Magellanic Clouds. A European Southern Observatory Presentation: Principal Prospects, Current Observational and Theoretical Approaches, and Prospects for Future Research, Based on the Symposium on the Magellanic Clouds, held in Santiago de Chile, March 1969, on the Occasion of the Dedication of the European Southern Observatory.* 1971, XII + 189 pp.
24. B. M. McCormac (ed.), *The Radiating Atmosphere. Proceedings of a Symposium Organized by the Summer Advanced Study Institute, held at Queen's University, Kingston, Ontario, August 3–14, 1970.* 1971, XI + 455 pp.
25. G. Fiocco (ed.), *Mesospheric Models and Related Experiments. Proceedings of the 4th ESRIN-ESLAB Symposium, held at Frascati, Italy, July 6–10, 1970.* 1971, VIII + 298 pp.
26. I. Atanasijević, *Selected Exercises in Galactic Astronomy.* 1971, XII + 144 pp.
27. C. J. Macris (ed.), *Physics of the Solar Corona. Proceedings of the NATO Advanced Study Institute on Physics of the Solar Corona, held at Cavouri-Vouliagmeni, Athens, Greece, 6–17 September 1970.* 1971, XII + 345 pp.
28. F. Delobeau, *The Environment of the Earth.* 1971, IX + 113 pp.
29. E. R. Dyer (general ed.), *Solar-Terrestrial Physics/1970. Proceedings of the International Symposium on Solar-Terrestrial Physics, held in Leningrad, U.S.S.R., 12–19 May 1970.* 1972, VIII + 938 pp.
30. V. Manno and J. Ring (eds.), *Infrared Detection Techniques for Space Research. Proceedings of the 5th ESLAB-ESRIN Symposium, held in Noordwijk, The Netherlands, June 8–11, 1971.* 1972, XII + 344 pp.
31. M. Lecar (ed.), *Gravitational N-Body Problem. Proceedings of IAU Colloquium No. 10, held in Cambridge, England, August 12–15, 1970.* 1972, XI + 441 pp.
32. B. M. McCormac (ed.), *Earth's Magnetospheric Processes. Proceedings of a Symposium Organized by the Summer Advanced Study Institute and Ninth ESRO Summer School, held in Cortina, Italy, August 30–September 10, 1971.* 1972, VIII + 417 pp.
33. Antonin Rükl, *Maps of Lunar Hemispheres.* 1972, V + 24 pp.
34. V. Kourganoff, *Introduction to the Physics of Stellar Interiors.* 1973, XI + 115 pp.
35. B. M. McCormac (ed.), *Physics and Chemistry of Upper Atmospheres. Proceedings of a Symposium Organized by the Summer Advanced Study Institute, held at the University of Orléans, France, July 31–August 11, 1972.* 1973, VIII + 389 pp.
36. J. D. Fernie (ed.), *Variable Stars in Globular Clusters and in Related Systems. Proceedings of the IAU Colloquium No. 21, held at the University of Toronto, Toronto, Canada, August 29–31, 1972.* 1973, IX + 234 pp.
37. R. J. L. Grard (ed.), *Photon and Particle Interaction with Surfaces in Space. Proceedings of the 6th ESLAB Symposium, held at Noordwijk, The Netherlands, 26–29 September, 1972.* 1973, XV + 577 pp.
38. Werner Israel (ed.), *Relativity, Astrophysics and Cosmology. Proceedings of the Summer School, held 14–26 August 1972, at the BANFF Centre, BANFF, Alberta, Canada.* 1973, IX + 323 pp.
39. B. D. Tapley and V. Szebehely (eds.), *Recent Advances in Dynamical Astronomy. Proceedings of the NATO Advanced Study Institute in Dynamical Astronomy, held in Cortina d'Ampezzo, Italy, August 9–12, 1972.* 1973, XIII + 468 pp.
40. A. G. W. Cameron (ed.), *Cosmochemistry. Proceedings of the Symposium on Cosmochemistry, held at the Smithsonian Astrophysical Observatory, Cambridge, Mass., August 14–16, 1972.* 1973, X + 173 pp.
41. M. Golay, *Introduction to Astronomical Photometry.* 1974, IX + 364 pp.
42. D. E. Page (ed.), *Correlated Interplanetary and Magnetospheric Observations. Proceedings of the 7th ESLAB Symposium, held at Saulgau, W. Germany, 22–25 May, 1973.* 1974, XIV + 662 pp.
43. Riccardo Giacconi and Herbert Gursky (eds.), *X-Ray Astronomy.* 1974, X + 450 pp.
44. B. M. McCormac (ed.), *Magnetospheric Physics. Proceedings of the Advanced Summer Institute, held in Sheffield, U.K., August 1973.* 1974, VII + 399 pp.
45. C. B. Cosmovici (ed.), *Supernovae and Supernova Remnants. Proceedings of the International Conference on Supernovae, held in Lecce, Italy, May 7–11, 1973.* 1974, XVII + 387 pp.
46. A. P. Mitra, *Ionospheric Effects of Solar Flares.* 1974, XI + 294 pp.
47. S.-I. Akasofu, *Physics of Magnetospheric Substorms.* 1977, XVIII + 599 pp.

48. H. Gursky and R. Ruffini (eds.), *Neutron Stars, Black Holes and Binary X-Ray Sources.* 1975, XII + 441 pp.
49. Z. Švestka and P. Simon (eds.), *Catalog of Solar Particle Events 1955–1969. Prepared under the Auspices of Working Group 2 of the Inter-Union Commission on Solar-Terrestrial Physics.* 1975, IX + 428 pp.
50. Zdeněk Kopal and Robert W. Carder, *Mapping of the Moon.* 1974, VIII + 237 pp.
51. B. M. McCormac (ed.), *Atmospheres of Earth and the Planets. Proceedings of the Summer Advanced Study Institute, held at the University of Liège, Belgium, July 29–August 8, 1974.* 1975, VII + 454 pp.
52. V. Formisano (ed.), *The Magnetospheres of the Earth and Jupiter. Proceedings of the Neil Brice Memorial Symposium, held in Frascati, May 28–June 1, 1974.* 1975, XI + 485 pp.
53. R. Grant Athay, *The Solar Chromosphere and Corona: Quiet Sun.* 1976, XI + 504 pp.
54. C. de Jager and H. Nieuwenhuijzen (eds.), *Image Processing Techniques in Astronomy. Proceedings of a Conference, held in Utrecht on March 25–27, 1975.* 1976, XI + 418 pp.
55. N. C. Wickramasinghe and D. J. Morgan (eds.), *Solid State Astrophysics. Proceedings of a Symposium, held at the University College, Cardiff, Wales, 9–12 July, 1974.* 1976, XII + 314 pp.
56. John Meaburn, *Detection and Spectrometry of Faint Light.* 1976, IX + 270 pp.
57. K. Knott and B. Battrick (eds.), *The Scientific Satellite Programme during the International Magnetospheric Study. Proceedings of the 10th ESLAB Symposium, held at Vienna, Austria, 10–13 June 1975.* 1976, XV + 464 pp.
58. B. M. McCormac (ed.), *Magnetospheric Particles and Fields. Proceedings of the Summer Advanced Study School, held in Graz, Austria, August 4–15, 1975.* 1976, VII + 331 pp.
59. B. S. P. Shen and M. Merker (eds.), *Spallation Nuclear Reactions and Their Applications.* 1976, VIII + 235 pp.
60. Walter S. Fitch (ed.), *Multiple Periodic Variable Stars. Proceedings of the International Astronomical Union Colloquium No. 29, held at Budapest, Hungary, 1–5 September 1976.* 1976, XIV + 348 pp.
61. J. J. Burger, A. Pedersen, and B. Battrick (eds.), *Atmospheric Physics from Spacelab. Proceedings of the 11th ESLAB Symposium, Organized by the Space Science Department of the European Space Agency, held at Frascati, Italy, 11–14 May 1976.* 1976, XX + 409 pp.
62. J. Derral Mulholland (ed.), *Scientific Applications of Lunar Laser Ranging. Proceedings of a Symposium held in Austin, Tex., U.S.A., 8–10 June, 1976.* 1977, XVII + 302 pp.
63. Giovanni G. Fazio (ed.), *Infrared and Submillimeter Astronomy. Proceedings of a Symposium held in Philadelphia, Penn., U.S.A., 8–10 June, 1976.* 1977, X + 226 pp.
64. C. Jaschek and G. A. Wilkins (eds.), *Compilation, Critical Evaluation and Distribution of Stellar Data. Proceedings of the International Astronomical Union Colloquium No. 35, held at Strasbourg, France, 19–21 August, 1976.* 1977, XIV + 316 pp.
65. M. Friedjung (ed.), *Novae and Related Stars. Proceedings of an International Conference held by the Institut d'Astrophysique, Paris, France, 7–9 September, 1976.* 1977, XIV + 228 pp.
66. David N. Schramm (ed.), *Supernovae. Proceedings of a Special IAU-Session on Supernovae held in Grenoble, France, 1 September, 1976.* 1977, X + 192 pp.
67. Jean Audouze (ed.), *CNO Isotopes in Astrophysics. Proceedings of a Special IAU Session held in Grenoble, France, 30 August, 1976.* 1977, XIII + 195 pp.
68. Z. Kopal, *Dynamics of Close Binary Systems.* XIII + 510 pp.
69. A. Bruzek and C. J. Durrant (eds.), *Illustrated Glossary for Solar and Solar-Terrestrial Physics.* 1977, XVIII + 204 pp.
70. H. van Woerden (ed.), *Topics in Interstellar Matter.* 1977, VIII + 295 pp.
71. M. A. Shea, D. F. Smart, and T. S. Wu (eds.), *Study of Travelling Interplanetary Phenomena.* 1977, XII + 439 pp.
72. V. Szebehely (ed.), *Dynamics of Planets and Satellites and Theories of Their Motion. Proceedings of IAU Colloquium No. 41, held in Cambridge, England, 17–19 August 1976.* 1978, XII + 375 pp.
73. James R. Wertz (ed.), *Spacecraft Attitude Determination and Control.* 1978, XVI + 858 pp.

74. Peter J. Palmadesso and K. Papadopoulos (eds.), *Wave Instabilities in Space Plasmas. Proceedings of a Symposium Organized Within the XIX URSI General Assembly held in Helsinki, Finland, July 31–August 8, 1978.* 1979, VII + 309 pp.
75. Bengt E. Westerlund (ed.), *Stars and Star Systems. Proceedings of the Fourth European Regional Meeting in Astronomy held in Uppsala, Sweden, 7–12 August, 1978.* 1979, XVIII + 264 pp.
76. Cornelis van Schooneveld (ed.), *Image Formation from Coherence Functions in Astronomy. Proceedings of IAU Colloquium No. 49 on the Formation of Images from Spatial Coherence Functions in Astronomy, held at Groningen, The Netherlands, 10–12 August 1978.* 1979, XII + 338 pp.
77. Zdeněk Kopal, *Language of the Stars. A Discourse on the Theory of the Light Changes of Eclipsing Variables.* 1979, VIII + 280 pp.
78. S.-I. Akasofu (ed.), *Dynamics of the Magnetosphere. Proceedings of the A.G.U. Chapman Conference 'Magnetospheric Substorms and Related Plasma Processes' held at Los Alamos Scientific Laboratory, N.M., U.S.A., October 9–13, 1978.* 1980, XII + 658 pp.
79. Paul S. Wesson, *Gravity, Particles, and Astrophysics. A Review of Modern Theories of Gravity and G-variability, and their Relation to Elementary Particle Physics and Astrophysics.* 1980, VIII + 188 pp.
80. Peter A. Shaver (ed.), *Radio Recombination Lines. Proceedings of a Workshop held in Ottawa, Ontario, Canada, August 24–25, 1979.* 1980, X + 284 pp.
81. Pier Luigi Bernacca and Remo Ruffini (eds.), *Astrophysics from Spacelab.* 1980, XI + 664 pp.
82. Hannes Alfvén, *Cosmic Plasma,* 1981, X + 160 pp.
83. Michael D. Papagiannis (ed.), *Strategies for the Search for Life in the Universe,* 1980, XVI + 254 pp.
84. H. Kikuchi (ed.), *Relation between Laboratory and Space Plasmas,* 1981, XII + 386 pp.
85. Peter van der Kamp, *Stellar Paths,* 1981, XXII + 155 pp.
86. E. M. Gaposchkin and B. Kołaczek (eds.), *Reference Coordinate Systems for Earth Dynamics,* 1981, XIV + 396 pp.
87. R. Giacconi (ed.), *X-Ray Astronomy with the Einstein Satellite. Proceedings of the High Energy Astrophysics Division of the American Astronomical Society Meeting on X-Ray Astronomy held at the Harvard-Smithsonian Center for Astrophysics, Cambridge, Mass., U.S.A., January 28–30, 1980.* 1981, VII + 330 pp.
88. Icko Iben Jr. and Alvio Renzini (eds.), *Physical Processes in Red Giants. Proceedings of the Second Workshop, helt at the Ettore Majorana Centre for Scientific Culture, Advanced School of Agronomy, in Erice, Sicily, Italy, September 3–13, 1980.* 1981, XV + 488 pp.
89. C. Chiosi and R. Stalio (eds.), *Effect of Mass Loss on Stellar Evolution. IAU Colloquium No. 59 held in Miramare, Trieste, Italy, September 15–19, 1980.* 1981, XXII + 532 pp.
90. C. Goudis, *The Orion Complex: A Case Study of Interstellar Matter.* 1982, XIV +306 pp.
91. F. D. Kahn (ed.), *Investigating the Universe. Papers Presented to Zdeněk Kopal on the Occasion of his retirement, September 1981.* 1981, X + 458 pp.
92. C. M. Humphries (ed.), *Instrumentation for Astronomy with Large Optical Telescopes, Proceedings of IAU Colloquium No. 67.* 1981, XVII + 321 pp.
93. R. S. Roger and P. E. Dewdney (eds.), *Regions of Recent Star Formation, Proceedings of the Symposium on "Neutral Clouds Near HII Regions - Dynamics and Photochemistry", held in Penticton, B.C., June 24–26, 1981.* 1982, XVI + 496 pp.
94. O. Calame (ed.), *High-Precision Earth Rotation and Earth-Moon Dynamics. Lunar Distances and Related Observations.* 1982, XX + 354 pp.
95. M. Friedjung and R. Viotti (eds.), *The Nature of Symbiotic Stars,* 1982, XX + 310 pp.
96. W. Fricke and G. Teleki (eds.), *Sun and Planetary System,* 1982, XIV + 538 pp.
97. C. Jaschek and W. Heintz (eds.), *Automated Data Retrieval in Astronomy,* 1982, XX + 324 pp.
98. Z. Kopal and J. Rahe (eds.), *Binary and Multiple Stars as Tracers of Stellar Evolution,* 1982, XXX + 503 pp.
99. A. W. Wolfendale (ed.), *Progress in Cosmology,* 1982, VIII + 360 pp.
100. W. L. H. Shuter (ed.), *Kinematics, Dynamics and Structure of the Milky Way,* 1983, XII+ 392 pp.